LOW BACK PAIN PATHOGENESIS AND TREATMENT

Edited by **Yoshihito Sakai**

Low Back Pain Pathogenesis and Treatment
Edited by Yoshihito Sakai

Published by InTech
Janeza Trdine 9, 51000 Rijeka, Croatia

Publishing Process Manager Romana Vukelic

Technical Editor Teodora Smiljanic

Cover Designer InTech Design Team

Additional hard copies can be obtained from orders@intechopen.com

Low Back Pain Pathogenesis and Treatment, Edited by Yoshihito Sakai
p. cm.
ISBN 978-953-51-0338-7

For EU product safety concerns:
IN TECH d.o.o., Prolaz Marije Krucifikse Kozulić 3, 51000 Rijeka, Croatia,
info@intechopen.com or visit our website at intechopen.com.

Meet the editor

Dr. Yoshihito Sakai, born 1968 in Nagoya, has been an Assistant Professor at the Department of Spine Surgery, National Center for Geriatrics and Gerontology, Obu, Japan since 2009. Dr. Sakai is also a lecturer at Nagoya University School of Medicine. He received his Ph.D. "vertebral osteophyte formation and gene polymorphism" in 2007 from Nagoya University School of Medicine, Department of Orthopaedic Surgery. His research interests cover the area of degenerative disease of the lumbar spine and low back pain. He has published research articles in over 100 national and international refereed journals, and he has also written 10 books which have been awarded. Dr. Sakai is also an accomplished musician.

Contents

Preface

Low back pain is a common pandemic disorder which affects the lumbar spine, and is associated with substantial morbidity for about 80% of the general population at some stages during their lives. Although low back pain is usually a self-limiting and benign disorder that trends to improve spontaneously over time, the etiology of low back pain is generally unknown and the diagnostic label, "nonspecific low back pain", is frequently given when no specific pathologic process or structure can be identified. Although low back pain resolves within a few weeks, approximately 10% of the patients develop chronic low back pain, which imposes large burden on society to the health care, and also absence from work, and lost productivity. The direct cost of low back pain was estimated to be US$ 91 billion, and individuals with low back pain incurred health care expenditures about 60% higher than individuals without low back pain.

Exercise therapy is a widely used treatment for low back pain. The response to managed care for low back pain has been emphasized on several active therapies including rehabilitative exercise programs in physical therapy, occupational therapy, and chiropractic therapy. Conservative treatment with exercise therapy has been reported its effectiveness for the patients with low back pain; however, chronic low back pain including degenerative spinal disease is difficult to treat because of the range of causative mechanisms that may be involved. This difficulty leads to the limited efficacy of current physical therapy and exemplified by the inconsistency of existing treatment patterns. Now that optimal pain management demands not only on pain relief, but also a beneficial effects on functionality and quality of life, other non-surgical intervention such as psychosocial care, pharmacotherapy, radiofrequency treatment, and surgical intervention.

This current book contains reviews and original articles with emphasis on pathogenesis and treatment of low back pain except for the rehabilitative aspect which may be carried in another volume. Consisting of three main sections, the first section of the book has a focus on pathogenesis of low back pain, while the second and third sections are on the treatment including conservative and surgical procedure, respectively. The authors of each chapter are experts in their respective fields, and this book volume is intended for all clinicians caring for the patients with low back pain, including physicians, therapists, orthopaedic surgeons, spine surgeons, fellows and residents in the disciplines of the lumbar spine medicine.

Yoshihito Sakai
Assistant Professor, Spine Surgery Department,
National Center for Geriatrics and Gerontology,
Obu,
Japan

Part 1

Pathogenesis

1

Spinal Alignment and Low Back Pain Indicating Spine Shape Parameters

Schroeder Jan and Mattes Klaus
University of Hamburg, Dpt. Human Movement and Training Science
Germany

1. Introduction

Low back pain is thought of as having no structural correlates in radiographic findings. But an associated deconditioning syndrome is assigned by back pain complaints accompanied by functional deficits, especially peak force and performance deficits of deep trunk muscles. We were aiming at investigating if there might be comparable relations between spinal mal-alignment and complaints in chronic low back pain patients. And if spine shape aberrations were in fact associated with low back pain, could they be used to determine exercise programs for an active low back pain therapy, as is generally known for diagnostic screening procedures and low back pain therapy monitoring based on muscle function deficits? Seeking for exercise induced adaptations, we intended to find statistical correlations indicating some kind of specificity for those individualized exercise programs which are based on initial findings in spinal alignment and trunk muscle function.

Our scientific approach involved two aspects that were important for both practical applications and scientific analysis methods in the field of low back pain treatment and research. First of all, our spine shape assessment was non-invasive, and therefore suitable for screening and monitoring without any risks for patients and volunteers. And secondly, indirect spine shape assessment by means of video raster stereography allowed an easy access to multivariate statistical analysis approaches. Therefore, variable interdependencies could be taken into account which might have covered significant effects in earlier investigations.

2. Background

From an economic point of view, low back pain (LBP) is one of the most emerging and cost-pushing health disorders in the western world, and for the majority of cases neither direct organic signs nor structural correlates can be identified (Waddell et al., 1980). According to McGill (2007, p. 5), more than 80% of all patients with back complaints suffer from non-specific low back pain. He suggests that, besides other factors, insufficient diagnosis procedures may contribute to the current uncertainty regarding the true incidence of specific low back pain issues.

Several influencing factors are discussed to be essential in the etiology of low back pain, such as psycho-social components (Waddell et al., 1980), and organic mechanisms in terms of spinal instability due to ligament function and deficits in neuromuscular coordination and compensation: neutral zone spinal instability hypothesis (Panjabi, 1992).

With respect to these biomechanical and social-medical findings, and being aware of muscular dysfunction in LBP patients compared to pain free volunteers (Cady et al., 1979; Denner, 1997; McNeil et al., 1980), reconditioning of muscle function and neuromuscular coordination patterns is supposed to be a successful intervention mode in the therapy of low back pain (Denner, 1997; McGill, 2007; Panjabi, 1992; Waddell et al., 1980), especially when segmental stabilization is taken into account (Ljunggren et al., 1997; O'Sullivan, Twomey & Allison, 1997; Richardson, Hodges & Hides, 2004).

Beside deficits in muscle function of LBP patients, there are anthropometric risk factors for the development and progredience of LBP which deal with spinal shape asymmetries in the frontal plane (Balagué, Troussier & Salminen, 1997) and the alignment of the lumbosacral transition in the sagittal plane (Adams, Mannion & Nolan, 1997; Lewit, 1991, p. 60). Video raster stereographic back shape reconstruction offers a valid and reliable and – in contrast to radiographic screening procedures – a non-invasive, non-aggressive high-resolution system for spine shape assessment in screening and monitoring (Drerup & Hierholzer, 1994).

Recent video raster stereographic investigations of the spinal form of male and female LBP patients and pain free volunteers revealed spine shape parameters indicating LBP by means of multivariate factor analyses: trunk imbalance and trunk inclination (Schröder, Stiller & Mattes, 2010). While a more extended trunk inclination should be considered to be due to the higher age of the patients (Gelb et al., 1995; Kobayashi et al, 2004; Takeda et al., 2009), trunk imbalance remained as a marker for low back pain. Additionally, there was some evidence for a flatter lumbar lordosis in male patients, revealed by means of discriminant analyses (Schröder, Strübing & Mattes, 2010). With female patients, too, pelvis torsion and pelvis tilt were found to be indicating low back pain (Schröder, Stiller & Mattes, 2011). It is highly probable that video raster stereography offers some possibilities in the process of differential diagnosis of sacroiliac disorders (Foley & Buschbacher, 2006).

Furthermore, there was some evidence for non-parametric signs in the spinal alignment of back pain patients with vertebral blockades (Schröder, Färber & Mattes, 2009) or a lumbar facet joint syndrome (Schröder, Strübing & Mattes, 2010). These findings and some specific kind of profile of spinal shape parameters should be helpful for diagnosis procedures in the field of orthopaedic practioneers. This work is in process.

The findings mentioned above might provide an opportunity to create therapeutic exercise programs based on spinal form deviation signs, comparable to individualized exercise programs based on muscle function deficits (Denner, 1997). So far, specific correlations between adaptations of muscle function and clinical out-come parameters could hardly be established (Mannion et al., 2001b; 2001c). Nevertheless, first results of a pilot study seemed to show specific adaptations following individualized exercise programs, e.g. trunk imbalance decreased mainly in patients who showed extraordinary values in the frontal plane before a short-term training period of ten weeks. This specific decrease correlated with pain reduction and was accompanied by increases in peak forces of trunk muscle strength (Schröder et al., 2009).

In general, spinal form adaptations are difficult to prove by means of statistical calculations (Kuo, Tully & Galea, 2009), because they depend on the degree of mal-alignment, and adaptations are varying considerably among individuals (Weiß, Dieckmann & Gerner, 2003; Weiß & Klein, 2006). Age and gender also seem to be influencing factors for the degree of spinal form adaptations in some parameters (Schröder & Mattes, 2010). Correlations between clinical out-come and muscle function increases are augmented, when spinal form adaptations are taken into account in multiple regression models.

3. Methods

3.1 Study design

First of all, a cross-sectional study was conducted to identify spine shape parameters associated with low back pain. Secondly, a pre-post-effect analysis was carried out, seeking for exercise induced adaptations in the process of reconditioning.

3.2 Subjects

At least 405 subjects could be examined, 213 patients suffering from low back pain (LBP) and 192 volunteers - most of them freshmen at the University of Hamburg - serving as controls (CON). The controls were included if there was no diagnosis dealing with back pain complaints, no serious back pain history for two years, and no back pain at all in the last six months.

Participants were divided into female and male subsamples. Due to the large sample size, the observed - relatively small - differences in anthropometric parameters between patients and controls were almost significant, except for the body weight of the males (tab. 1).

	age [y]	height [m]	weight [kg]	BMI [kg/m²]
LBP females	50,5	1,68	67,9	24,2
SD (n=129)	14,2	0,06	6,0	1,6
LBP males	47,6	1,83	82,4	24,6
SD (n=84)	15,3	0,06	6,0	1,4
CON females	26,5	1,70	65,7	22,8
SD (n=79)	4,7***	0,06*	6,5*	1,4***
CON males	27,6	1,85	82,2	24,0
SD (n=113)	4,4***	0,05**	5,5	1,2***

Table 1. Anthropometric data of low back pain patients (LBP) and pain free controls (CON) (mean ± standard deviation; LBP vs. CON: * $p \leq 0,05$; ** $p \leq 0,01$; *** $p \leq 0,001$ Student's t-test)

Female patients were significantly older ($t = -17,636$; $p < 0,000$), had a slightly smaller body height ($t = 2,475$; $p = 0,014$), a slightly larger body weight ($t = -2,517$; $p = 0,013$) than the female controls and also showed a slightly higher body mass index ($t = -6,353$; $p < 0,000$).

Male patients were significantly older ($t = -11,668$; $p < 0,000$), had a slightly smaller body height ($t = 2,395$; $p = 0,018$), a nearly identical mean body weight ($t = -0,330$; $p = 0,742$), and showed a slightly higher body mass index ($t = -3,298$; $p < 0,001$), too (tab. 1).

Patients were included after clinical and radiographic examinations by an orthopaedic physician (Buchholz & Partner, Hamburg, Germany), who qualified the pain syndrome as chronic unspecific back pain (LBP), when no correlation to structural signs could be established and when patients suffered from low back pain for a time period of six months minimum. In fact, back pain history varied from six months to more than nine years (average: 8 months) and most of the patients had gone through several treatment trials before. Specific signs, such as vertebral fractures, spinal surgery, severe scoliosis or acute sciatic symptoms were exclusion criteria, as well as a back pain state of more than 5 points in the CR10 pain scale reacing from zero to ten points (Borg, 1998) at examination time.

107 of those patients mentioned above went through an exercise therapy program and were re-examined in a post-test. Treatment effects could be analysed for 61 female patients (57%), and for 46 males (43%). Females were 48,7 ± 14,1 years of age, body height was 1,70 ± 0,07 m, body weight was 67,8 ± 10,7 kg, and their body mass index (BMI) was 23,6 ± 3,3 kg/m^2. Males were of the same age (49,6 ± 14,3 years), but naturally higher (1,80 ± 0,07 m) and heavier (81,4 ± 12,9 kg), while the body mass index was comparable (24,9 ± 2,9 kg/m^2) to the females, and not indicating obesity.

3.3 Spine shape assessment

Spine shape parameters were calculated by means of video raster stereography (Formetric®-System[1]), a high resolution back shape reconstruction device (reconstruction error 0,2 to 0,5 mm; resolution 10 pts./cm^2) (Drerup & Hierholzer, 1994). Reproducibility of back shape reconstruction was proved. Reliability coefficients (ICC: Intra Class Correlation) were ranging between 0,99 and 0,91 for the sagittal plane, and between 0,82 and 0,69 for the frontal plane. For the coronal plane, reliability was 0,81 (Mohukum et al., 2009; Schröder & Mattes, 2009; Schröder, Reer & Mattes, 2009) (tab. 2).

Specific back surface landmarks - like the vertebra prominens (VP), the beginning of the rima ani representing the sacrum point (SP), and the right and left lumbar dimple (DR, resp. DL) representing the position of spinae iliaca posterior superior (SIPS) of the pelvis - were recognized automatically to build up a Cartesian coordinate system. This coordinate system served as calibration reference frame for a three-dimensional surface reconstruction using triangulation equations that ensured a valid correlation between back shape reconstructions and radiographic assessments of the anatomy of spine and pelvis characters [2] (Drerup & Hierholzer, 1985; 1987a; 1987b) (fig. 1).

[1] Diers International, Schlangenbad, Germany

[2] Using stereography, the three-dimensional coordinates of every point on a given surface might be calculated by two cameras. In video raster stereography one camera is substituted by a projector – quasi like an inverse camera (fig. 1). If the geometry of projector and camera is known and invariant, triangulation equations enable the system not only to detect every point on the back surface, but also to reconstruct invariant back shape characters based on two phenomenons: First of all, the surface around every point spreads into two directions. The curvature of these planes may be calculated from the three-dimensional coordinates of any reconstruction point. As a consequence, the surface of the reconstructed body may show nothing but a convex, a concave or a saddle-shaped curvature as an invariant representation of the back shape (fig. 1), not depending on the position of the reconstructed body. Additionally, every point on the surface has an orientation determined by structures beneath the skin surface, which can be expressed mathematically by the surface normal. For back shape reconstruction, the spinous processes and the lumbar dimples representing pelvic processes are of a certain interest (fig. 1) (Drerup et al., 2001).

Spine shape parameter	Short/ ICC	Explication
Trunk imbalance [mm]	Tr-Imb ICC=0,82	Plumb deviation from vertebra prominens to midpoint between dimples in the frontal plane (fig. 2)
Trunk inclination [mm]	Tr-Inc ICC=0,91	Plumb deviation from vertebra prominens to pelvis position/ midpoint between dimples in the sagittal plane (fig. 2)
Pelvis tilt [mm]	P-Tilt ICC=0,81	Deviation of the axis of lumbar dimples to the floor line in the frontal plane (fig. 2)
Pelvis torsion [°]	P-Tors ICC=0,69	Relative torsion between left and right side pelvis bones (os ilium) in the frontal-transversal plane
Vertebral side deviation [mm]	Side-rms ICC=0,71	Average deviation of vertebral bodies in the frontal plane (rms from vertebra prominens to midpoint between dimples)
Vertebral rotation (rms) [°]	Rot-rms ICC=0,81	Average rotation of vertebral bodies in the transversal plane (rms from vertebra prominens to midpoint between dimples)
Kyphosis angle (ICT-ITL) [°]	KA-max ICC=0,91	Maximum thoracic angle calculated from ICT and ITL triangles (fig. 2)
Lordosis angle (ITL-ILS) [°]	LA-max ICC=0,99	Maximum lumbar angle calculated from ITL and ILS triangles (fig. 2)

Table 2. Spine shape parameters, short-cuts with Intra Class Correlation coefficient (ICC), and a description of anatomy and corresponding geometry

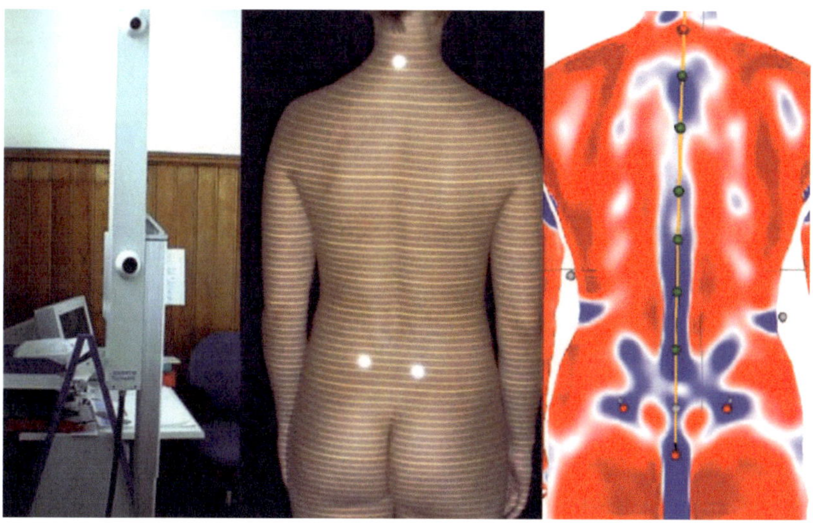

Fig. 1. Video raster stereography with camera and projector system (left), projection lines on the back surface with vertebra prominens (VP) and lumbar dimples (DL+DR) high-lighted - here with optical markers only for demonstration (middle), and video raster stereography back surface reconstruction with landmarks recognized automatically (red dots) and plane curvatures representing convex (red areas) or concave (blue areas) back shape profiles (right) (modified from: Schröder, Förster & Mattes, 2008, p. 46)

For a better understanding of geometry and corresponding anatomical landmarks, spine shape parameters were illustrated in an animation, especially for the sagittal plane (fig. 2).

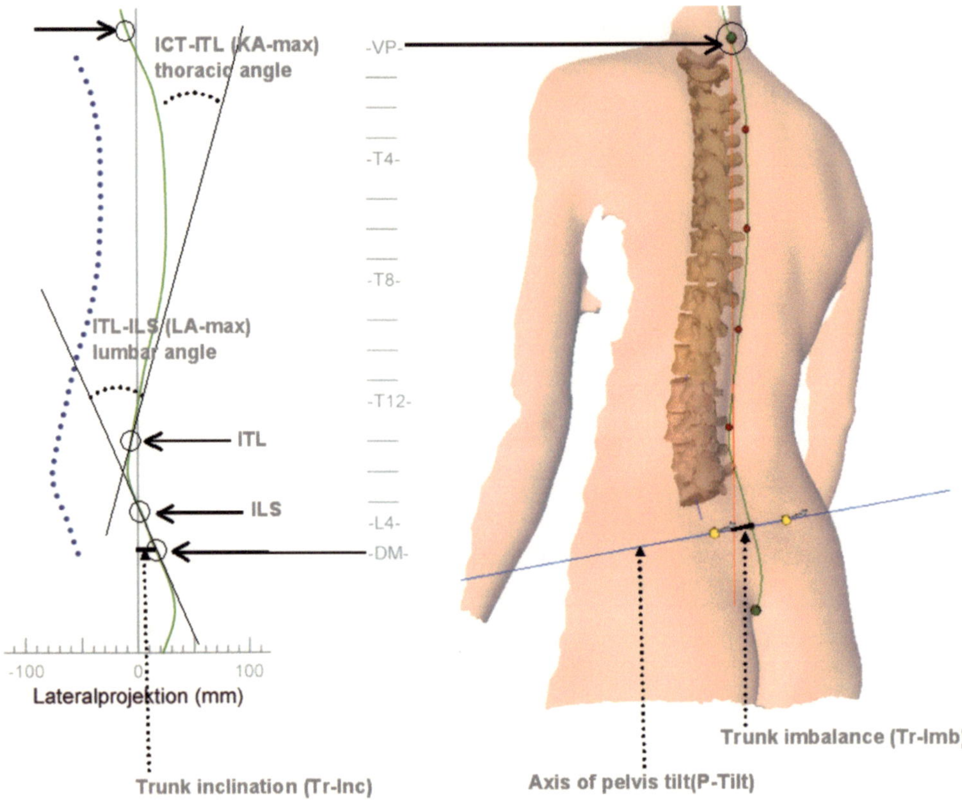

Fig. 2. Spine shape in the sagittal plane: kyphosis angle (KA-max) and lordosis angle (LA-max) with inflectional points of the curvature from cervical to thoracic spine (ICT), from thoracic to lumbar spine (ITL) and from lumbar to sacral spine (ILS) and three dimensional animation of back surface with lumbar dimples (yellow dots - with arrows representing the direction of the mathematical normal on each dimple's plane) and spinous processes like vertebra prominens (VP) and sacrum point (SP) marking the beginning of the rima ani (green dots) (Formetric®-System) (modified from: Schröder, Stiller & Mattes, 2010, p. 92)

3.4 Trunk muscle peak force assessment

Torques of the superficial trunk muscles were assessed by means of isometric peak forces (sensor sample rate 100 Hz, sensibility 0,85 mV/V, signal smoothing by a sliding average over 0,3 sec) in a test chair that allowed data acquisition in all three dimensions (extension-flexion, lateral flexion, axial rotation) (Myoline®)[3], while patients or volunteers had to be fixed only once for all test contractions in a universal standard position. Reproducibility was verified, and reliability coefficients were ranging between 0,85 and 0,94 for trunk muscle testing in all three dimensions (Schröder, Reer & Mattes, 2009).

[3] Diers International, Schlangenbad, Germany

3.5 Pain documentation

Pain was described by means of the CR10 pain scale questionnaire, an instrument for self-rated pain and exertion, evaluated by Gunnar Borg (1998). The CR10 pain scale (0=nothing at all, 0,5=extremely weak, 1=very weak, 2=weak, 3=moderate, 5=strong, 7=very strong and 10=extremely strong) combined categorical and rational aspects of the phenomenon pain - for a valid assessment with respect to the non-linear relation between pain state and semantic expressions for its description[4]. Reliability had been verified earlier, and coefficients ranged between 0,78 to 0,99 (Borg, 1998, pp. 41-43).

3.6 Treatment

About 50% of all low back pain patients (n=107) went through an individualized exercise program for a time period of 10 to 12 weeks from pre- to post-testing. There were 18 training sessions altogether, normally two sessions per week. Every session took 60 minutes and followed a fixed schedule of seven phases: a systematic ergometer warm-up (5 min), functional strengthening (2 to 4 exercises) and stretching (4 to 6 exercises), as well as physiotherapist pulley and weight training (4 to 6 exercises) using standard training devices. But the exercise program was dominated by Segmental Stabilization Training (SST), which was learned and re-learned in every session (2 to 3 min) in a basic exercise (fig. 3), and which was applied in several static (2 to 4 exercises) and dynamic (2 to 3 exercises) tasks with an emphasis on the special SST-coordination[5] pattern.

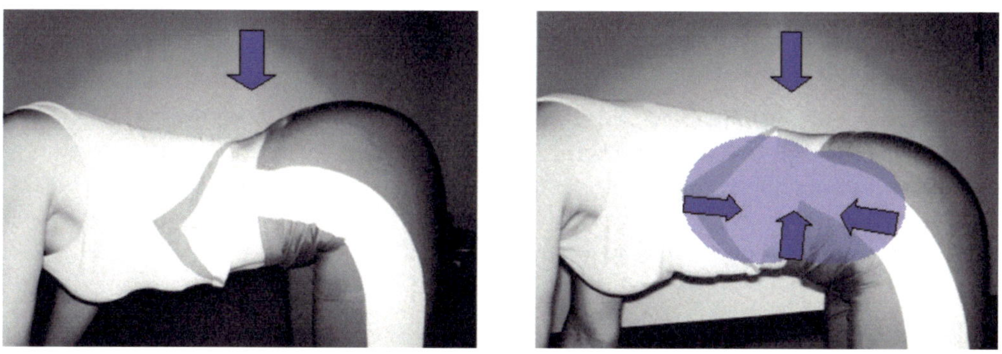

Fig. 3. Coordination pattern of Segmental Stabilization Training (SST) (from: Schröder, Förster & Mattes, 2008, p. 48)

---

[4] As there is no linear relation between increasing pain and expressions for its description, the CR10 pain scale has a higher rational resolution for an almost weak pain state and includes more steps on the rational pain scale for stronger pain states, which matches the character of pain and the possibility for a valid assessment better than an ordinary visual analogue scale (VAS) (Borg, 1998).

[5] Segmental stabilisation means a special coordination pattern to involve deep trunk and lumbar back muscles. A slight tension of the pelvic floor, accompanied by a draw-in task for the belly button - submaximal activity of the musculus transversus abd. - and breathing slightly against the diaphragm is meant to increase the activity of deep back muscles, such as musculus multifidus. Using this coordination pattern, stability of lumbar vertebral segments and the transition to sacroiliac joints were found to be improved (Richardson, Hodges & Hides, 2004). Therefore, Segmental Stabilization Training is meant to represent that kind of specific exercise therapy, which was requested for the treatment of low back pain (Panjabi, 1992; Waddell et al. 1980).

All exercises were performed for one to three sets, with an intensity that allowed 10 to 15 repetitions or 20 to 30 seconds of static resistance, respectively. Number of sets and reps (volume and intensity and the choice of exercise itself (content) were determined by individual findings in the pre-test and anamnesis information right before starting the intervention. Training took place in the field of out-patient rehabilitation in groups of three to five patients and was conducted and controlled by at least one physiotherapist (Schröder & Färber, 2010).

3.6 Statistics

Data were described as mean ± standard deviation (SD), mean ± CI (95% confidence interval) for figure 4, and mean ± SEM (68% confidence interval meaning the Standard Error of the Mean) for figure 5. Normal distribution was proved using the Kolmogorov-Smirnov-test.

For the cross-sectional study, a factor analysis (SPSS 12: principle components extraction, Kaiser-normalisation with varimax rotation) was conducted to explore a spine shape structure model of almost independent factors, determined by video raster stereography spine shape parameters, seeking for differences between low back pain patients and pain free controls. In a second multivariate approach, discriminant analyses (SPSS 12) were calculated for males and females to reveal spine shape parameters being able to separate low back pain patients from pain free volunteers. At least, these extracted parameters were analysed for significant differences between patients and controls by means of univariate procedures (Student's t-test), and a spine shape profile was illustrated for males and females with or without low back pain.

For the analysis of treatment effects in the sample of patients who went through an exercise program, three-way ANOVAs (SPSS 12: within-subjects factor: pre vs. post exercise program, between-subjects factor for gender: female vs. male and between-subjects factor for age: under 60 years vs. over 60 years) were calculated. Bivariate Pearson correlations and linear multiple regression models based on pre-post-differences were calculated to analyse interdependencies of variables monitored in the process of reconditioning.

Significance was accepted for p-values of $p \leq 0{,}05$ *. Differences showing p-values of $p \leq 0{,}01$ ** or $p \leq 0{,}001$ *** were deemed very significant.

4. Results

4.1 Cross-sectional study

4.1.1 Factor analysis

A factor analysis revealed components describing almost independent spine shape characters determined by video raster stereography parameters, with respect to the interdependency of theses parameters. Different models for the controls (CON) and for the low back pain patients (LBP) indicated low back pain markers (tab. 3).

Trunk inclination (Tr-Inc), trunk imbalance (Tr-Imb), pelvis tilt (P-Tilt), pelvis torsion (P-Tors), thoracic kyphosis angle (KA-max), lumbar lordosis angle (LA-max), mean (root-mean-square) vertebral side deviation (Side-rms), and mean (root-mean-square) vertebral

	Components (CON) n=192				Components (LBP) n=213			
	1	2	3	4	1	2	3	4
Tr-Inc	**-0,673**	-0,083	-0,033	0,156	-0,189	-0,099	0,045	**0,756**
Tr-Imb	-0,031	0,065	0,046	**0,946**	-0,093	0,199	**0,759**	-0,067
P-Tilt	0,088	0,029	**0,829**	-0,070	0,131	0,217	-0,047	**0,700**
P-Tors	-0,148	-0,025	**0,758**	0,114	0,232	-0,138	**0,702**	0,068
KA-max	**0,721**	-0,098	-0,055	0,318	**0,798**	-0,034	0,191	0,049
LA-max	**0,739**	0,001	-0,055	-0,064	**0,861**	0,023	-0,067	-0,110
Rot-rms	0,220	**0,805**	0,130	0,010	-0,036	**0,768**	0,210	0,167
Side-rms	-0,199	**0,806**	-0,123	0,051	0,019	**0,833**	-0,118	-0,042

Table 3. Factor analysis – principle components extraction – for controls (CON: n=192) and low back pain patients (LBP: n=213) (factor loading coefficients over 0,65 printed in bold)

rotation (Rot-rms) served as variables (tab. 2). In the rotated component matrix, factor loading coefficients higher than 0,650 were enhanced to mark relevance (tab. 3). A factor analysis for the pain free controls revealed four components with an Eigen value greater than one, explaining 66% of the total variance. The table showed factor loading coefficients constituting independent factors for a summarizing description of human spinal alignment. Factors could be named as 'sagittal spine shape' (factor 1: LA-max 0,739 x KA-max 0,721 x Tr-Inc -0,673), 'vertebral deviations' (factor 2: Rot-rms 0,805 x Side-rms 0,806), 'pelvis parameters' (factor 3: P-Tilt 0,829 x P-Tors 0,758) and 'trunk deviation' (factor 4: Tr-Imb 0,946) (tab. 3). For low back pain patients, a component model of all four components explaining a total variance of 64,8 % could be revealed. In the first and most important component 'sagittal spine shape' the trunk inclination lost its influence (factor 1: LA-max 0,861 x KA-max 0,798). The second component 'vertebral deviations' did not differ from the controls (factor 2: Rot-rms 0,768 x Side-rms 0,833). Compared to the controls, there were some significant changes for the pelvis parameters. For low back pain patients, pelvis torsion was associated with trunk imbalance (factor 3: Tr-Imb 0,759 x P-Tors 0,702), and pelvis tilt was associated with trunk inclination (factor 4: Tr-Inc 0,756 x P-Tilt 0,700). So, the pelvis parameters were influencing the upper body position in the frontal and sagittal plane in back pain patients (tab. 3).

Summarizing the factor analyses, there were four independent components to describe spinal alignment for pain free persons: 'sagittal spine shape', 'vertebral deviations', 'pelvis parameters', and 'trunk deviation'. Low back pain was indicated by changes of the evidence of pelvis parameters compared to pain free controls. They were no longer an independent component, but were influencing the upper body position or deviation in low back pain patients.

4.1.2 Discriminant analysis

Discriminant analyses for male and female patients and controls included all spine shape parameters used before for the factor analysis.

For males, there was a relatively poor canonical correlation (eta^2 = 0,399), but the discriminant function led to a high significant solution for a group separation (Chi^2 = 32,810; $p \leq 0,001$; Wilks' Lambda = 0,841). For the males, there was a correctly predicted group membership of 70% using the discriminant function, 72% for the controls and 68% for the low back patients, respectively. Trunk imbalance offered the best capability to separate groups by means of the canonical discriminant function coefficients (Tr-Imb: 0,743) for males (tab. 4).

	Canonical discriminant coefficients	
	males (n=197)	females (n=208)
Tr-Inc	0,336	**0,610**
Tr-Imb	**0,743**	**0,441**
P-Tilt	0,157	0,006
P-Tors	0,066	**0,470**
KA-max	-0,146	-0,183
LA-max	-0,350	0,105
Rot-rms	-0,342	-0,374
Side-rms	0,067	0,340

Table 4. Canonical discriminant coefficients for males (controls n=113 and low back pain patients n=84) and females (controls n=79 and low back pain patients n=129) (relevant coefficients printed in bold)

For females, the canonical correlation was a little higher (eta^2 = 0,448; Wilks' Lambda = 0,799) than for males, and the discriminant function also led to a high significant solution for a group separation (Chi^2 = 44,570; $p \leq 0{,}001$). The prediction of correct group membership showed a ratio of 69%, 74% for the controls and 65% for the female low back patients, respectively. Trunk inclination and a little less trunk imbalance and pelvis torsion offered the best capability to separate groups by means of the canonical discriminant function coefficients for females (tab. 4).

Summarizing the results of the discriminant analyses, we found poor but acceptable discriminating functions for males and females, where group membership (LBP vs. CON) could be predicted correctly for approximately 70 % of all cases. Trunk imbalance in males and trunk imbalance with trunk inclination and pelvis torsion in females were the most appropriate spine shape variables to separate groups using a multivariate discriminant analysis function.

Evaluating both factor analysis and discriminant analysis, there were video raster stereography spine shape parameters that could be established to be associated with low back pain: trunk inclination and trunk imbalance with pelvis parameters mainly found in females, trunk inclination with trunk imbalance and the lumbar lordosis angle mainly found in males. Univariate analyses confirmed these multivariate findings.

4.1.3 Univariate analysis

Univariate comparisons revealed statistically significant mean differences between low back pain patients and controls for both men and women in their video raster stereography spinal alignment (tab. 5), illustrated as spine shape profiles (fig. 4).

	Tr-Inc [mm]	Tr-Imb [mm]	P-Tilt [mm]	P-Tors [dgr]	KA-max [dgr]	LA-max [dgr]	Rot-rms [dgr]	Side-rms [mm]
CON females	8,6	7,8	4,7	2,0	47,8	42,2	3,7	6,1
± SD	15,1	5,3	4,9	1,5	9,7	8,3	1,8	4,0
LBP females	21,6 ***	11,3 ***	4,9	3,1 ***	47,5	41,7	3,4	7,1
± SD	20,1	7,7	4,0	2,8	9,3	9,0	1,9	3,7
CON males	10,9	7,2	4,9	3,0	48,4	35,6	3,4	6,8
± SD	16,8	6,2	3,8	2,4	9,0	6,5	1,6	3,6
LBP males	18,7 *	11,6 ***	5,3	3,4	46,4	32,6 **	3,1	6,6
± SD	24,2	8,1	4,2	2,5	8,1	6,8	1,7	4,2

Table 5. Spine shape parameters for female and male controls vs. low back pain patients (Student's t-test: p≤0,05 *, p≤0,01 **, p≤0,001 ***)

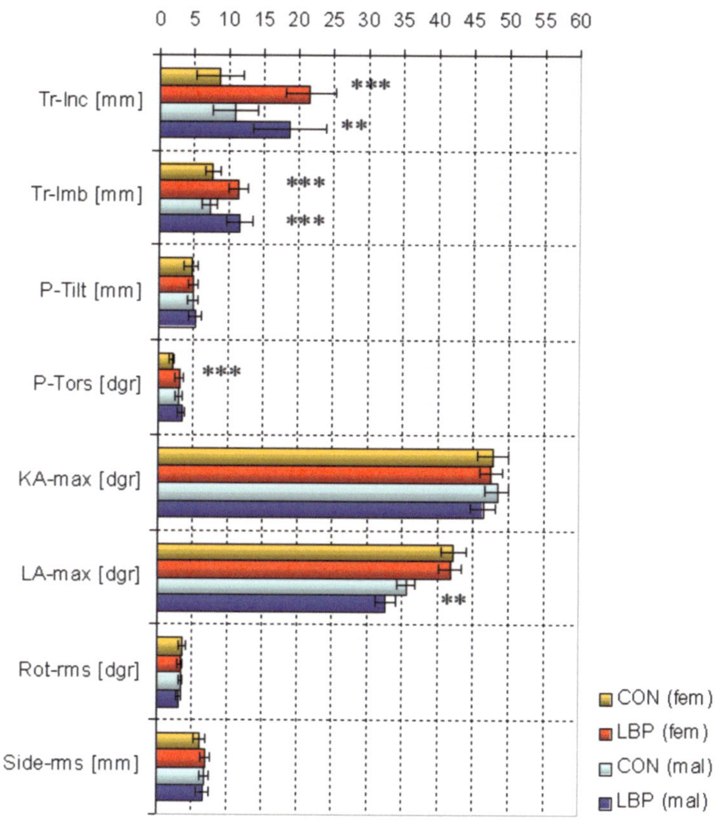

(Mean ± 95% CI; Student's t-test: p≤0,05 *, p≤0,01 **, p≤0,001 ***)

Fig. 4. Video raster stereography spine shape profiles for males (light and dark blue: CON n=113 and LBP n=84) and females (light and dark red: CON n=79 and LBP n=129)

First of all, trunk inclination (Tr-Inc) (females' mean difference: 13,0 mm; t=-4,959; p≤0,001; males' mean difference: 7,8 mm; t=-2,534; p=0,012) and trunk imbalance (Tr-Imb) (females' mean difference: 3,6 mm; t=-3,993; p≤0,001; males' mean difference: 4,4 mm; t=-4,211; p≤0,001) differed significantly between low back pain patients and pain free controls. As for women, there was a significant difference in the parameter pelvis torsion (P-Tors) (mean difference: 1,2°; t=-3,811; p≤0,001), and for men in the parameter maximum lumbar lordosis angle (LA-max) (mean difference: 3,1°; t=3,204; p=0,002), respectively (tab. 5).

4.2 Effect analysis

The effect analysis dealt with changes in low back pain, trunk muscle function, and spinal alignment following a ten-week exercise program, where adaptations were assigned to a process called reconditioning. Effects were analysed using a three-way ANOVA to verify within-subjects effects and interactions with independent factors, like gender and age.

4.2.1 Parameters of low back pain reconditioning

Development of pain state (CR10) and trunk muscle peak forces following the exercise program were described as mean and standard deviation, and the within-subjects effect showed at least very significant increases of peak forces, and a decrease of pain, respectively. Trunk muscle torque was expressed as corresponding masses in kilogram for more transparency (tab. 6).

As for the low back pain state, there were neither significant differences between males and females (F=0,371; p=0,544) nor between younger and older patients (F=0,647; p=0,423) (between subjects factors), and there were no interactions for gender (F=2,910; p=0,091) or age (F=0,941; p=0,334) with the treatment effect, which in itself was very significant (F=60,603; p≤0,001) (within-subject factor) (tab. 6).

	CR10 [pts.]	Ext. [kg]	Flex. [kg]	Lat-lt. [kg]	Lat-rt. [kg]	Rot-lt [kg]	Rot-rt [kg]
LBP t1 total	3,8	56,7	30,5	29,4	29,7	30,6	32,1
± SD	2,3	25,7	17,0	11,8	12,5	16,1	15,3
LBP t2 total	2,3	69,9	35,7	34,3	34,1	36,5	38,0
± SD	1,8	28,8	17,3	13,5	13,7	19,6	19,5
Mean-diff.	-1,5 ***	+13,2 ***	+5,2 ***	+4,9 ***	+4,4 **	+5,9 ***	+5,9 ***
F=	60,603	30,563	16,969	16,727	9,280	17,515	23,426
p≤	0,001	0,001	0,001	0,001	0,003	0,001	0,001

Table 6. Pain (CR10 points) and trunk muscle peak forces for back extension (Ext), trunk flexion (Flex), lateral flexion to the right (Lat-rt) and to the left (Lat-lt), as well as axial rotation to the right (Rot-rt) and to the left (Rot-lt) before (t1) and after (t2) the exercise program for the total (n=107) of low back pain patients (within-subjects effect: p≤0,05 *, p≤0,01 **, p≤0,001 ***)

Of course, trunk muscle peak forces differed between males and females (extension: F=42,351; p≤0,001; flexion: F=23,482; p≤0,001; lateral-left: F=44,251; p≤0,001; lateral-right: F=33,686; p≤0,001; rotation-left: F=40,841; p≤0,001; rotation-right: F=47,507; p≤0,001), and also between younger and older patients (extension: F=7,745; p=0,006; flexion: F=21,945; p≤0,001; lateral-left: F=20,271; p≤0,001; lateral-right: F=6,923; p=0,010; rotation-left: F=7,821; p=0,006; rotation-right: F=4,441; p=0,038), but there were no significant interactions at all between grouping variables and the within-subjects factor (p>0,05), while the treatment effect itself was very significant in any dimension (extension: F=30,563; p≤0,001; flexion: F=16,969; p≤0,001; lateral-left: F=16,727; p≤0,001; lateral-right: F=9,280; p=0,003; rotation-left: F=17,515; p≤0,001; rotation-right: F=23,462; p≤0,001) (tab. 6).

With respect to references in the field of low back pain research, relative peak force increases were illustrated for both males and females separately (fig. 5). Relative increases ranged between approximately 20% to about 40%. Increases were higher in the back extension (approx. 35%) and trunk flexion (from about 30 to 45%) than in the lateral flexion (approx. 25%) and the axial rotation (from about 20 to 30%) (fig. 5).

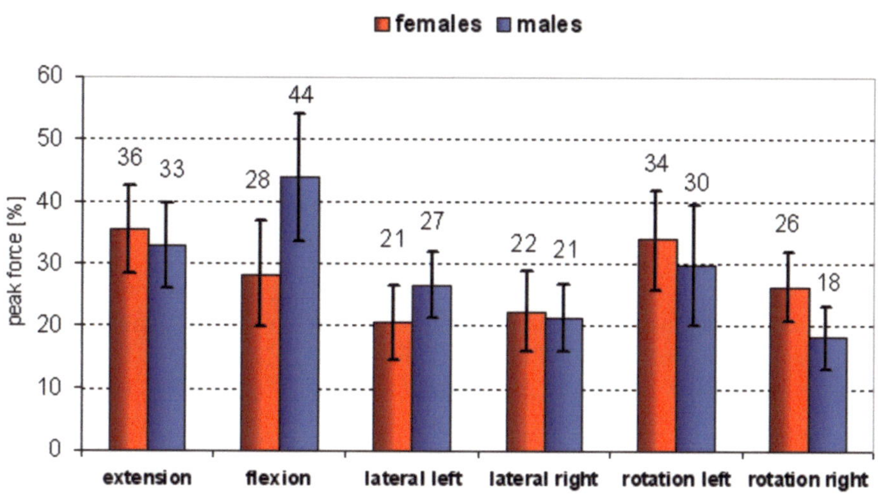

Fig. 5. Relative increases of trunk muscle peak forces for males (n=46) and females (n=61) (Mean±SEM)

Investigating the relation between clinical out-come and muscle function, multiple regression models for the estimation of low back pain decreases by means of relative peak force increases led to a multiple regression coefficient of R=0,292 (R^2=9%). The only predictor showing a tendency for a significant contribution to explain pain decrease was the relative increase of trunk flexion (β=0,216; p=0,055).

4.2.2 Parameters of spinal alignment

Three-way ANOVAs revealed significant within-subjects effects for only a few video raster stereography parameters of the spinal alignment. Pelvis torsion (P-Tors) and lumbar lordosis angle (LA-max) showed significant changes for the total of the low back pain patients - a group statistically verified - manifesting themselves in pelvis position correction

(-0,6°; F=5,145; p=0,025) and lumbar spinal erection (-0,7°; F=6,548; p=0,012), respectively (tab. 7).

	Tr-Inc [mm]	Tr-Imb [mm]	P-Tilt [mm]	P-Tors [dgr]	KA-max [dgr]	LA-max [dgr]	Rot-rms [dgr]	Side-rms [mm]
LBP t1 total	19,4	12,4	4,6	3,2	47,1	37,2	3,1	6,5
± SD	19,2	9,5	3,8	2,6	9,2	9,0	1,6	3,1
LBP t2 total	21,2	11,2	4,3	2,6	46,5	36,5	3,5	6,8
± SD	22,7	8,4	3,3	1,9	9,7	8,8	1,8	3,4
Mean-diff.	+1,8	-1,2	-0,3	-0,6 *	-0,6	-0,7 *	+0,4	+0,3
F=	2,122	1,671	2,524	5,145	1,698	6,548	3,029	0,131
p=	0,148	0,199	0,115	0,025	0,196	0,012	0,085	0,718

Table 7. Spine shape parameters before (t1) and after (t2) the exercise program for the total (n=107) of low back pain patients (within-subjects effect: p≤0,05 *, p≤0,01 **, p≤0,001 ***)

Changes in the sagittal plane were depending on gender (interaction: F=6,651; p=0,011), but not on age (interaction: F=2,596; p=0,110). Naturally, there were differences in the lumbar lordosis angle (LA-max) between males (t1: 31,6° ± 7,2°; t2: 31,8° ± 7,6°) and females (t1: 41,3° ± 8,0°; t2: 40,0° ± 8,0°) (between-subjects effect: F=25,305; p≤0,001), but there was no significant difference between younger and older patients (between-subjects effect: F=2,420; p=0,123).

Changes of pelvis torsion (P-Tors) were neither depending on gender (interaction: F=0,041; p=0,840) nor on age (interaction: F=0,582; p=0,447). There were no significant differences between males and females in the pelvis torsion (between-subjects effect: F=0,353; p=0,554), and also not between younger and older patients (between-subjects effect: F=0,642; p=0,425).

Differences from pre- to post-test for the total of the examined low back pain patients (n=107) in any other spine shape parameter did not reach significance levels (within-subjects effects: p>0,05) (tab. 7). And there were no significant between-subjects effects for gender (p>0,05) or age (p>0,05), except for the trunk inclination, where older people showed significantly larger values than younger persons (F=13,063; p≤0,001). Furthermore, there were no significant interactions between the within-subjects factor (treatment) and the between-subjects factors (gender and age), neither for trunk inclination (Tr-Inc), trunk imbalance (Tr-Imb), pelvis tilt (P-Tilt), and thoracic kyphosis angle (KA-max) nor for the vertebral side deviation (Side-rms) or the vertebral rotation (Rot-rms) (p>0,05).

Looking for specific adaptations of spinal alignment, bivariate correlations of alterations – maybe corrections – of spine shape parameters with extra-ordinary deviations (out-layers of the standard deviation interval before the start of the exercise program) in the frontal plane revealed significant correlation coefficients for trunk imbalance (r=0,40; p=0,021; n=33), pelvis tilt (r=0,43; p=0,038; n=23), and pelvis torsion (r=0,72; p≤0,001; n=26). There were no significant correlation coefficients for any other spine shape parameter, neither for the sagittal plane nor for the coronal plane, in this specific pre-post-analysis investigating parameter changes depending on the initial state prior to the exercise intervention.

Taking account of the alterations of spine shape parameters additional to the peak force increases, a linear multiple regression model explained the total variance of pain decrease ($R=0,399$; $R^2=16\%$) better than using only peak force increases as predictors ($R=0,292$; $R^2=9\%$). Only trunk imbalance contributed significantly as a predictor ($\beta=0,248$; $p=0,036$) to explain pain decrease.

5. Discussion

5.1 Cross-sectional findings

A literature review from the beginning of the 21st century did not come to a conclusive position of evidence (Bernard, 2002). Are there any correlations between posture or spinal mal-alignment and muscle function deficits connective with low back complaints? Univariate investigations - using video raster stereography or not - could not confirm these expectations (Heckmann et al., 2008; Nourbakhsh, Arabloo & Salavati, 2006). But in the field of physiotherapy or manipulative medicine and respective treatment as well as diagnostic procedures of low back pain (LBP) this assumption is considered to be a major guide line for therapy interventions (Lewit, 1991; Seeger et al., 1997).

There is some evidence for the relevance of psychosocial factors influencing the development and the progredience of low back pain. Furthermore, chronification and behavioral aspects of individual coping strategies could be established to be predictive factors for a treatment success (Hildebrandt et al., 1997). But with respect to organic signs, low back pain is considered to be unspecific. Pain is not assigned to structural correlates. Radiographic findings indicate the cause of low back pain only accidentally (Waddell et al., 1980). From an organic point of view, spinal instability seems to be a major risk factor, and probably might be a criterion for diagnosis procedures and therapy interventions (Panjabi, 1992).

According to this instability hypothesis, significant associations could be verified between low back pain and functional deficits of trunk muscle peak force (Cady et al., 1979; Denner, 1997; McNeill et al., 1980) and neuromuscular coordination patterns (Richardson, Hodges & Hides, 2004). Resulting deconditioning syndromes might not only be accompanied by functional disorders, but also by spinal mal-alignment and postural abberations (Müller, 1999).

Some epidemiological reviews or radiographic cross-sectional and follow-up studies extracted frontal plane asymmetries and a flatter lumbosacral transition as anthropometric risk factors for the development and progredience of low back pain (Adams, Mannion & Dolan, 1999; Balagué, Troussier & Salminen, 1999; During et al., 1985; Harrison et al., 1998; Masset, Piette & Malchaire, 1998; Nissinen et al., 1994).

As a main result, the present investigations could confirm these findings from the literature by means of multivariate analysis approaches and with the help of a non-invasive spine shape reconstruction device. Using video raster stereography, particular spine shape parameters were identified to be associated with low back pain (tab. 3 and tab. 4). Patients with chronic low back pain showed larger values for trunk imbalance (Tr-Imb: $p<0,01$) and trunk inclination (Tr-Inc: $p<0,001$) compared to pain free volunteers (tab. 5). Trunk inclination should be considered to be due to the higher age of the patients sample (Gelb et al., 1995; Kobayashi et al., 2004; Takeda et al., 2009), but trunk imbalance remained as an indicator variable to identify low back pain (Schröder, Stiller & Mattes, 2010; 2011). Additionally, female patients showed higher values in the parameter pelvis torsion (P-tors: $p<0,001$), and male patients had a flatter lumbar lordosis (LA-max: $p<0,01$), respectively (fig. 4) (Schröder, Strübing & Mattes, 2010). These findings were in a line with earlier studies based on radiological methods or mathematical models, respectively (During et al., 1985; Harrison et al., 1998).

Those recent results provide the idea that spinal mal-alignment should be associated with low back pain. Spine shape abberations might be one organic risk factor for the development of low back pain, but - on the other hand - it might also be a symptom of deconditioning processes in chronic low back complaints, as is well known for deficits of muscle function (Cady et al., 1979; Denner, 1997; 1999; McNeill et al., 1980).

5.2 Reconditioning and spinal alignment

Referring to systematic associations between spinal mal-alignment or aberrations of 'normal' spine shape and back complaints in chronic low back pain patients described above, we conducted a longitudinal study to analyse adaptations of an individualized exercise program. The exercise program was determinded by individual spine shape parameter findings, muscle function findings, and anamnestic data related to individual back complaints - comparable to programs based only on functional profiles of trunk muscle performance, evaluated earlier (Denner, 1997). Patients were meant to face individually composed tasks to generate almost individual adaptations - with an idea of treatment specificity.

In the present study, clinical outcome variables and muscular function parameters increased like they did in comparable studies using intensive muscle activation (Denner, 1999; Mannion et al., 2001a; 2001b; 2001c; Uhlig, 1999). Low back pain patients started the exercise therapy with a pain state of 3,8 (±2,3) points, in terms of Borg's CR10 scale meaning a back pain level from moderate to strong. Pain decreased to 2,3 (±1,8) points, meaning a pain level from very weak to moderate. These decreases were accompanied by peak force increases ranging from about 20% to approximately 40% (fig. 5), assigning that kind of reconditioning process described elsewhere for low back patients who went through an active rehabilitation program (Denner, 1997; 1999; Mannion et al., 2001a; 2001b; 2001c; Schröder et al., 2009). Multivariate analysis procedures seeking for a direct correlation between pain decrease and muscle function increases could not reveal significant coefficients (R=0, 292). These findings were in a line with earlier investigations, where a correlation coefficient of $r=0,20$ ($p=0,60$) could be established, which was also not suitable to support an assumption of a direct dependency between clinical out-come and muscle function state (Mannion et al., 2001b). Psychological factors, like awareness of increased muscle function and re-established self-confidence, were assumed to be reasonable mediators between decreases of pain or increases of health state parameters and increased muscle function and performance parameters (Mannion et al., 2001c).

Additionally, systematic and significant spine shape alterations - apparent in lumbar erection and correction of pelvis asymmetries - could be verified (tab. 7), comparable to earlier investigations (Schröder et al., 2009). With respect to the knowledge of interindividual spine shape variability and intra-individual variations in repeated measurements of spinal alignment (Jackson et al., 2000) known as 'margin error' in pre-post-analyses (Weiß, Dieckman & Gerner, 2003; Weiß & Klein, 2006) these small changes of pelvis torsion (P-tors: -0,6° ; p=0,025) and lumbar lordosis angle (LA-max: -0,7°; p=0,012) were interpreted as relevant and statistically significant effects, following an active exercise program based on individual findings and using specific treatment elements, like a reasonably high training intensity (Dalichau et al., 2005; Denner, 1997; 1999; Uhlig, 1999) and the special coordination patterns for deep trunk muscles known as Segmental Stabilization Training (Richardson, Hodges & Hides, 2004).

Unfortunately, the evidence of specificity of those exercise induced adaptations was still lacking. On the one hand, adaptations of spine shape parameters in the frontal plane (trunk imbalance, pelvis tilt, pelvis torsion) were greater the more abnormal these values were before the treatment (r=0,40 to 0,72; p≤0,05), but on the other hand, pain reduction could not be explained sufficiently, neither by increases of muscle function (R=0,292) nor by corrections of spinal mal-alignment (R=0,256), nor by the total of all parameters, muscle function and spinal alignment (R=0, 399).

Since correlations between clinical out-come variables and functional adaptations of trunk muscle peak force had rarely been investigated, correlations between pain decrease and alterations in the spinal alignment - with a focus on the monitoring of low back pain intervention and using video raster stereography - had as yet not been investigated anywhere else, apart from our own pilot study, where decreasing values of trunk imbalance were associated with pain decrease in those patients who showed sacroiliac symptoms (Schröder et al., 2009). Dalichau et al. (2005) used an ultra sound topometry device (Zebris®, Isny, Germany) to detect a thoracic erection following three modes of muscle activation exercise programs. Spinal erection was accompanied by trunk muscle peak force increases, adaptations in the performance of the Matthiass-Test (at the end of a 30-second test period) and pain decreases. Dalichau et al. (2005) found high correlation coefficients, but not directly between spine shape and peak force or pain changes. They correlated the degree of deviation of the thoracic kyphosis angle at the end of the Matthiass-Test with back pain intensity (r=0,91) and functional deficits (r=0,89). So, the results of Mannion and collaborates (2001b; 2001c), mentioned above, might serve as the only reference remaining for directly calculated correlations in a longitudinal study between peak force increases and pain decreases (r=0,20; p=0,60), but not taking into account exercise induced spine shape alterations.

6. Additional applications of spine shape analysis

Although the majority of all low back pain cases are of unknown etiology, new diagnosis procedures, such as video raster stereography, might be able to find structural or functional correlates of some specific origin for back pain complaints (McGill, 2007, p. 5).

For example, video raster stereography (Formetric®-system) is able to detect local changes of the convexity of the spinal curvature [6]. A sensitivity study of n=21 volunteers suffering from

[6] Kyphosis or lordosis describes an angle referring to geometric relations of the human anatomy, but there are changes of convexity also in the microstructure of the alignment of the spinous processes. If

accidental vertebral blockades provided the idea of automatically detectable structural deviations in the alignment of spinous processes in terms of overreaching the midline in the curve of the second mathematical differentiation of the lateral projection of the spine (fig. 6).

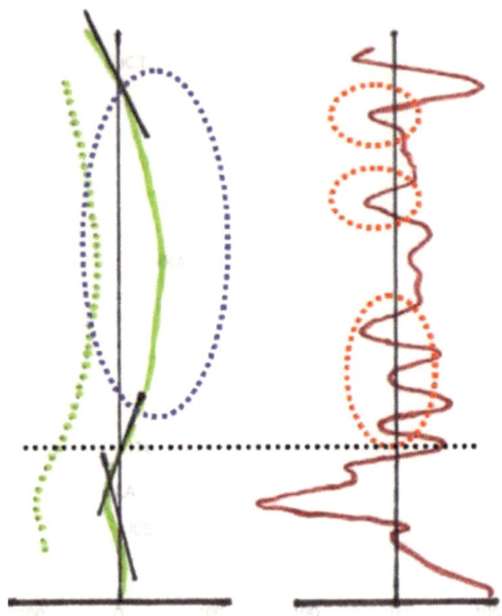

Fig. 6. Lateral projection of spinal alignment (left) with back surface (drawn green line) and calculated line of vertebral centres (dotted green line) with a focus on the thoracic spine (blue dotted oval) and the second mathematical differentiation (right) with the curve of local changes of angles at a given point (drawn red line) with an emphasis on curve areas reaching or overreaching the midline (red dotted ovals) indicating structural deviations in the normal spinal alignment of the thoracic spine (area above the black dotted line) (modified from Schröder, Stiller & Mattes, 2011, p. 165)

But video raster stereographic signals indicated signs for a vertebral blockade much more often than a manual examination by an expert did. Sensitivity of video raster stereography was almost poor (23%) (Schröder, Färber & Mattes, 2009; Schröder, Stiller & Mattes, 2011).

Furthermore, there was some evidence for the possibility to get helpful additional diagnostic information to identify sacroiliac joint (SIJ) pain origins in patients with single localized low back pain. Problems concerning the sacroiliac joints are supposed to be the cause for about 20% of all low back complaints, but diagnosis is difficult (Foley & Buschbacher, 2006). In a cross-sectional study, women with single localized low back pain corresponding to the area of sacroiliac joints (n=23) showed significantly higher values for trunk imbalance (mean-diff.: 4,9 mm; $p \leq 0{,}001$), for pelvis tilt (mean-diff.: 2,8 mm; $p = 0{,}007$)

---

the direction of the curvature at a given segmental position changes completely from a right-sided convexity to a left-sided convexity, the curve of the second mathematical differentiation of the lateral projection of the spinal alignment reaches or overreaches the midline (fig. 6). Those changes of local convexity assign structural deviations of the normal spinal alignment, such as scoliosis curvatures or vertebral blockings.

and for pelvis torsion (mean-diff.: 1,1°; p=0,014) than pain free women (n=89). This was indicating deviations in the frontal plane like in low back pain patients, but enhancing the role of exceeded pelvis parameters. Maybe due to the normal differences between shape and geometry of male and female pelvis anatomy, these sacroiliac signs could not be confirmed statistically for male patients with comparable single localized pain (Schröder, Stiller & Mattes, 2011).

In the field of specific low back complaints, we could identify signals in the spinal alignment of the lumbar lordosis that referred to structural abberations of specific vertebral segments in low back pain patients suffering from a facet joint syndrome (fig. 7) (Schröder, Strübing & Mattes, 2010).

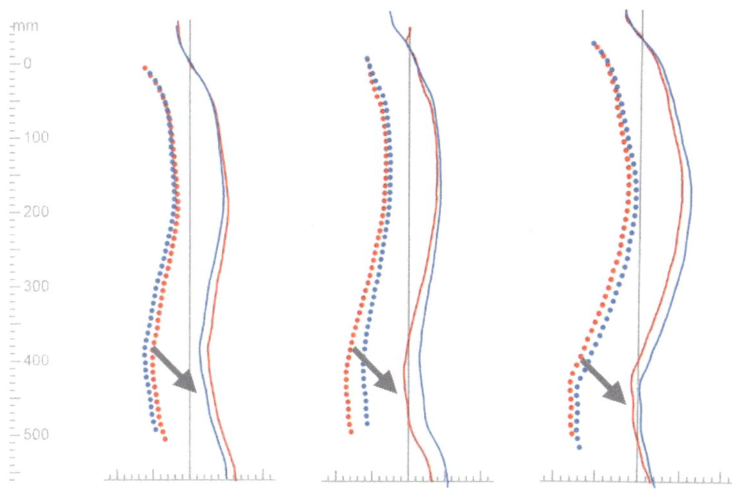

Fig. 7. Spinal alignment of three low back pain patients with different types of spine shape suffering from lumbar facet syndrome in repeated measurements (back surface [drawn] and calculated vertebral centres [dotted] before [red] and after [blue] treatment) with signals for structural changes of vertebral elements [arrows]

A functional diagnosis procedure to quantify leg length differences and to try out the best fitting correction had been evaluated earlier (Drerup et al., 2001). A functional test protocol for the quantification of spinal flexibility - especially for back extension limitations - by means of video raster stereography is currently performed (fig. 8), as the evidence of lumbar hypermobility or flexibility deficits is well known as a cause or a symptom of low back pain.

With regard to technical limitations of the high resolution Formetric®-system - anticipation of problems dealing with an automatic recognition of the vertebra prominens without manually fixed extra markers, while the upper body was hyperextended maximally and the camera was looking at it from above - the test protocol had to include three test positions. Data acquisition had been performed in a normal position, serving as a native reference to qualify the individual's spinal alignment. But pictures had also to be taken in a position with a forced hyper kyphosis as a basic reference for the following test position with the same artificial hyper kyphosis performed in a maximally extended spine position (fig. 8). Spinal flexibility for the backward hyperextension could be quantified in terms of changes of the lumbar lordosis angle, which was not affected by the artificial hyper kyphosis test position.

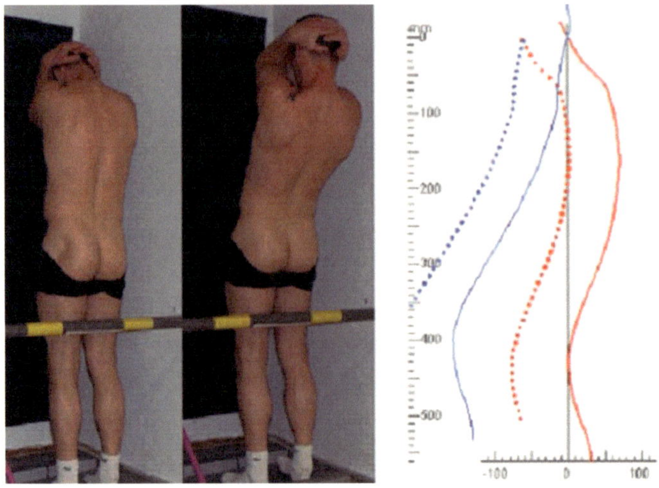

Fig. 8. Test position with artificial hyper kyphosis in a basic (left) and a maximally hyper-extended position (middle) and the video raster stereographic representation of spinal mobility (right) for the back extension task

7. Conclusion

A single cross-sectional study does not allow to draw any conclusions, whether spine shape alterations are the cause of low back pain or the symptoms following a process of deconditioning. But exercise induced adaptations of spinal alignment suggest the assumption that there is the possibility for a correction of mal-alignment. These alterations should be considered to be due to a functional restoration, comparable to increases of trunk muscle peak forces observed in the process of reconditioning.

Finally, the role of video raster stereography for quality management should be emphasized. The indirect and non-invasive assessment of the spinal curvature and pelvis position parameters offered valid, reliable and helpful information throughout the screening and monitoring processes for out-patient low back pain rehabilitation.

Further investigations, if possible with clustered samples of the degree of chronification or personal strategies of behavioral coping and - if possible - distinguished specific back pain complaints, are necessary to learn more about the role of spinal mal-alignment in patients with low back pain, and probably more about specific effects of different exercise treatment modes.

8. Acknowledgment

We would like to thank our colleagues and collaborates at the Department of Human Movement Science at the University of Hamburg, Germany, especially Prof. R. Reer and Prof. K.-M. Braumann from the Institute of Sports Medicine, as well as our project partners of the industry (Diers International, Schlangenbad, Germany) and of the out-patient rehabilitation (Buchholz & Partners, Hamburg, Germany). Furthermore, we would like to encourage all colleagues in the field of low back pain treatment and research to reproduce and enlarge our results to support validity and evidence of our findings.

9. References

Adams, M.; Mannion, A. & Dolan, P. (1999). Personal risk factors for first time low back pain. *Spine,* Vol.24, No.23, (December 1999), pp. 2497-2505, ISSN 0362-2436

Balagué, F.; Troussier, B. & Salminen, J. (1999). Non-specific low back pain in children and adolescents: risk factors. *European Spine Journal,* Vol.8, No.6, (June 1999), pp. 429-438 , ISSN 0940-6719

Bernard, M. (2002). The influence of muscular condition and physical training on adult posture : a systematic literatur review (germ.). *Zeitschrift für Physiotherapeuten,* Vol.54, No.7, (July 2002), pp. 1070-1087, ISSN 1614-0397

Borg, G. (1998). *Borg's perceived exertion and pain scales.*, Human Kinetics, ISBN 0-88011-623-4, Champaign, Illinois, USA

Cady, L.; Bischoff, D.; O'Connel, E.; Thomas, P. & Allan, J. (1979). Strength fitness and subsequent back injuries in firefighters. *Journal of Occupational Medicine,* Vol.21, No.4, (April 1979), pp. 269-272, ISSN 1076-2752

Dalichau, S.; Stein, B.; Schäfer, K.; Buhlmann, J. & Menken, P. (2005). Quantification of spinal configuration and postural capacity by ultrasound topometry for evaluation of different muscle strengthening programs in the therapy of back pain (germ.). *Zeitschrift für Orthopädie,* Vol.143, No. , pp. 79-85, ISSN 0044-3220

Denner, A. (1997). [Die wirbelsäulenstabilisierende Muskulatur chronischer Rückenpatienten. Dekonditionierung versus Rekonditionierung] (germ.). *Manuelle Medizin,* Vol.35, No.2 (May 1997), pp. 94-102, ISSN 0025-2514

Denner, A. (1999). [Die Trainierbarkeit der Rump-, Nacken- und Halsmuskulatur von dekonditionierten Rückenschmerzpatienten] (germ.). *Manuelle Medizin,* Vol.37, No.1, (January 2011), pp. 34-39, ISSN 0025-2514

Drerup, B.; Ellger, B.; Meyer zu Bentrup, F. & Hierholzer, E. (2001). Functional examinations with rasterstereography. A new method for the biomechanical analysis of skeletal geometry (germ.). *Der Orthopäde,* Vol.30, No.4, (April 2001), pp. 242-250, ISSN 0085-4530

Drerup, B. & Hierholzer, E. (1985). Objective determination of anatomical landmarks on the body surface: Measurement of the vertebra prominens from surface curvature. *Journal of Biomechanics,* Vol.18, No.6, (June 1985), pp. 467-474, ISSN 0021-9290

Drerup, B. & Hierholzer, E. (1987a). Automatic localization of anatomical landmarks on the back surface and construction of a body-fixed coordinate system. *Journal of Biomechanics,* Vol.20, No.10, (October 1987), pp. 961-970, ISSN 0021-9290

Drerup, B. & Hierholzer, E. (1987b). Movement of the human pelvis and displacement of related anatomical landmarks on the body surface. *Journal of Biomechanics,* Vol.20, No.10, (October 1987), pp. 971-977, ISSN 0021-9290

Drerup, B. & Hierholzer, E. (1994). Back shape measurement using video rasterstereography and three-dimensional reconstruction of spinal shape. *Clinical Biomechanics,* Vol.9, No.1, (January 1994), pp. 28-36, ISSN 0268-0033

During, J.; Goudfrooij, H.; Keesen, W.; Beeker T. & Crowe, A. (1985). Towards standards for posture: postural characteristics of the lower back system in normal and pathologic conditions. *Spine,* Vol.10, No.1, (January 1985), pp. 83-87, ISSN 0362-2436

Foley, B. & Buschbacher, R. (2006). Sacroiliac joint pain: anatomy, biomechanics, diagnosis, and treatment. *American Journal of Physical Medicine and Rehabilitation,* Vol.85, No.12, (December 2006), pp. 997-1006, ISSN 0894-9115

Gelb, D.; Lenke, L.; Bridwell, K.; Blanke, K. & McEnery, K. (1995). An analysis of sagittal spinal alignment in 100 asymptomatic middle and older aged volunteers. *Spine,* Vol.20, No.12, (June 1995), pp. 1351-1358, ISSN 0362-2436

Harrison, D.D.; Cailliet, R.; Janik, T.; Troyanovich, S.; Harrison, D.E. & Holland C. (1998). Elliptical modeling of the sagittale lumbar lordosis and segmental rotation angles as a method to discriminate between normal and low back pain subjects. *Journal of Spinal Disorders,* Vol.11, No.5, (May 1998), pp. 430-439, ISSN 0895-0385

Heckmann, T.; Tschan, H.; Kinzlbauer, M.; Guschelbauer, R. & Bachl, N. (2008). [Analyse der Körperhaltung bei Jugendlichen mit Hilfe der Videorasterstereographie unter Berücksichtigung der Prävalenz des Rückenschmerzes und der körperlichen Aktivität] (germ.). *Österreichisches Journal für Sportmedizin,* Vol.38, No.1, (January 2008), pp. 25-36, ISSN 1012-3156

Hildebrandt, J.; Pfingsten, M.: Saur, P. & Jansen, J. (1997). Prediction of success from a multidisciplinary treatment program for chronic low back pain. *Spine,* Vol.22, No.9, (September 1997), pp. 990-1001, ISSN 0362-2436

Jackson, R.; Kanemura, T.; Kawakami, N. & Hales, C. (2000). Lumbopelvic lordosis and pelvic balance on repeated standing lateral radiographs of adult volunteers and untreated patients with constant low back pain. *Spine,* Vol.25, No.5, (May 2000), pp. 575-586, ISSN 0362-2436

Käser, L.; Mannion, A.; Rhyner, A.; Weber, E.; Dvorak, J. & Müntener, M. (2001). Active therapy for chronic low back pain: part 2. Effects on paraspinal muscle cross-sectional area, fibre type size, and distribution. *Spine,* Vol.26, No.8, (April 2001), pp. 909-919, ISSN 0362-2436

Kobayashi, T.; Atsuta, Y.; Matsuno, T. & Takeda, N. (2004). A longitudinal study of congruent sagittal spinal alignment in an adult cohort. *Spine,* Vol.29, No.6, (March 2004), pp. 671-676, ISSN 0362-2436

Kuo, Y.; Tully, E. & Galea, M. (2009). Sagittal spinal posture after Pilates-based exercise in healthy older adults. *Spine,* Vol.34, No.10, (May 2009), pp. 1046-1051, ISSN 0362-2436

Lewit, K. (1991). *Manipulative Therapy in Rehabilitation of the Locomotor System. 2nd Ed.,* Butterworths, ISBN 0-750-51123-5, London & Oxford, United Kingdom

Ljunggren, A.; Weber, H.; Kogstad, O.; Thom, E. & Kirkesola, G. (1997). Effect of exercise on sick leave due to low back pain - A randomized, comparative, long-term study. *Spine,* Vol.22, No.14, (July 1997), pp. 1610-1617, ISSN 0362-2436

Mannion, A; Taimela, S.; Müntener, M. & Dvorak, J. (2001a). Active therapy for chronic low back pain: part 1. Effects on back muscle activation, fatigability, and strength. *Spine,* Vol.26, No.8, (April 2001), pp. 897-908, ISSN 0362-2436

Mannion, A; Junge, A.; Taimela, S.; Müntener, M.; Lorenzo, K. & Dvorak, J. (2001b). Active therapy for chronic low back pain: part 3. Factors influencing self-rated disability and its change following therapy. *Spine,* Vol.26, No.8, (April 2001), pp. 920-929, ISSN 0362-2436

Mannion, A.; Dvorak, J.; Taimela, S. & Müntener, M. (2001c). Increase in strength after active therapy in chronic low back pain (CLBP) patients: muscular adaptations and clinical relevance (germ.). *Der Schmerz,* Vol.15, No.6, (December 2001), pp. 468-473, ISSN 0932-433X

Masset, D.; Piette, A. & Malchaire, J. (1998). Relation between functional characteristics of the trunk and the occurrence of low back pain. Associated risk factors. *Spine*, Vol.23, No.3, (February 1998),pp. 359-365, ISSN 0362-2436

McGill, S. (2007). *Low back disorders: evidence based prevention and rehabiltation. 2nd Ed.*, Human Kinetics, ISBN 0-7360-6692-6, Champaign, Illinois, USA

McNeill, T.; Warwick, D.; Andersson, G. & Schultz, A. (1980). Trunk strengths in attempted flexion, extension, and lateral bending in healthy subjects and patients with low-back disorders. *Spine*, Vol.5, No.6, (November December 1980), pp. 529-538, ISSN 0362-2436

Mohukum, M.; Wolf, U.; Mendoza, S.; Paletta, J.; Sitter, H. & Skwara, A. (2009). Reproducibility of rasterstereography for kyphotic and lordotic angles and for trunk length and trunk inclination - A reliability study. *European Spine Journal*, Vol.18, No.11, (November 2009), p. 1746, ISSN 0940-6719

Müller, G. (1999). [Zur Evaluation von Funktionsstörungen an der Wirbelsäule. Strukturdiagnostik versus Funktionsdiagnostik] (germ.). *Manuelle Medizin*, Vol.37, No.1, (January 1999), pp. 18-25, ISSN 0025-2514

Nissinen, M.; Heliövaara, M.; Seitsamo, J.; Alaranta, H. & Poussa, M. (1994). Anthropometric Measurements and the incidence of low back pain in a cohort of pubertal children. *Spine*, Vol.19, No.12, (June 1994), pp. 1367-1370, ISSN 0362-2436

Nourbakhsh, M.; Arabloo, A. & Salavati, M. (2006). The relationship between pelvic cross syndrome and chronic low back pain. *Journal of Back and Musculoskeletal Rehabilitation*, Vol.19, (2006) pp. 119-128, ISSN 1053-8127

O'Sullivan, P.; Twomey, L. & Allison, G. (1997). Evaluation of specific stabilization exercise in the treatment of chronic low-back-pain with radiologic diagnosis of spondylolysis or spondylolisthesis. *Spine*, Vol.22, No.24, (December 1997), pp. 2959-2967, ISSN 0362-2436

Panjabi, M. (1992). The stabilizing system of the spine. Part I. Function, dysfunction, adaptation, and enhancement. *Journal of Spinal Disorders*, Vol.5, No.4, (August 1992), pp. 383-389, ISSN 0895-0385

Richardson, C.; Hodges, P. & Hides, J. (2004). *Therapeutic exercise for lumbopelvic stabilization. 2nd Ed.*, Churchill Livingstone, ISBN 0-443-07293-0, Edinburgh, United Kingdom

Schröder, J. & Färber, I. (2010). *[Segmentales Stabilisierungstraining als Baustein einer evidenzbasierten Bewegungstherapie bei Rückenbeschwerden]* (germ.), Feldhaus Edition Czwalina, ISBN 978-3-88020-558-1, Hamburg, Germany

Schröder, J.; Färber, I. & Mattes, K. (2009). First results for the comparison of manual techniques and a biomechanical device (Formetric-System) in the identification of vertebral blockades, *Proceedings of 14th Congress ECSS*, p. 411, ISBN 978-82-502-0420-1, Oslo, Norway, June 24-27, 2009

Schröder, J.; Färber, I.; Meyer-Hofmann, K.; Schaar, H. & Mattes, K (2009). [Kraft- und Haltungsdiagnostik als Basis einer individualisierten Bewegungstherapie bei Rückenschmerzpatienten] (germ.). *Zeitschrift für Physiotherapeuten*, Vol.61, No.5, (Mai 2009), pp. 420-432, ISSN 1614-0397

Schröder, J.; Förster, J. & Mattes, K. (2008). A pragmatic variant of segmental stabilisation training (SST) for sports practice on the basis of conspicuous findings concerning the spinal form of volleyballers (germ.). *Leistungssport*, Vol.38, No.4, (July 2008), pp. 45-51, ISSN 0341-7387

Schröder, J. & Mattes, K. (2010). Spine shape changes following individualized exercise programs in back pain patients over 60 years of age, *Proceedings of 15th Congress ECSS*, p. 207, ISBN 978-605-61427-0-3, Antalya, Turkey, June 23-26, 2010

Schröder, J. & Mattes, K. (2009). Posture analysis: variations and reliability of biomechanical parameters in bipedal standing by means of Formetric-System, *Proceedings of 14th Congress ECSS*, p. 618, ISBN 978-82-502-0420-1, Oslo, Norway, June 24-27, 2009

Schröder, J.; Reer, R. & Mattes, K. (2009). Biomechanical diagnosis procedures in the clinical routine: reliability of trunk muscle testing and posture analysis in the treatment of back pain patients (germ). *Orthopädische Praxis*, Vol.45, No.6, (June 2009), pp. 288-294, ISSN 0030-588X

Schröder, J.; Stiller, T. & Mattes, K. (2010). Spine shape parameters as indicators for low back pain disorders (germ.). *Deutsche Zeitschrift für Sportmedizin*, Vol.61, No.4, (April 2010), pp. 91-96, ISSN 0344-5925

Schröder, J.; Stiller, T. & Mattes, K. (2011). Reference data for spine shape analysis (germ.). *Manuelle Medizin*, Vol.49, No.3, (July 2011), pp. 161-166, ISSN 0025-2514

Schröder, J.; Strübing, K. & Mattes K. (2010). Back complaints and spinal form (germ.). *Manuelle Medizin*, Vol.48, No.6, (December 2010), pp. 454-459, ISSN 0025-2514

Seeger, D.; Koch, D.; Heinemann, R; Saur, P. & Hildebrandt, J. (1997). Physiotherapeutic examination in view of the ambulatory rehabilitation of patients with chronic low back pain. Part 1: Diagnosis (germ.). *Krankengymnastik/ Zeitschrift für Physiotherapeuten*, Vol.49, No.1, (January 1997), pp. 7-34, ISSN 1614-0397

Takeda, N.; Kobayashi, T.; Atsuta, Y.; Matsuno, T.; Shirado, O. & Minami, A. (2009). Changes in the sagittal spinal alignment of the elderly without vertebral fractures: a minimum 10-year longitudinal study. *Journal of Orthopaedic Science*, Vol.14, No.6, (November 2009), pp. 748-753, ISSN 0949-2658

Uhlig, H. (1999). [Die Rekonditionierbarkeit chronischer Rückenpatienten mit muskulärer Insuffizienz] (germ.). *Manuelle Medizin*, Vol.37, No.1, (January 2011), pp. 40-45, ISSN 0025-2514

Waddell, G.; McCulloch, J.; Kummel, E. & Venner, R. (1980). Nonorganic physical signs in low-back pain. *Spine*, Vol.5, No.2, (March/April 1980), pp. 117-125, ISSN 0362-2436

Weiß, H.; Dieckmann, J. & Gerner, H. (2003). The practical use of surface topography: following up patients with Scheuermann's disease. *Pediatric Rehabilitation*, Vol.6, No.1, (January 2003), pp. 39-45, ISSN 1363-8491

Weiß, H. & Klein, R. (2006). Improving excellence in scoliosis rehabilitation. *Pediatric Rehabilitation*, Vol.9, No.3, (July 2006), pp. 190-200, ISSN 1363-8491

2

Osteophyte Formation in the Lumber Spine and Relevance to Low Back Pain

Yoshihito Sakai
Department of Orthopaedic Surgery
National Center for Geriatrics and Gerontology
Japan

1. Introduction

Vertebral osteophyte formation is a well-documented phenomenon that is associated with degeneration and altered mechanics of the spine, both of which have been considered to be the result of aging, a purely physiologic response to load bearing, or intrinsic spinal disease as etiologic factors. (Lane et al., 1993, O'Neill et al., 1999) They are recognized radiologically as hyperostosis at the region of the attachment of the annular fibers to the vertebral body and localized increases in bone mineral density. (Nathan et al., 1994) As the etiologic factors, the compressive forces on the vertebral endplates (Nathan et al., 1962), bone mineral density (Kinoshita et al., 1998), obesity (O'Neill et al., 1999) and genetic factors (Sambrook et al., 1999) have been reported as causes, although the absence of a single definitive factor causing spinal degeneration has led to a suggestion that several factors including both genetic and nongenetic ones contribute to the development of osteophyte formation. (Harada et al., 1998, Liu et al., 1997)

On the other hand, low back pain (LBP) is one of the most common musculoskeletal disorders of the elderly, of which risk factor seems to be related to lumbar disc degeneration. (van Tulder et al., 1997) While data from many studies suggest an association with lumbar disc degeneration and LBP (Lawrence, 1969, Simmons et al., 1991, Jayson, 1994), asymptomatic lumbar disc degeneration is common (Powell et al., 1986, Borenstein et al., 2001), and the correlation between LBP and disc degeneration observed in radiographs is only moderate or poor (Witt et al., 1984). Osteophyte formation in the lumbar spine is a characteristic feature of intervertebral disc degeneration, however, the relationship between osteophytes and LBP is less clear (van Tulder et al., 1997). Symmons et al. reported that osteophytes were no more common among women with recurrent back pain compared to those without (Simmons et al., 1991), while O'Neill et al. concluded that osteophytes affecting the lumbar spine are associated with LBP in men (O'Neill et al., 1999). Meanwhile, we often encounter prominent osteophytes in the absence of intervertebral disc degeneration as supporting a lack of association between the two factors reported by Oishi *et al* (Oishi et al, 2003). They concluded intervertebral disc degeneration and osteophyte formation of the vertebral bodies seemed to represent different factors affecting the lumbar spine. Thus, there are some doubts as to the relationship between osteophyte formation and disc degeneration. We investigated the factors influencing osteophyte formation of the lumbar spine without disc degeneration and estimate the implications of osteophytes from the viewpoint of LBP and gene polymorphism.

2. Methods

2.1 Study subjects

The subjects consisted of Japanese volunteers who attended "a basic health checkup" supported by a local government. A total of 387 elderly persons from 60 to 81 years (average 68.0±6.4 years, 153 males, 234 females), most of whom were engaged in farming and fishing, were invited to participate in the study with a written informed consent form as well as a sheet describing the study outline. Patients with rheumatoid arthrosis, vertebral fracture, or disorders known to affect bone metabolism, including diabetes mellitus and other endocrinologic diseases were excluded from this study.

2.2 Evaluation of chronic LBP

Two spine surgeons performed a brief interview including visual analogue scale (VAS) for low back pain (0-100) and physical examination regarding LBP after taking blood samples and radiograms of the lumbar spine. LBP was defined as more than 20 in VAS and lasting for recent 3 months. Body weight, height, body mass index (BMI), body fat ratio, bone stiffness (QUS), back muscle strength, smoking status and alcohol intake were evaluated. Bone stiffness was measured with a quantitative ultrasound (QUS) densitometry device (A-1000PlusII; LUNAR, WI, USA) to calculate the Stiffness Index on the calcaneus recognized by the American Food and Drug Administration (FDA) and which has become the world standard. Back muscle strength was determined as the maximal isometric strength of the trunk muscles in standing posture with 30° lumbar flexion using a digital back muscle strength meter (T.K.K.5402, TAKEI Co., Japan).

2.3 Radiographic evaluation

The participants were instructed to stand on both feet shoulder-width apart while maintaining a level gaze. Film-focus distance was unified at 150 cm, and a film was correctly put along a gravity plum line. According to Miyakoshi (Miyakoshi et al., 2003), the degree of disc height narrowing was scored as 0 (0-20% reduction in disc height, as compared with the L1/2 disc), 1 (20-50% reduction), or 2 (more than a 50% reduction), and the total score from the L2/3 to the L5/S1 disc was defined as the disc score. The disc score of 0 was defined as "no disc degeneration". Osteophyte formation was assessed according to Nathan's classification (0-4) (Nathan et al., 1962), and a total number from L1/2 to L5/S1 (Osteophyte score) of more than 6 was defined as osteophyte (+).

2.4 Classification of osteophyte formation

The cases with osteophyte formation were classified by the presence of disc height narrowing into two groups: Group A; osteophyte (+) with disc height narrowing, Group B; osteophyte (+) without disc height narrowing. Group C was defined as the cases without osteophyte formation.

2.5 Selected polymorphisms

The gene polymorphism examinations were conducted in accordance with "Ethical Guidelines for Human Genome and Gene Research" (approved: March 29, 2001, implemented: April 1, 2001), with adequate explanation provided to the subjects. The genotypes of the alcohol sensitivity related polymorphisms (alcohol dehydrogenase 2 (ADH2 Arg47His), aldehyde dehydrogenase 2 (ALDH2 Glu487Lys)), tobacco sensitivity related polymorphisms (NADH quinone oxidoreductase 1 (NQO1 C609T), glutathione S transferase M1 (GSTM1), glutathione S transferase T1 (GSTT1)), inflammation related polymorphisms (interleukin 1β (IL-1B), tumor necrosis factor α (TNF-A)), longevity-associated polymorphism of mitochondrial DNA (mt5179), allergy-associated polymorphism of interleukin-4 (IL-4), immunity-associated polymorphism of CD14, vitamin D receptor (VDR) and transforming growth factor β (TGFB1) were characterized by a polymerase chain reaction with the confronting two-pair primers (PCR-CTPP) method (Hamajima et al., 2000). This is a new genotyping method invented independently, recently found to be based on the same logic as bi-directional PCR amplification of specific alleles. Twenty-eight subjects were excluded because of inadequate blood samples. A total of 197 subjects in group A, 93 in group B and 65 in group C were recruited for the polymorphism study. To characterize the features of osteophyte formation without disc degeneration, group B was compared with combined groups A and C (n=262).

3. Results

Disc height narrowing and presence of osteophytes were observed in 245 cases (63.3%) and 316 cases (81.6%), respectively. Vacuum phenomenon and degenerative spondylolithesis were seen in 107 cases (27.6%) and 73 cases (18.9%), respectively. Osteophye formation and vacuum phenomenon were significantly seen in elderly person. ($p<0.01$) (Fig.1) Whereas disc height narrowing was almost seen in L4/5 and L5/s1, osteophyte formation was extensively presented from L1/2 to L5/s1.

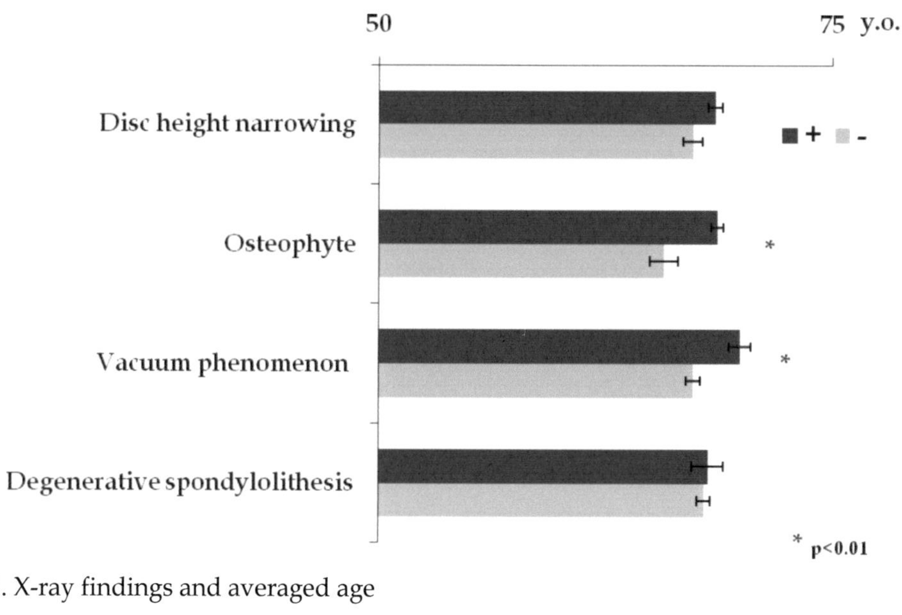

Fig. 1. X-ray findings and averaged age

3.1 Disc score and osteophyte score

Averaged disc score and osteophyte score were 1.4±1.5 and 8.3±3.1, respectively. (Fig.2) There were no significant correlation between Disc score and Osteophyte score with Spearman's rank test. (p=0.084, r=0.293) Multiple regression analysis revealed that Disc score was associated with age, and Osteophyte score was associated with age, gender (male > female). (Table 1) Subjects who present vacuum phenomenon had significantly higher Disc score and Osteophyte score. (p<0.01) (Fig. 3)

According to the classification of osteophyte formation (2.4), Group A (osteophyte (+) with disc height narrowing) and Group B (osteophyte (+) without disc height narrowing) were seen in 217 and 99 cases, respectively. Group C (without osteophyte formation) was seen in 71 cases. Reduction of disc height was significantly associated with the presence of osteophyte. (p<0.01) (Table 2)

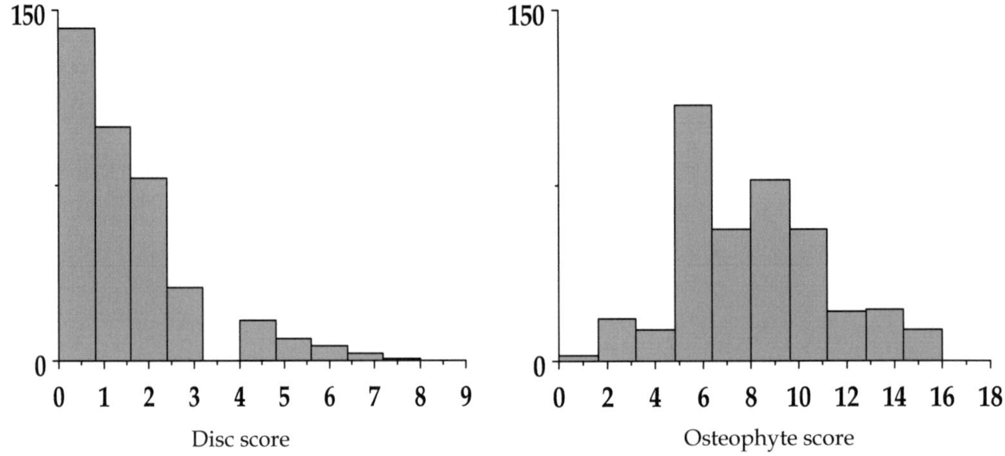

Fig. 2. Frequency distribution of disc score and osteophyte score

	Disc score		Osteophyte score	
	r	*p*	*r*	*p*
Age	0.150	0.006	0.122	0.021
Gender	0.061	0.568	0.400	<0.001
Height	0.181	0.664	0.462	0.223
Weight	0.312	0.619	0.593	0.283
BMI	0.271	0.592	0.224	0.617
Body fat ratio	0.015	0.891	0.142	0.156
Brinkman index	0.106	0.089	0.061	0.276
Alcohol consumptio	0.044	0.461	0.002	0.970
Bone stiffness	0.046	0.392	0.046	0.340

Data are correlation coefficients (*r*) by Spearman's rank test and statistical significance (*p* value)

Table 1. Correlations between the parameters of physical factors and Disc score, Osteophyte score

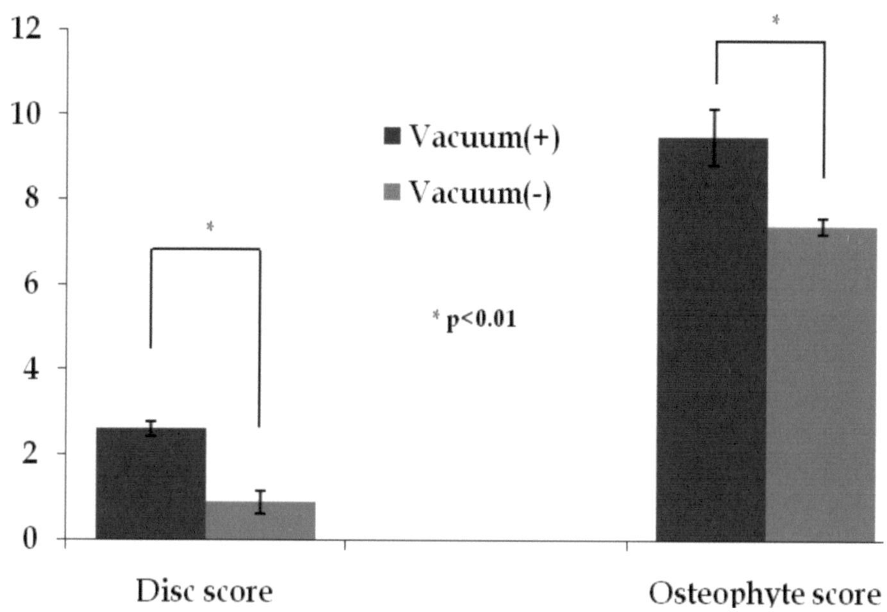

Fig. 3. Disc score and Osteophyte score according to the presence of vacuum phenomenon.

	Reduction of Disc height (+)	Reduction of Disc height (-)	Total
Osteophte (+)	217 case * (Group A)	99 cases (Group B)	316 cases
Osteophte (-)	28 cases	43 cases	71 cases (Group C)
Total	245 cases	142 cases	387 cases

* $p<0.01$

Table 2. Presence of disc height loss and osteophyte formation

3.2 Osteophyte and gene polymorphism (Sakai et al., 2007)

There were no significant differences in the disc score, osteophyte score or the ratio of group B in all polymorphisms, though Arg/Arg polymorphism in ADH2 tended to be less frequent. ($p=0.051$) (Table 3) Results of the logistic regression model to select gene polymorphism factors associated with the presence of osteophyte formation without disc height narrowing were shown in Table 4. In the polymorphism of alcohol dehydrogenase

	Genotypes			*P* value
	His/His (n=212)	Arg/His (n=124)	Arg/Arg (n=19)	
Disc score	1.33 ± 1.56	1.42 ± 1.47	1.30 ± 1.83	0.812
Osteophyte score	8.12 ± 3.12	7.91 ± 3.05	6.75 ± 2.90	0.160
No. of Group A+C/B	147/65	98/26	17/2	0.051

Group A=the cases with osteophyte formation with disc height narrowing
Group B=the cases with osteophyte formation without disc height narrowing
Group C=the cases with no osteophyte formation

Table 3. Disc score and Osteophyte score for ADH2 polymorphysm

(ADH2; Arg47His), the prevalence of osteophyte formation without disc height narrowing (group B) was less in His/Arg (OR=0.57, 95%CI=0.33-0.97,p=0.041) and Arg/Arg (OR=0.41, 95%CI=0.1-1.5,p=0.18) than His/His. In the other polymorphisms, there were no significant differences in osteophyte formation without disc height narrowing.

	GroupA+C (n=262)	Group B (n=93)	Genotypes (Arg/Arg, Arg/His, His/His)		
	Alelle Frequency (Arg, His)		Arg/Arg	Arg/His versus His/His	Arg/Arg + Arg/His
ADH2					
Arg	0.25	0.16			
His	0.75	0.84			
P value			0.183	0.035	0.027
Odds ratio (95% CI)			0.41 (0.10-1.50)	0.57 (0.33-0.97)	0.55 (0.32-0.93)

Allele frequencies were estimated by the gene counting method. *P* values were adjusted for age, gender, BMI, body fat ratio, bone stiffness, smoking habit and alcohol consumption.

Table 4. Multivariate logoistic regression analysis of the effect of ADH2polymorphism on the prevalence of osteophyte without disc height narrowing

3.3 LBP and osteophyte

The prevalence of LBP was 40.4% (156 cases) with average VAS scale of 34.9±28.5 (10-100). Back muscle strength was significantly lower in the LBP group than in the non-LBP group. ($p<0.05$) Disc score was significantly higher in the LBP group than in the non-LBP group ($p<0.01$), whereas there was no significant difference in the osteophyte score between the two groups. (Table 5) Characteristics of the groups A, B, and C were shown in Table 6. In group C, male subjects, Brinkman index and drinkers were significantly fewer than in group

	LBP group (n=156)	non-LBP group (n=224)
Age (years)	68.7 ± 6.8	67.7 ± 5.9
Gender (male/female)	59/97	93/131
Height (cm)	154.0 ± 8.3	155.5 ± 8.0
Body weight (kg)	58.0 ±9.0	58.6 ± 9.5
BMI (kg/m^2)	24.3 ± 3.0	24.2 ±3.0
Body fat ratio (%)	28.2 ±7.2	27.4 ±6.9
Brinkman index	277.3 ± 475.4	226.6 ±425.3
Drinker (no. (%))	41(26.2)	53 (23.7)
VAS	34.9 ± 28.5*	3.5 ± 8.0*
Back muscle strength (kg)	63.6 ± 24.2**	72.6 ± 28.4**
Bone stiffness	97.0 ± 18.6	96.7 ± 21.0
Disc score	1.72 ± 1.71*	1.09 ± 1.33*
Osteophyte score	8.28 ± 3.10	7.71 ± 3.11

BMI = body mass index, VAS = visual analogue scale, * $p<0.01$, **$p<0.05$

Table 5. Characteristics of the LBP group and the non-LBP group

A and the group B. Although vacuum phenomenon was more frequent in the group A ($p<0.01$), the presence of vertebral fracture and degenerative spondylolisthesis were equivalent to the group B and C. Disc score was significantly higher in the group A than in the group C. ($p<0.01$) Osteophyet score was significantly higher in the group A than in the group B. ($p<0.05$) (Fig.4) Both VAS scale and the prevalence of LBP were significantly greater in group A than group B and group C. In group B, VAS scale and numbers of LBP were equivalent to those in group B, but significantly less than those in group A. (Fig. 5)

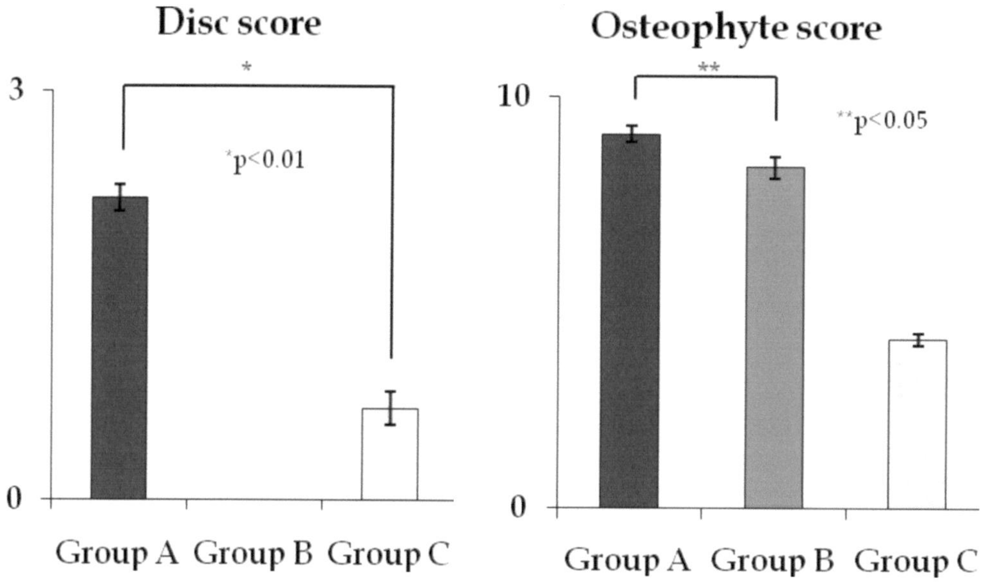

Fig. 4. Disc score and Osteophyte score in the group A, B and C.

	Group A (n=217)	Group B (n=99)	Group C (n=71)
Age (years)	68.9± 6.2	67.8± 6.2	65.6±6.6
Gender (male/female)	98/119	48/51	7/64 *, **
Height (cm)	155.0 ± 8.2	156.5± 8.8	151.9±6.1
Body weight (kg)	58.5 ± 9.4	60.2 ±9.9	55.2±7.2
BMI (kg/m^2)	24.3 ± 3.1	24.5 ± 3.0	23.9±2.8
Body fat ratio (%)	27.0 ± 7.4	27.3 ± 6.2	30.7±5.8
Brinkman index	232.2 ± 427.1	384.1 ± 532.8	95.6±276.0 **,***
Drinker (no. (%))	52 (30.0)	35 (35.3)	8 (11.3) *,**
Back muscle strength (kg)	70.5 ± 28.6	71.6 ± 28.5	60.5 ± 17.3
Bone stiffness	98.5 ± 19.6	95.8 ± 20.5	93.4 ± 20.1
Vertebral fracture (%)	3.26	3.06	2.85
Vacuum phenomenon (%)	45.53 **	7.37	4.35
Degenerative spondylolisthesis	21.96	18.37	28.17

Group A=the cases with osteophyte formation with disc height narrowing
Group B=the cases with osteophyte formation without disc height narrowing,
Group C=the cases with no osteophyte formation
* $p<0.01$ vs Group A; ** $p<0.01$ vs Group B; *** $p<0.05$ vs Group A

Table 6. Characteristics of the groups A, B, and C

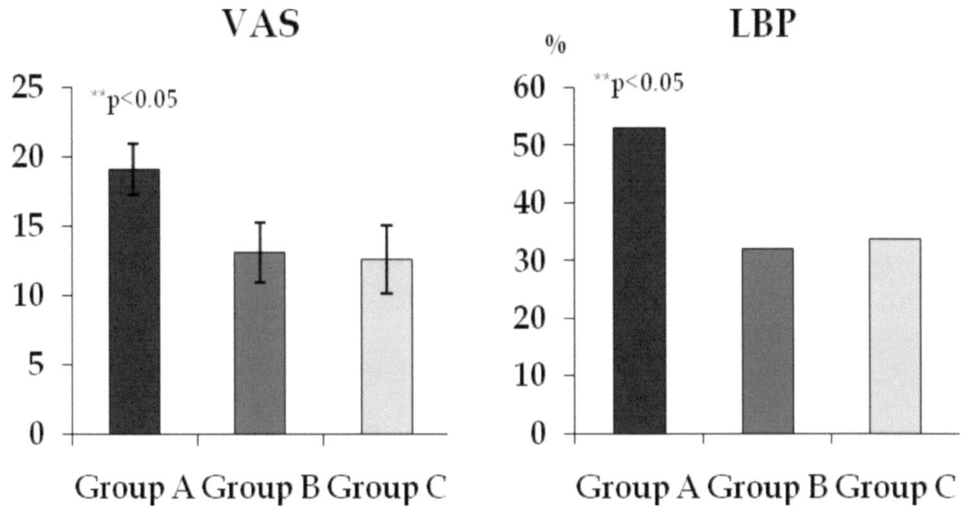

Fig. 5. Visual analogue scale and low back pain in the group A, B and C.

4. Discussion

It is commonly recognized that the degenerative changes that occur in the intervertebral discs are the point of departure of osteophyte formation. During the degeneration process, the discs undergo progressive structural changes in the form of dehydration of the nucleus and disintegration of the annulus fibrosus resulting in decreased disc height (Buckwalter et al., 1995), and lead to an increase in the compression stiffness and reduction in disc fiber strain (Kim et al., 1991). Biomechanically, the nucleus has lost some of its proteoglycan and water contents and increased its collagen content (Andersson, 1998). With progressive matrix alterations of the nucleus, changes in disc morphology such as a reduction in disc height become visible in plane radiographs. Degenerative changes within may result in an alteration of its mechanical properties, increased flexibility and decreased disc height, which in turn contribute to changes in the local stress within the disc (An et al., 2004). There is a general agreement that changes induced by aging lead to alternations in the thickness of the disc, but some differences are seen in the account of the effect of aging on the thickness of the lumbar disc. Vernon-Roberts et al. stressed that reduction of the disc height with age is inevitable (Vernon-Roberts et al., 1977), however, an increase in disc height with age has been reported. (Twomery et al., 1987, Amonoo-Kuofi, 1991, Roberts et al., 1997) Shao et al. demonstrated that the vertebral endplates became more concave with age, resulting the lumbar disc height increase (Shao et al., 2002). The effect of aging on the disc height has not been well understood, and the term of disc degeneration is imprecisely defined.

On the other hands, osteophytes form as a specific tissue reaction to these stresses and strains (Bick, 1995), and are attributed to higher stress more frequently anteriorly than posteriorly. Schmorl et al. formulated a pathogenic hypothesis that as a result of tears in the attachment of the annulus fibrosus into the margibnal ring of the vertebral body, the nucleus protrudes forward against the anterior longitudinal ligament. The increased strain causes the formation of spurs in the area of its attachment to the periosteum covering the cortex of the vertebral bodies (Schmorl et al., 1932). (Schmorl's rim lesion theory) Colins formulated a theory of osteophyte formation that associates degeneration of the entire intervertebral disc (collapsed disc) and the resultant anterior protrusion with subsequent osteophyte formation. The protrusion of the disc lifts the periosteum lateral to anterior longitudinal ligament and stimulates new subperiosteal bone (Collons, 1949). (Collins' bulging disc theory) According to Macnab's theory, osteophytes form as a result of instability between adjacent vertebral bodies (Macnab, 1971)(Macnab's instability theory). Traction spur, which projects horizontally and never curve toward the disc, differentiated from claw osteophytes (Nathan et al., 1962). Nathan concluded that osteophytes form as a natural physiologic response to compressive loads, serving to stabilize the spine (Nathan et al., 1962). In any case, there is wide agreement about the close association of disc degeneration with osteophyte formation and precedence in disc degeneration over vertebral deformities (Nathan et al., 1962, Vernon-Roberts et al., 1977, Lipson et al, 1980, Milgram, 1982). Our study date results provide further evidence substantiating that osteophyte formation and disc height narrowing are not always closely correlated, as identified by the prevalence of osteophytes without disc height narrowing in about 30% and the lack of correlation between disc height reduction and osteophytes. This finding stresses that these two features of spinal "degenerative" changes represent different factors affecting lumbar spine and the potential for osteophyte formation caused by factors other than spinal degeneration. Oishi et al. showed that intervertebral disc degeneration and osteophyte formation of the vertebral bodies represented different factors affecting the lumbar spine in postmenopausal Japanese women; however, the difference of osteophytes with or without disc degeneration was not mentioned. There are no detailed studies concerning osteophytes not accompanied with disc degeneration. We considered it informative to investigate the features of such osteophytes that are often observed clinically (Oishi et al., 2003).

There are few epidemiological data about osteophyte formation on the lumbar spine compared to the number of studies about osteoarthritis of the knee and hip. Nathan reported with regard to the frequency and degree of development of anterior osteophytes that the prevalence of osteophytes was greater in whites of both sexes than in Negroes with no statistically significant differences, with the frequency being much higher in the males in both races (Nathan et al., 1962). Our results revealed a significant influence of gender, smoking and alcohol consumption on osteophyte formation irrespective of the presence of disc height narrowing, however, showed no epidemiological differences between osteophyte formation with disc height narrowing and without narrowing, namely the differences in osteophytes depending on intervertebral disc degeneration. The present study illustrated that the prevalence of LBP in group B was significantly lower than in group A, and this suggests that lumbar disc narrowing may have a propensity for LBP, indicating osteophytes may prevent the clinical manifestation of pain. While data from many studies suggest an association between LBP and osteophyte formation (Frymoyer et al., 1984, Biering-Sorensen et al., 1985, Symmons et al., 1991, Pye et al., 2004), several studies indicate that osteophyte formation do not have an independent association with LBP (van Tulder et al., 1997, O'Neill et al., 1999, Schepper et al. 2004). Whether the stabilization of osteophytes or low frequency of disc degeneration decreased the prevalence of LBP is not clear. However, when osteophyte formation occurs before disc degeneration advances as a physiologic response to stabilize the spine, LBP may be evitable. While many studies have focused on LBP in relation to lumbar disc degeneration (Parkkola et al., 1993, Paajanen et al., 1997, Luoma et al., 2000, Jarvik et al., 2001, Videman et al., 2003), there are no reports regarding the association of LBP with osteophytes without lumbar disc degeneration.

Discal degeneration is generally considered as the primary source of LBP. In addition to nociceptive nerve fibers in the annulus and nucleus that can be sensitized by the cytokines and neuropeptides present in the degenerated disc, other sources of notiception can be found in the spinal unit including muscles, ligaments and facet joints (Freemont et al., 1997, Benoist, 2003). Nociception coming from these various tissues makes it difficult to distinguish from osteophytes in spinal pain. Thus, the cause of the decreased LBP should not be determined to be osteophyte formation before disc degeneration, although, it would be intriguing to investigate the genetic predisposition in cases with osteophytes without disc degeneration.

Several studies on factors associated with genetic susceptibility to spinal osteophyte formation, such as VDR (Videman et al., 2001, Jordan et al., 2005) and TGFB1 (Yamada et al., 2000) referred to osteophytes with spinal degeneration. Our results did not show any relationship between these polymorphisms and osteophyte formation without disc degeneration. Alcohol dehydrogenase β subunit is an enzyme that converts ethanol to acetaldehyde, whose gene, ADH2 located in 4q22, has a functional polymorphism Arg47His (Matsuo et al., 1989). Both ADH2 and ALDH2 (aldehyde dehydrogenase 2) are polymorphic, and genetic polymorphisms have been shown to functionally affect alcohol detoxification. The enzyme activity is higher in the 47His allele (ADH2*2) than in the 47Arg allele (ADH2*1) (Yin et al., 1984), and the former leads to a higher rate of oxidation of ethanol, resulting in an arginine/histidine exchange in the protein. In particular, an association of the 47His allele with flushing has been reported (Takeshita et al., 1996), and results of a number of studies seem to indicate that the 47His allele protects against alcohol abuse and alcoholism in Asians (Muramatsu et al., 1995, Shen et al., 1997) and Caucasians (Whitfield et al., 1998, Borras et al., 2000). In Japanese, the incidence of 47Arg allele is low, different from Caucasians (Sherman et al., DI, 1993, Higuchi et al., 1994). On the other hand, most alcoholics exhibit radiographic evidence of osteopenia (Bilke et al., 1985), leading to a hypothesis that reduced osteoblast activity resulting in underfilling of resorptive lacunae is primary responsible for alcohol-induced bone loss (Turner et al., 2001). Ethanol has been shown to increase bone resorption (Callaci et al., 2004) and to decrease trabecular bone volume (Rico et al., 1987). Additionally, administration of ethanol to healthy volunteers results in an acute decrease in serum osteocalcin levels (Rico et al., 1987, Nielsen et al., 1990). The present study demonstrated that carriers of 47Arg allele might suppress osteophyte formation unaffected by intervertebral disc degeneration, and this could be supported by these studies showing that ethanol contributes to decreased bone formation. Further research will be required to investigate the osteophyte development and molecular characterization, however, our study would encourage further studies on the mechanisms underlying osteophyte formation.

5. Conclusion

Osteophyte formation of the lumbar spine without disc degeneration was investigated, and estimated the implications of osteophytes from the viewpoint of LBP and gene polymorphism. The 47His polymorphism in the ADH2 may act to suppress osteophyte formation unaffected by disc degeneration. The subjects with osteophyte development preceding intervertebral disc degeneration had a lower risk of LBP compared with those without osteophytes.

6. References

Amonoo-Kuofi, HS. (1991) Morphometric changes in the height and anteropsterior diameters of the lumbar intervertebral discs with age. *J Anat* Vol. 175, pp. 159-168, ISSN 0021-8782

An, HS.; Anderson, PA.; Haughton, VM., Iatridis, JC.; Kang, JD.; Lotz, JC.; Natarajan, RN.; Oegema, TR Jr.; Roughley, P.; Setton, LA.; Videman, T.; Andersson, GB. & Weinstein, JN. (2004) Introduction: disc degeneration: summary. *Spine* Vol.29, No. 23, pp. 2677-2678, ISSN 0887-9869

Andersson, GBJ. (1998) What are the age-related changes in the spine? *Baillière's Clinical Rheumatology,* Vo. 12, No. 1, pp. 161-173, ISBN 0-7020-2380-9

Benoist, M. (2003) Natural history of the aging spine. *Eur Spine J* Vol. 12, Suppl 2, S86-89, ISSN 0940-6719

Bick, EM. (1995) Vertebral osteophytosis: Pathological basis of its roentgenology. *Am J Roentgenol Radium Ther Nucl* , Vol. 73, No. 6, pp. 979-983, ISSN 0002-9580

Biering-Sørensen, F.; Hanson, FR.; Schroll, M. & Runeborg, O. (1985) The relation of spinal x-ray to low back pain and physical activity among 60 year old men and women. *Spine* Vol. 10, No. 5, pp.445-451, ISSN 0887-9869

Bilke, DD.; Genant, HK.; Cann, C.; Recker, RR.; Halloran, BP. & Strewler, GJ. (1985) Bone disease in alcohol abuse. *Ann Intern Med* Vol. 103, No. 1, pp. 42-48, ISSN 0003-4819

Borenstein, DG.; O'Mara, JW Jr.; Boden, SD.; Lauerman, WC.; Jacobson, A.; Platenberg, C.; Schellinger, D. & Wiesel, SW. (2001) The value of magnetic resonance imaging of the lumbar spine to predict low-back pain in asymptomatic subjects: a seven-year follow-up study. *J Bone Joint Surg [Am]* Vo. 83, No.9, pp. 1306-1311, ISSN 0021-9355

Borras, E.; Coutelle, C.; Rosell, A.; Fernández-Muixi, F.; Broch, M.; Crosas, B.; Hjelmqvuist, L.; Lorenzo, A.; Gutiérrez, C.; Santos, M.; Szczepanek, M.; Heilig, M.; Quattrocchi, P.; Farrés, J.; Vidal, F.; Richart, C.; Mach, T.; Bogdal, J.; Jörnvall, H.; Seitz, HK.; Couzigou, P. & Parés, X. (2000) Genetic polymorphism of alcohol dehydrogenase in Europeans. ADH2 allele decreases the risk of alcoholism and is associated with ADH3*1. *Hepatology* Vol. 31, No. 4, pp. 984-989, ISSN 0270-9139

Buckwalter, JA. (1995) Aging and degeneration of the human intervertebral disc. *Spine* Vol. 20, No. 11, pp. 1307-1314, ISSN 0887-9869

Callaci, JJ.; Junknelis, D.; Patwardhan, A.; Sartori, M.; Frost, N. & Wezeman, FH. (2004) The effects of binge alcohol exposure on bone resorption and biomechanical and structural properties are offset by concurrent bisphosphonate treatment. *Alcohol Clin Exp Res* Vol. 28, No. 1, pp. 589-594, ISSN 0145-6008

Collins , DH. (1949) *The pathology of articular and spinal diseases.* Edward-Arnold and Co., London de Schepper, EI.; Damen, J.; van Meurs, JB.; Ginai, AZ.; Popham, M.; Hofman, A.; Koes, BW. & Bierma-Zeinstra, SM. (2010) The association between lumbar disc degeneration and low back pain: the influence of age, gender, and individual radiographic features. *Spine* Vol. 35, No. 5, pp. 531-536, ISSN 0887-9869

Freemont, AJ.; Peacock, TE.; Goupille, P.; Hoyland, JA.; O'Brien, J. & Jayson, MI. (1997) Nerve ingrowth into diseased intervertebral disc in chronic back pain. *Lancet* Vol. 350, No. 9072, pp. 178-181, ISSN 0140-6736

Frymoyer, JW.; Newberg, A.; Pope, MH.; Wilder, DG.; Clements, J. & McPherson, B. (1984) Spine radiographs in patients with low back pain. An epidemiological study in men. *J Bone J Surg [Am]* Vol. 66, No. 7, pp. 1048-1055, ISSN 0021-9355

Hamajima, N.; Saito, T.; Matsuo, K.; Kozaki, K.; Takahashi, T. & Tajima, K. (2000) Polymerase chain reaction with confronting two-pair primers for polymorphism genotyping. *Jpn J. Canser Res* Vol.91, No. 9, pp. 865-868, ISSN 0021-4922

Harada, A.; Okuizumi, H.; Miyagi, N. & Genda, E. (1998) Correlation between bone mineral density and intervertebral disc degeneration. *Spine* Vol. 23, No. 8, pp. 857-861, ISSN 0887-9869

Higuchi, S. (1994) Polymorphisms of ethanol metabolizing enzyme genes and alcoholism. *Alcohol Alcohol* Suppl.2, pp. 29-34, ISSN 0735-0414

Jordan, KM.; Syddall, H.; Dennison, EM.; Cooper, C. & Aeden, NK. (2005) Birthweight, vitamin D receptor gene polymorphism, and risk of lumbar spine osteoarthritis. *J Rheumatol* Vol. 32, No.4, pp. 678-683, ISSN 0315-162X

Jarvik, JJ.; Hollingworth, W.; Heagerty, P.; Haynor, DR. & Deyo, RA. (2001) The longitudinal assessment of imaging and disability of the back study: Baseline data. *Spine* Vol. 26, No. 10, pp. 1158-1166, ISSN 0887-9869

Jayson, MI. (1994) Mechanisms underlying chronic back pain. *Br Med J* Vo. 309, No. 6956, pp. 681-682, ISSN 0007-1447

Kim, YE.; Goel, VK.; Weinstein, JN. & Lim, TH. (1991) Effects of disc degeneration of one level on the adjacent level in axial mode. *Spine* Vol. 16, No. 3, pp. 331-335, ISSN 0887-9869

Kinoshita, H.; Tamaki, T.; Hashimoto, T. & Kasagi, F. (1998) Factors influencing lumbar spine bone mineral density assessment by dual-energy X-ray absorptiometry: Comparison with lumbar spine radiogram. *J Orthop Sci* Vol. 3, No.1, pp. 3-9, ISSN 0949-2658

Lane, NE.; Nevitt, MC.; Genant, HK. & Hchberg, MC. (1993) Reliability of new indices of radiographic osteoarthritis of the hand and hip and lumbar disc degeneration. *J Rheumatol* Vol. 20, No.11, pp. 1911-1918, ISSN 0315-162X

Lawrence, JS. (1969) Disc degeneration. Its frequency and relationship to symptoms. *Ann Rheum Dis* Vol. 28, No. 2, pp.121-138, ISSN 0003-4967

Lipson, SJ. & Muir, H. (1980) Vertebral osteophyte formation in experimental disc degeneration. Morphologic and proteoglycan changes over time. *Arthritis Rheum* Vol. 23, No. 3, pp. 319-24, ISSN 0004-3591

Liu, G.; Peacock, M.; Eilam, O.; Dorulla, G.; Braunstein, E. & Johnston, EE. (1997) Effect of osteoarthritis in the lumbar spine and hip on bone mineral density and diagnosis of osteoporosis in elderly men and women. *Osteoporos Int* Vol. 7, No.6, pp. 564-569, ISSN 0937-941X

Luoma, K.; Riihimaki, H.; Luukkonen, R.; Raininko, R.; Viikari-Juntura, E. & Lamminen, A. (2000) Low back pain in relation to lumbar disc degeneration. *Spine* Vol. 25, No. 4, pp. 487-492, ISSN 0887-9869

Macnab, I. (1971) The traction spur: An indicator of spinal instability. *J Bone Joint Surg [Am]*, Vol. 53, No. 4, pp. 663-670, ISSN 0021-9355

Matsuo, Y.; Yokoyama, R. & Yokoyama, S. (1989) The genes for human alcohol dehydrogenases β1 and β2 differ by only one nucleotide. *Eur J Biochemistry* Vol. 183, NO. 2, pp. 317-320, ISSN 0014-2956

Milgram, JW. (1982) Osteoarthritic changes at the severely degenerative disc in humans. *Spine* Vol. 7, No.5, pp. 498-505, ISSN 0887-9869

Miyakoshi, N.; Itoi, E.; Murai, H.; Wakabayashi, I.; Ito, H. & Minato T. (2003) Inversed relation between osteoporosis and spondylosis in postmenopausal women as evaluated by bone mineral density and semiquantitative scoring of spinal degeneration. *Spine* Vol. 28, No. 5, pp. 492-495, ISSN 0887-9869

Muramatsu, T.; Wang, ZC.; Fang, YR.; Hu, KB.; Yan, H.; Yamada, K.; Higuchi, S.; Harada, S. & Kono, H. (1995) Alcohol and aldehyde dehydrogenase genotypes and drinking behavior of Chinese living in Shanghai. *Hum Genet* Vol. 96, No. 2, pp. 151-154, ISSN 0340-6717

Nathan, H. & Islael, J. (1962) Osteophyte of the vertebral column. An anatomical study oftheir development according to age, race, and sex with considerations as to their etiology and significance. *J Bone Joint Surg[Am]* Vol. 44, No. 2, pp. 243-268, ISSN 0021-9355

Nathan, M.; Pope, MH. & Grobler, LJ. (1994) Osteophyte formation in the vertebral column: A review of the etiologic factors- Part 1. *Contemporary Orthopaedics* Vol. 29, No. 1, pp. 31-37, ISSN 0194-8458

Nielsen, HK.; Lundby, L.; Rasmussen, K.; Charles, P. & Hansen. (1990) Alcohol decreases serum osteocalcin in a dose-dependent way in normal subjects. *Calcif Tissue Int* Vol. 46, No. 3, pp. 173-178, ISSN 0171-967X

Oishi, Y.; Shimizu, K.; Katoh, T.; Nakao, H.; Yamaura, M.; Fukuko T.; Narusawa, K. & Nakamura T. (2003) Lack of association between lumbar disc degeneration and osteophyte formation in elderly Japanese women with back pain. *Bone* Vol. 32, No. 4, pp. 401-411, ISSN 8756-3282

O'Neill, TW.; McCloskey, EV.; Kanis, JA.; Bhalla, AK.; Reeve, J.; Reid, DM.; Todd, C.; Woof, AD. & Silman, AJ. (1999) The distribution, determinants, and clinical correlates of vertebral osteophytosis: a population based survey. *J Rheumatol* Vol. 26, No.4, pp. 842-848, ISSN 0315-162X

Paajanen, H.; Erkintalo, M.; Parkkola, R.; Salminen, J. & Kormano, M. (1997) Age-dependant correlation of low-back pain and lumbar disc degeneration. *Arch Orthop Trauma Surg* Vol. 116, No. 1, pp. 106-107, ISSN 0344-8444

Parkkola, R.; Rytokoshi, U. & Kormano, M. (1993) Magnetic resonance imaging of the discs and trunk muscles in patients with chronic low back pain and healthy control subjects. *Spine* Vol. 18, No. 7, pp. 830-836, ISSN 0887-9869

Powell, MC.; Wilson, M.; Szypryt, P.& Summonds, EM. (1986) Prevalence of lumbar disc degeneration observed by magnetic resonance imaging in symptomless women. *Lancet* Vol. 13, No. 2, pp.1366-1367, ISSN 0140-6736

Pye, SR.; Reid, DM.; Smith, R.; Adams, JE.; Nelson, K.; Silman, AJ. & O'Neill, TW. (2004) Radiographic features of lumbar disc degeneration and self-reported back pain. *J Rheumatol* Vol. 31, No. 4, pp. 753-758, ISSN 0315-162X

Rico, H.; Cabranes, JA.; Cabello, J.; Gomez-Castresana, F. & Hermandez, ER. (1987) Low serum osteocalcin in acute alcohol intoxication: A direct toxic effect of alcohol on osteoblast. *Bone Miner* Vol. 2, No. 3, pp. 221-225, ISSN 0268-3369

Roberts, N; Gratin, C. & Whitehouse, GH. (1997) MRI analysis of lumbar intervertebral disc height in young and older populations. *J Magn Reson Imaging* Vol. 7, No. 5, pp. 880-886, ISSN 1053-1807

Sakai, Y.; Matsuyama, Y.; Hasegawa, Y.; Yoshihara, H.; Nakamura, H.; Katayama, Y.; Imagama, S.; Ito, Z.; Ishiguro, N. & Hamajima, N. (2007) Association of gene polymorphisms with intervertebral disc degeneration and vertebral osteophyte formation. *Spine* Vol.32, No. 12, pp.1279-1286, ISSN 0887-9869

Sambrook, PN.; MacGregor, AJ. & Spector, TD. (1999) Genetic influences on cervical and lumbar disc degeneration. Magnetic resonance imaging study in twins. *Arthritis Rheum* Vol. 42, No.2, pp. 366-372, ISSN 0004-3591

Schmorl, G. & Junghanns, H. (1932) *The Human Spine in Health and Disease.* Grune and Stratton, New York Shao, Z.; Rompe, G. & Schiltenwolf, M. (2002) Radiographical changes in the lumbar intervertebral discs and lumbar vertebrae with age. *Spine* Vo. 27, No. 3, pp. 263-268, ISSN 0887-9869

Shen, YC.; Fan, JH.; Edenberg, HJ.; Li, TK.; Cui, YH.; Wang, YF.; Tian, CH.; Zhou, CF.; Zhou, RL.; Wang, J.; Zhao, ZL. & Xia, GY. (1997) Polymorphism of ADH and ALDH genes among four ethnic groups in China and effects upon the risk for alcoholism. *Alcohol Clin Exp Res* Vol. 21, No. 7, pp. 1272-1277, ISSN 0145-6008

Sherman, DI.; Ward, RJ.; Warren-Perry, M.; Williams, R. & Peters, TJ. (1993) Association ofrestriction fragment length polymorphism in alcohol dehydrogenase 2 gene with alcohol induced liver damage. *BrMed J* Vol. 307, No. 6916, pp. 1388-90, ISSN 0007-1447

Symmons, DP.; van Hemert, AM.; Vandenbroucke, JP. &Valkenburg, HA. (1991)A longitudinal study of back pain and radiological changes in the lumbar spines of middle aged women. II. Rsdiologic findings. *Ann Rheum Dis* Vol.50, No. 3, pp. 162-166, ISSN 0003-4967

Takeshita, T.; Mao, XQ. & Moritomo, K. (1996) The contribution of polymorphism in thealcohol dehydrogenase β subunit to alcohol sensitivity in a Japanese population. *Hum Genet* Vol. 97, No. 4, pp. 409-413, ISSN 0340-6717

Turner, RT.; Kidder, LS.; Kennedy, A.; Evans, GL. & Sibonga, JD. (2001) Moderate alcoholconsumption suppresses bone turnover in adult female rats. *J Bone Miner Res* Vol. 16, No. 3, pp. 589-594, ISSN 0884-0431

Twomery, LT. & Taylor, JR. (1987) Age changes in lumbar vertebrae and intervertebral discs. *Clin Orthop Relat Res* Vol. 224, pp. 97-104, ISSN 0009-921X

van Tulder, MW.; Assendelft, WL.; Koes, BW. & Boulter LM. (1997) Spinal radiographicfindings and nonspecific low back pain. A systematic review of observational studies. *Spine* Vol. 15, No. 22, pp. 427-434, ISSN 0887-9869

Vernon-Roberts, B. & Pirie, CJ. (1977) Degenerative changes in the intervertebral discs of thelumbar spine and their sequelae. *Rheumatol Rehabil* Vol. 16, No. 1, pp. 13-21, ISSN 0300-3396

Videman, T.; Gibbons, KE.; Battie, MC.; Maravilla, K.; Vannine, E.; Leppavuori, J.; Kaprio, J.& Peltonen, L. (2001) The relative roles of intragenic polymorphisms of the Vitamin D Receptor gene in lumbar spine degeneration and bone density. *Spine* Vol. 26, No. 3, pp.7-12, ISSN 0887-9869

Videman, T.; Battie, MC.; Gibbons, LE.; Maravilla, K.; Manninen, H. & Kaprio, J. (2003)Associations between back pain history and lumbar MRI findings. *Spine* Vol. 28, No. 6, pp. 582-588, ISSN 0887-9869

Whitfield, JB.; Nightingale, BM.; Bucholz, KK.; Madden, PAF.; Heath, AC. & Marin, NG.(1998) ADH genotypes and alcohol use and dependence in Europeans. *Alcoho Clin Exp Res* Vol. 22, No. 7, pp. 1463-1469, ISSN 0145-6008

Witt, I.; Vestergaard, A. & Rosenklint, A. (1984) A comparative analysis of x-ray findings ofthe lumbar spine in patients with and without lumbar pain. Spine Vol. 9, No. 3, pp. 298-300, ISSN 0887-9869

Yamada, Y.; Okuizumi, H.; Miyauchi, A.; Takagi, Y.; Ikeda, K. & Harada, A. (2000)Association of transforming growth factor β1 genotype with spinal osteophytosis in Japanese women. *Arthritis Rheum* Vol. 43, No. 2, pp. 452-460, ISSN 0004-3591

Yin, SJ.; Bosron, WF.; Magnes, LJ. & Li, TK. (1984) Human liver alcohol dehydrogenase: purification and kinetic characterization of β2β2, β2β1, αβ2, and β2γ1 "oriental" enzymes. *Biochemistry* Vol. 23, No. 24, pp. 5847-5853, ISSN 0006-2960

3

Low Back Pain and Injury in Athletes

Wayne Hoskins
Department of Orthopaedic Surgery, Royal Melbourne Hospital, Grattan St, Parkville Victoria, Australia

1. Introduction

Low back pain is an extremely common entity in the general population. Athletes are no different in their affliction for suffering low back pain and injuries, particularly in sports that carry specific low back demands. Whilst traditionally low back pain in the non-athletic population has been thought of in terms of being acute or chronic in nature, recent long-term epidemiological studies have suggested there is a need to revise views regarding the natural history of low back pain. Low back pain is not simply either acute or chronic but fluctuates over time with frequent recurrences or exacerbations and should not be considered self-limiting. The natural history of low back pain in athletes is most probably no different. The very nature of athletic preparation requires mechanical overload. Athletic manoeuvres produce significant compressive forces directed at the lumbar spine. A trade-off is likely to exist between athletic demands and injury, with greater duration of training, training intensity and a lack of relative rest occurring at the expense of tissue overload and ongoing injury. This may explain why some athletes tend to have more persistent, chronic and recurrent low back symptoms, frequently associated with early degenerative joint disease.

Although most low back pain in both the athletic and non-athletic population is non-specific and mechanical in nature, athletes are often at special risk of more serious causes of back pain that are often sport specific in their aetiology. This is a result of the repetitive mechanical loading and often specific and unique motion imposed on the spines of athletes through various sporting requirements in training and competition. Furthermore, the paediatric sporting population carries a special risk for injury given they have less musculoskeletal maturity and they may be at a heightened risk for more severe and permanent skeletal damage, structural abnormalities and chronic pain.

The initial differential diagnosis list for athletic low back pain should be broad. Diagnosis should include a thorough history excluding red flag conditions, examination and a focussed evidence based approach to imaging. Attention should be paid to the mechanism of injury or the inciting event to assist in predicting the potential injury, implementing preventative measures and in developing a management and rehabilitation program. Consideration of the athlete's age and an understanding of the sports specific biomechanics of an athlete is required. It is unclear about the relevance of yellow flags in the development of low back injuries and chronic pain in athletic populations. A lack of research exists investigating the management of low back pain in athletic populations. Elite level competitors are likely more willing to train and compete with pain and injury as a result of the financial commitments they receive from competition as well as their drive for competitive success, making the management of athletes with low back injury a challenge for the sports clinician. It is likely that management should mirror published guidelines designed for the non-athletic population and incorporate a period of relative rest, avoiding aggravating activities, changes to training and technique along with appropriate rehabilitation therapy.

The coming chapter will discuss the prevalence of low back pain and injury in sport, identify risk factors that athletes have for low back pain, highlight some of the consequences that back pain carries, discuss the diagnosis and management of back pain in athletes and identify areas for future research.

2. Prevalence of low back pain in sporting populations

The anatomical boundaries of the low back being a shaded area between the last ribs and the gluteal folds (Figure 1) has been found to be the most commonly used in a review of methodologically sound low back pain prevalence studies (Walker, 2000). The prevalence of low back pain in the general, non-sporting population has been well described with numerous well designed, long term epidemiological studies and systematic reviews existing

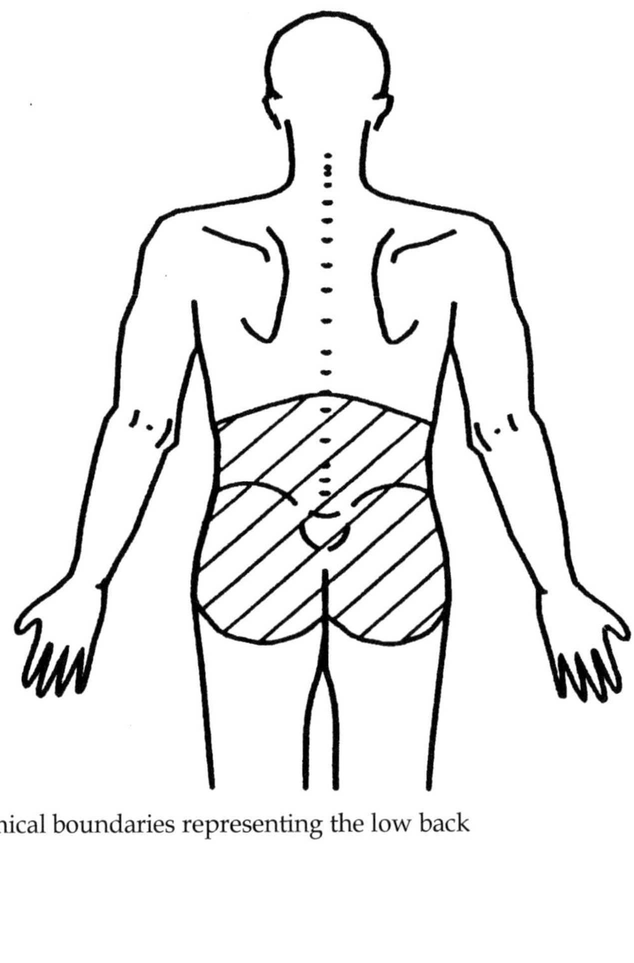

Fig. 1. The anatomical boundaries representing the low back

in the published scientific literature databases (Lebouef-Yde & Lauritsen 1995, Walker 2000). Evidence from this literature has clearly documented that low back is a very common entity and is responsible for substantial economic burden to society (Druss et al., 2002). However, most low back pain that people experience is low-intensity and low-disability in nature (Walker et al., 2004). Figures documenting lifetime prevalence have been as high as 84% with a point-prevalence between 12% and 33% recorded in a systematic review (Walker, 2000).

Despite the amount and quality of literature investigating low back pain in the general, non-athletic population, less interest has been afforded to investigating the prevalence, severity and epidemiology of low back pain in athletic populations. In particular there are very few large, long term epidemiological studies assessing low back pain amongst active competing athletes, especially at the elite and professional level of competition. Of the literature that exists, studies have documented that low back pain prevalence and severity can vary between sports, with, not surprisingly, an increase in pain noted in those sports that carry with them significant low back demands (Sward et al., 1990; Bahr et al., 2004). Noteworthy is the reported lack of significant difference in low back injury rates between contact and non-contact sports (Greene et al., 2001), suggesting that other factors may be more important in the development of most cases of low back pain and injury. However, the true prevalence, severity and natural history of low back pain in sporting populations remains unclear due to a lack of well designed, large-scale prospective and longitudinal scientific literature.

When comparing the literature that exists, it is not entirely clear whether competing athletes are at a risk of a higher prevalence or increased severity of low back pain compared with the non-athletic population. This is largely due to a lack of homogeneity in study design and methodology. It also has not been investigated whether low back pain prevalence or severity varies at different levels of athletic competition. Evidence suggests that sporting participation in the general population, regardless of activity, contributes to less frequent low back pain (Jacob et al., 2004). However, once low back pain is established, participation in sporting activities may indirectly contribute to increased severity of pain (Jacob et al., 2004). Bahr et al. analysed low back pain prevalence between elite athletes competing in endurance based sports: cross-country skiing (n=257), rowing (n=199), orienteering (n=278) as well as a non-athletic group (n=197) (Bahr et al., 2004). Low back pain lifetime (51-65%), year (48-63%) and seven day prevalence (20-25%) was similar between groups although lower in non-athletes. As far as the author is aware, despite smaller studies existing (Sward et al., 1991; Kujala et al., 1996), no other large study has used homogeneity in study design and methodology to make direct comparisons between active athletes and non-athletes. One difficultly in measuring low back pain in an athletic population is the lack of validated questionnaires to quantify the functional disability associated with low back pain. Whilst the validated questionnaires measuring pain severity and quality are likely to be useful, the validated questionnaires in use asking about functional limitations are unlikely to be useful as the parameters asked about are not created for sporting populations and questions asked are likely to be irrelevant to the high functional demands of athletes. The development of a validated sports specific, functional low back pain questionnaire is encouraged.

Much of the current sporting literature on low back pain and injury has tended to focus on sports with specific low back demands such as rowing (O'Kane et al., 2003; Teitz et al., 2003; Bahr et al., 2004), skiing (Mahlamaki et al., 1988; Eriksson et al., 1996; Ogon et al., 2001; Bahr et al., 2004), gymnastics (Sward et al., 1990; Hutchinson, 1999; Cupisti et al., 2004), diving (Baranto et al., 2006), wrestling (Lundin et al., 2001; Iwai et al., 2004), golf (McHardy & Pollard, 2005), cricket fast bowling (Elliott & Khangure, 2002; Ranson et al., 2010), tennis (Lundin et al., 2001) and American football (Iwamoto et al., 2004). Elite sporting activity is these sports is known to produce significant compressive forces directed at the lumbar spine (Hosea et al., 1989). The repetitive mechanical loading on the spines of athletes in these sports, often in positions involving end range of motion and the increased volume of training required for elite athletic performance is likely to result in tissue overload and subsequent injury. This, combined with a lack of full recovery between episodes of pain and injury due to many athletes not wanting to miss time off training or competition, may explain why athletes may have more persistent, chronic and recurrent low back symptoms, frequently associated with degenerative joint disease (Ong et al., 2003).

The knowledge surrounding the prevalence and magnitude of low back pain in sports that are not known for having specific low back demands, including the various highly popular football codes, remains largely unknown. Research has tended not to focus on low back pain as an area of interest in these sport, likely for a variety of reasons. Firstly, there are other well known more common and more serious injuries that tend to impact the functional demands of these athletes, resulting in loss of competition match play. Secondly, unlike other injuries that athletes experience, it is uncommon that low back pain is severe enough to prevent a professional footballer from competing or from relinquishing his place in team selection. This is particularly true when medical management frequently incorporates epidural steroid injections (Bono, 2004) and local anaesthetic agents (Orchard, 2004a), considered 'part of the game' in professional football (Orchard, 2001). Despite this, injury surveillances have documented that low back injury if present can be severe and have high recurrence rates. In one study on elite soccer, low back pain was reported as the most common overuse injury (Walden et al., 2005). In elite rugby league, 'back injuries' have been shown to have the highest rates of recurrence for all injuries (Orchard, 2004b), whilst in retired elite rugby league players, chronic low back pain has been the third most common complaint, reported by 39 % (Meir et al., 1997). In elite Australian Rules football, the Australian Football League's (AFL) long running injury surveillance has documented that five per cent of all players will miss a match each season with a 'lumbar or thoracic spine' injury, causing them to miss on average four weeks or matches per injury (Orchard & Seward, 2002). In amateur Australian Rules football players, 27% of player report a long term or recurrent back problem (McManus et al., 2004). In school children playing rugby union, low back pain has been shown to afflict over 40% of participants (Iwamoto et al., 2005).

Whilst there are many potential pain generators for low back pain, in reality most pain that both the general public and sporting population will experience, despite the use of advanced imaging techniques, can not be attributed to a tissue diagnosis and remain 'non-specific' and mechanical in diagnosis. However, there are several examples of where it is apparent that certain sports and activities have a clear association between the development of certain injuries and the mechanical demands associated with these sports and activities. Examples of this include spondylolisthesis in cricket bowlers (Ranson et al., 2010) and gymnasts (Toueg et al., 2010), herniated discs in weight lifters (Mundt et al., 1993) and traumatic injuries in body contact sports (Tewes et al. 1995). This will be discussed in further detail later in the chapter.

2.1 Prevalence of low back pain in adolescent sporting populations

There has been an increased awareness of low back pain in children and adolescents with several studies showing that low back pain is highly prevalent in the early years of life (Burton et al., 1996; Balague et al., 2003). Low back pain is known to increase with age during the first decades of life (Salminen et al., 1995), with prevalence increasing significantly following sexual maturity (LeResche et al., 2005). It has been theorized that low back pain in childhood may have important consequences for chronic low back pain in adulthood (Watson et al., 2002). This theory has more recently been validated with clear correlations now existing between low back pain in childhood and adolescence and in adulthood (Hestbaek et al., 2006). Hestbaek et al. in a large longitudinal study found low back pain in adolescence to be a significant risk factor for low back pain in adulthood with odds ratios as high as four (Hestbaek et al., 2006). A dose-response association was also demonstrated: the more days with low back pain the adolescent experienced, the higher the risk of future low back pain that they were more likely to experience. These findings are supported by other well conducted, long term research which has demonstrated that 90% of schoolchildren with low back pain will suffer from low back pain 25 years later (Harreby et al., 1996).

Questions have been raised regarding low back pain at the junior level of sporting competition, given that participation in adolescent sports has been found to be a risk factor for low back pain in one large, well conducted study (Kujala et al., 1997). Furthermore, sporting participation at an adolescent level has also been linked with higher low back pain prevalence than in adolescents who are non-athletes (Kujala et al., 1996). This is particularly true in the male sporting population (Burton et al., 1996). It is believed that adolescent athletes with less musculoskeletal maturity may be at a heightened risk for more severe and permanent skeletal damage and structural abnormalities, particularly when exposed to years of intense athletic training (Wojtys et al., 2000). However, there is a paucity of research documenting the true prevalence and severity of back pain in junior athletes and whether low back pain at a junior level predisposes increased prevalence of back pain later in a career. Like the adult literature, of the literature that does exist, it is extremely difficult to compare results due to a lack of homogeneity in study design. There is also a lack of literature comparing the prevalence and severity of back pain at varying levels of adolescent competition.

3. Risk factors for low back pain in sporting populations

Risk factors for the development of low back pain in the general population have been extensively researched in the published literature. Epidemiological studies into the prevalence of low back pain have identified that there are many individual, psychosocial and occupational risk factors for the onset of low back pain (Manek & MacGregor, 2005). A growing body of literature also exists implicating the role of genetic factors in back pain, in particular the development of disc injuries (Videman et al., 2005). Of the occupational factors there is evidence for a causal relationship between low back injuries and exposure to forceful exertions, awkward postures and vibration (Keyserling, 2000). Although not specifically targeted in research of athletic populations, it is probable that a combination of these 'occupational' factors is responsible for the development of most low back pain in athletic populations given many of the sports with low back demands are well known for their awkward posturing, forceful exertions and high mechanical loading of the lumbar spine (Hosea et al., 1989; Hosea & Boland, 1989; Cholewicki et al., 1991; Gatt et al., 1997).

Regardless of the sport in question, as Bono states, the low back is an important but under-recognized source of great dynamic power during a golf or baseball swing, a gymnast's landing, a power lifter's heavy squat, or a boxer's knockout punch. In static mode, it functions to help maintain an infielder's stand, a cyclist's tuck, or a ballerina's arabesque (Bono, 2006). These same sources of power and static control are likely to fail with fatigue, excess force and repetitive micro-trauma and result in low back injury. There are a few examples where a specific action or activity has been implicated in back injuries such as the fast bowling action in cricket, hyper-extension in gymnastics, prolonged flexion in skiing and cycling and repetitive lumbar flexion and loading in weight lifting pursuits. Despite this, there is a lack of literature investigating risk factors for the development of low back pain in athletic populations. Laboratory based studies exist demonstrating the high mechanical forces directed at the lumbar spine during a golfer's swing (Hosea et al., 1989), the rowing action (Hosea & Boland, 1989), American football blocking (Gatt et al., 1997) and weight lifting (Cholewicki et al., 1991). Low back pain is also likely be related to the type, intensity, duration and/or amount of athletic activity performed. In endurance based sports with low back demands a dose response relationship appears to exist with low back pain (Bahr et al., 2004). Causes of low back injury have also received much discussion in the large body of literature documenting changes in lumbar-pelvic muscle activation and recruitment due to low back pain, producing altered neuromuscular control strategies (Hungerford et al. 2003). This will be discussed later in the chapter.

What is less clear in athletic populations is the role that psychosocial factors have in both the development of low back injuries and also in the transition from acute to chronic pain. Multiple systematic reviews of the general population have shown that psychological factors have an important role in the transition from acute to chronic pain (Manek & MacGregor, 2005). In a recent systematic review of the literature, depression, psychological distress, passive coping strategies and fear-avoidance beliefs were sometimes found to be independently linked with poor outcome, whereas most social and socio-occupational factors were not (Ramond et al., 2011).

How this literature relates to athletic populations is unclear. Psychosocial factors may be more important for the professional and semi-professional athlete who has financial, contractual and performance concerns. These athletes generate a meaningful income and employment from their sporting endeavours. It has been suggested that a well motivated athlete may under-report pain in order to improve performance, their chances of team selection and for a positive mind frame (Lundin et al., 2001). Alternatively, pain may be over-reported as it may be provoked easily by intense training and competition requirements and hinder athletic performance (Lundin et al., 2001). The athlete may therefore place a greater impact on pain than may be appreciated. This situation is potentially more of a concern as exaggeration of self-reported low back pain and disability may be a predictor for low back pain chronicity (Gatchel et al., 1995). However, previous research on amateur athletes has found psychosocial issues such as level of satisfaction with coaches or team-mates not to be related to the development of low back pain (Greene et al., 2001). Despite this, it has been shown that low back pain in former elite athletes is predicted by psychosocial issues such as life dissatisfaction, neuroticism, hostility, extroversion and poor sleep quality (Videman et al., 1995). Future research is required to more broadly investigate psychosocial factors in athletes and their impact and relevance, if any, to the development of low back injuries and chronic pain during play and after a career has ended.

3.1 Risk factors for low back pain in adolescent sporting populations

As opposed to the adult population, literature investigating risk factors for the development of low back pain in adolescent populations is not as conclusive in it its findings. A recent systematic review of the literature included five studies (Hill & Keating, 2010). The included studies varied considerably in the methods used to gather data, definitions of low back pain, and recall periods for an episode of low back pain. Inconsistency in definitions of low back pain, pre-defined recall periods, and methods used to collect and analyse data limit conclusions that can be drawn about factors that identify children at risk of developing low back pain. As no risk factor has been validated in independent investigation, the authors concluded that there is no certainty that any factor places children at risk of developing low back pain (Hill & Keating, 2010).

Looking at studies investigating risk factors for low back pain in adolescent sporting populations, a large cross sectional survey has found that adolescents are at a greater risk of low back pain if they have low isometric muscle endurance in the back extensors, with no associations found for aerobic fitness, functional strength, flexibility, or physical activity level after adjustment for muscle endurance (Bo Anderson et al., 2006). It may be that the junior sporting population is initially protected from low back pain due to their increased physical fitness, but this could be lost following excessive spinal loading (Kujala et al., 1996) and high training duration (Kujala et al., 1992) that many become exposed to. This is more likely to be the case with the advanced professionalism and training commitments junior athletes face when they reach the transition to increased sporting specialization in elite junior and adult professional sporting competitions. This would be particularly the case if the athlete is allowed to progress with poor techniques that would predispose injury. Junior athletes at the elite level of competition also face pressure to play and train with low back pain (and other injuries) given non-participation or obvious injury history can affect future selection to professional adult level competition. This again makes management difficult.

Another potential reason for an increased incidence of low back pain in elite level adolescent athletes includes the likely increased prevalence of weight lifting training into the typical training programs of most athletes. The effects of the mechanical loading that weight lifting may have on the developing spine, particularly when poor lifting techniques and sub-optimal training programs focusing on body building exercises rather than more functional exercises, combined with the effects of increased loading and training volume has been discussed by other authors (Wotjys et al., 2000).

4. Consequences of low back injury

The development of low back injury when occurring in athletes has several potential consequences. This includes the development of future, recurrent and repeated episodes of low back pain and injury which may be related to the neuro-physiological changes to lumbar-pelvic stability that is known to occur secondary to low back pain, issues associated with current and future playing performance, potential associations with the occurrence of other injuries and pain and disability to the player in the post career stage.

4.1 Recurrent pain and neuro-physiological changes of back pain

Without a doubt the biggest risk factor for future occurrences of low back pain in athletes are a previous or a current history of low back pain (Greene et al., 2001; O'Kane et al., 2003). It may be reasonable to conjecture that regardless of the aetiology of the initial low back pain, that once an athlete has experienced significant low back injury, that they remain susceptible to future pain and aggravation or exacerbation of pain. This fits with the natural history of low back pain in the non-athletic population (Hestbaek et al., 2006).

Low back pain is known to result in clinical instability of the lumbar-pelvic spine (Kaigle et al., 1995). Panjabi states that clinical instability occurs when segmental control around the physiological neutral zone cannot be accomplished (Panjabi 1992a). It results in a loss of the normal pattern of spinal motion as the neural control system alters the timing of muscular contraction patterns and reflex responses (Panjabi, 1992b; O'Sullivan et al. 1997a). With this loss of segmental stability, there is evidence to support the concept of increased compensatory substitution of the global system (Edgerton et al. 1996; O'Sullivan et al., 1997b), including earlier activation of various muscles involved with lumbar-pelvic motor control (Hungerford et al., 2003).

The lumbar-pelvic spine is preferably supported by an intricate arrangement of deep local muscles, including the multifidus and transversus abdominus, which provide a stabilising base on which the global muscles can act. The local muscles support the individual spinal segments during continuous full-body movements and allow the powerful activation of more global muscles acting across larger joints without spinal injury occurring (Wilke et al., 1995). Coordination of local muscle contraction to provide ongoing spinal stability and prevent injury is a complicated neurological process. Proprioceptive sensory feedback is necessary to permit the correct series, quantity and timing of muscular contraction (O'Sullivan et al., 1997a, Panjabi, 2003), a property lost with lumbar-pelvic pain and dysfunction (O'Sullivan et al., 1997b).

Several authors have suggested that of the local lumbar-pelvic stabilisation muscles, the multifidus and transversus abdominus are key stabilisers (Wilke et al., 1995, Hodges & Richardson, 1996). The multifidus muscles are the deepest of the posterior stabilising muscles having predominantly vertebrae-vertebrae attachments, attaching to the zygapophyseal joint capsules and being segmentally innervated (Macintosh et al., 1986). They function to finely control lumbar vertebral movements about the neutral zone, with their anatomical arrangement, joint attachments and neurological innervation making them the principal muscle for this function (McGill, 1991; Wilke et al., 1995). The transversus abdominus is the deepest of the abdominal muscles. It has extensive attachments to the thoracolumbar fascia and with its advantageous line of attachment, is the most capable of all muscles in tensioning the thoracolumbar fascia, thereby having a major effect on lumbar-pelvic stability by restricting vertebral displacement (Hodges & Richardson, 1996) and controlling rotational and lateral stability of the spine via the thoracolumbar fascia (Cresswell, 1993). In normal participants both the multifidus and transversus abdominus have a large cross sectional area of type one, or slow twitch muscle fibres, which allows them to provide a tonic contraction to assist with their responsibility of providing constant lumbar-pelvic stability (Jorgensen et al., 1993). Activity of the multifidus and transversus abdominus should occur in advance of the muscles required to provide body movement and action in a feed forward mechanism (Hodges & Richardson, 1996). This occurs regardless of the direction of reactive forces (Hodges & Richardson, 1996).

In those with low back pain, significant changes to the multifidus and transversus abdominus have been recognised to occur which changes lumbar-pelvic stabilisation strategies (Biedermann et al., 1991; Hides et al., 1996; Hodges & Richardson, 1998; Hodges et al., 2003). This, in addition to the compensatory action of the global system, may produce altered muscle response patterns required for lumbar-pelvic stabilisation during sudden trunk loading in athletes following their clinical recovery from low back pain (Cholewicki et al., 2002). After the first episode of low back pain, selective atrophy of the multifidus can occur rapidly within days of pain occurrence, which can be as high as 31% in 24 hours (Hides et al., 1996), a temporal pattern suggestive of a neurogenic mechanism. This atrophy may not be restored following pain remission, which has been linked to a high rate of recurrent low back pain (Hides et al., 1996). In biomechanical research models, loss of even one segment of multifidus muscular control has been shown to significantly reduce the overall stability of the spine, particularly in controlling buckling when load on the spine is increased (Crisco & Panjabi, 1991). Multifidus also shows less endurance and greater fatigability after pain syndromes (Biedermann et al., 1991). This loss in endurance has enabled significant identification of athletes with existing low back pain (Roy et al., 1990). Changes to the internal structure of the type one fibres of the multifidus including a decrease in fibres can also occur following the onset of low back pain (Ford et al., 1983). This may result in reduction of neuromuscular control in fatigue situations and subsequent lumbar-pelvic clinical instability, as the multifidus cannot hold the contraction or the repetitive nature of contractions for the required time frame.

Low back pain is known to increase the threshold of transversus abdominus activation and cause a loss of its tonic activity so that it becomes phasic (Hodges et al., 2003). This suggests that the background stabilisation property provided by transversus abdominus is lost. In participants with a chronic history of low back pain, whilst in remission of pain, a delay in the activity of transversus abdominus has been found, regardless of the direction of imposed force (Hodges & Richardson, 1998). Importantly, this demonstrates that even with the absence of pain, there are alterations to the coordinated firing pattern, which predisposes injury. A lack of feed forward activation will have joints unprepared to take load at the point of loading so there is a higher risk of injury. Importantly this can occur in the absence of pain and may be related to performance deficit in the athlete. Other research has shown that imbalanced patterns of erector spinae activity and reduced trunk extension strength which results from low back pain remains present if low back pain does not resolve (Renkawitz et al., 2006).

Although likely to be multi-factorial, one explanation for recurrent low back pain in athletes could be that athletes who demonstrate neuromuscular control alterations to sudden trunk loading have an increased risk of sustaining a low back injury (Cholewicki et al., 2005). Previously it has been shown that athletes with a recent acute low back injury exhibit altered neuromuscular control strategies for sudden trunk loading (Cholewicki et al., 2002). These findings are relevant to the unexpected and expected contact nature of sports such as the various body contact football codes and related other sports but also for the agility, change of direction and sudden stop-start nature of many running based ball sports.

Lumbar muscle activity during gait functions to control trunk movements (Carlson et al., 1988). In a non-athletic population, low back pain has been shown to produce poorly coordinated activity of the lumbar muscles during gait (Lamonth et al., 2005). This situation occurring in athletes in running based sports may lead to forces being directed at unprotected spinal structures producing subsequent mechanical stress and injury. Greater and more frequent mechanical spinal loading could contribute to both injury and delayed healing response. Similar to the non-athletic population, a situation may exist where low back pain fluctuates over time with recurrences or exacerbations and temporary remissions (Hestbaek et al., 2003; van Tulder et al., 2002). In support of this mechanism for repetitive and recurrent injury, Green et al. documented that athletes with a history of low back injury with current low back pain have a six times greater risk for future injury (Greene et al., 2001). For athletes with a previous history of low back injury who are now asymptomatic, approximately a three times greater risk of injury exists (Greene et al., 2001; Cholewicki et al., 2005).

4.2 Consequences to athletic performance

Low back injury in athletes may be of further significance as Nadler et al. documented that athletes with resolved low back pain from a history of low back injury demonstrate significantly diminished athletic performance in a 20m shuttle run test compared with a healthy group (Nadler et al., 2002). Despite this study, there is very little scientific literature investigating the consequences of current or resolved low back pain and injury on athletic performance. Research findings have demonstrated that weak hip extensors have been associated with the presence of low back pain, and in female athletes, the presence of hip weakness identified at the time of the pre-participation physical has been shown to be predictive of the subsequent development of low back pain (Kankaanpaa et al., 1998; Leinonen et al. 2000; Nadler et al., 2001; Nadler et al., 2002). Gluteus maximus should be the primary hip extensor during sprinting (Simonsen et al. 1985). During sprinting, the hamstrings should act as a transducer of power between the knee and hip joint and contribute little to hip extension (Jacobs et al., 1996). This transfer of power is essential in the execution of explosive movements like sprinting (Gregoire et al., 1984).

Significant alterations to hip extensor recruitment have been shown to occur with chronic low back pain during walking, causing the gluteus maximus to be inhibited and hamstrings overactive (Vogt et al., 2003). Hypothetically, gluteus maximus inhibition during sprinting may impact power development and sprinting performance and may require the hamstrings to contribute more force to hip extension rather than acting in its transducer role, potentially predisposing hamstring injury. This fits with the often talked about, but poorly researched syndrome proposed by Janda, the lower crossed syndrome, where decreased hip joint range of motion leads to hypermobility of the lumbosacral region (Janda, 1996), which may be another potential mechanism for low back pain. It also fits with models of overactivity of the global muscle system and a compromise in the local spinal muscle system, predisposing excess force directed at unprotected spinal structures and further back injury.

Given the consequences of low back pain on the lumbar-pelvic muscular system discussed earlier in this chapter, it is highly likely that other measures of athletic ability may be reduced in athletes with a history of low back injury. If objective deficits are documented in research, further research is also required to document that rehabilitation and management protocols are successfully able to reverse the decline in athletic performance, not just resolve symptoms.

4.3 Association of low back pain with other injuries

Given that evidence exists documenting that low back pain produces changes in the neuromuscular control of the lumbopelvis (Demoulin et al., 2007) and in athletes, it produces altered muscle response patterns required for lumbar-pelvic stabilization during sudden trunk loading following clinical recovery from low back pain (Cholewicki et al., 2002), it is reasonable to hypothesize that low back pain could increase the risk an athlete has of suffering other injuries. A prospective study by Nadler et al. showed a correlation between the prevalence of low back pain in athletes and lower extremity overuse syndromes, through an unclear mechanism (Nadler et al., 1998). In community level Australian Rules footballers, a history of low back pain has been shown to be a risk factor for other injuries, producing a 19% increased risk of overall injury rates (McManus et al., 2004). Changes in lumbar-pelvic stabilisation and neuromuscular control could explain the high rates of injuries such as hamstring injuries, groin injuries and other lower limb muscle strains which occur in the various football codes, cricket and track and field to name a few sports.

Using magnetic resonance imaging to confirm diagnosis of hamstring injury, 14% to 19% of all hamstring injuries are without muscle damage (Verrall et al., 2001; Verrall et al., 2003; Woods et al., 2004), suggesting no local muscle pathology. A recent study found this figure could be as high as 45% (Gibbs et al., 2004). Injury in such cases could possibly be related to altered functional biomechanics or pain referral that does not appear on cross sectional imaging. It is known that referred myotomal pain from lumbar-pelvic structures, the sciatic nerve and the gluteal or piriformis muscles can mimic hamstring strains (Verrall et al., 2001). The term 'back related hamstring injury' has been coined and is used to classify injuries as having both local hamstring signs and positive lumbar signs (Bennell et al., 1998, Orchard, 2001).

The association between low back ailments and hamstring injuries has been recognised for some time (Baquie & Reid, 1999). However, this relationship has not received as much recognition in the scientific literature as what is suggested anecdotally. Verrall et al. have performed a prospective study which showed that that a past history of low back injury approached significance for being a predictor for hamstring injury ($p=0.06$), without reaching the statistically significant level (Verrall et al., 2001). Further specific research has not followed on from this 114 participant study. A strong correlation between common lower limb soft tissue injuries, including hamstring and calf injuries, that involve L5 and S1 nerve supply, with increasing player age has been clearly demonstrated in the AFL's injury survey (Orchard & Seward, 2002). Orchard et al. suggest that on the basis that low back injuries are very common in elite athletes with increased levels of lumbar degenerative changes at the L4/L5 and L5/S1 levels (Ong et al., 2003), that subtle pathology may be present, which increases with age and which predisposes hamstring and calf injury

(Orchard et al., 2004). The association between low back injury and pathology and hamstring injury has extended into treatment approaches with authors documenting the use of mobilisation (Baquie & Reid, 1999) and slump stretching protocols (Kornberg & Lew, 1989; Turl & George, 1998) in the management of hamstring injured athletes with signs of lumbar injury.

4.4 Post career low back pain

Questions need to be raised regarding whether low back pain normalizes following a career of athletic participation. It is known that former elite athletes are more likely to receive hospital care suffering from musculoskeletal complaints in general (Kujala et al., 1996). However, in the largest study performed using self-reported questionnaires, it appears that low back pain is less common in former elite athletes (29.3% of 937) than in non-athletes (44% of 620) (Videman et al., 1995). This is despite an increase in degenerative radiological findings in former elite athletes (Videman et al., 1995; Lundin et al., 2001). It is unclear whether participation in certain sports will affect post career pain or the intensity of low back pain experienced (Lundin et al., 2001).

5. Diagnosis

Although most low back pain is non-specific and mechanical in nature (Burton et al., 1996), athletes presenting to a sports clinician with back pain may have a pathological cause. It is important to initially consider a broad differential diagnosis list. A sports clinician looking after athletic patients is responsible for performing a diagnostic triage to rule out red flag conditions, diagnose the condition and either referring out, or being responsible themselves for treating symptomatic tissues and recognising and evaluating functional deficiencies and aetiological factors responsible for factors causing the low back injury. Dealing with an athlete can often be a challenge when compared with the general population. A sports clinician must assimilate a large body of clinical information unique to the diagnosis and management of the special needs of those who participate in sport. This includes being highly familiar with the vast array of sports and the potential injury mechanisms for low back pain that could occur in a particular sport.

Whilst most back pain will be mechanical in nature it is important to exclude other diagnoses such rheumatological or inflammatory conditions, infection, fracture and neoplasm. This is particularly the case when adolescent athletes present with low back pain as they are more likely to potentially have a pathologic cause for their symptoms (Micheli et al., 1995). For this reason, it is important for those caring for younger athletes to maintain a high index of suspicion for some of the more common pathologic causes of low back pain in this population. Sports-related diagnoses that have been said to be considered include disc-related back pain, atypical Scheuermann's kyphosis, spondylolysis, spondylolisthesis and other stress fractures of the pelvis, especially in female athletes (Waicus & Smith, 2002). Other research has documented that junior athletes with chronic low back pain form a population of adolescents who have degenerative disc disease identified on magnetic resonance imaging (Dimar et al., 2007). For adolescent athletes with degenerative disc disease, the relative risk of reporting recurrent low back pain up to the age of 23 years is 16 compared with those having no disc degeneration (McManus et al., 2004). Furthermore, disc protrusion and Scheuermann-type changes also contribute to the risk of persistently recurrent low back pain at a later age (Salminen et al., 1999). How this should alter management approaches remains unclear as there is also a large proportion of adolescent athletes with signs of degeneration present on imaging who remain symptom free. Low back pain in adolescent athletes is a problem that should not be ignored but instead fully evaluated.

The challenge with diagnosis of back pain is that the tissue diagnosis model is mostly not relevant, despite advances in imaging techniques. Whether a sporting population or not, history must identify and eliminate potential red flag conditions that may be present that would indicate more serious pathology. Red flags are clinical indicators of possible serious underlying conditions requiring further medical intervention. Red flags were designed for use in acute low back pain, but the underlying concept can be applied more broadly in the search for serious underlying pathology in any pain presentation. Red flag conditions are listed in Table 1, and should be enquired about in all patients. The presence of red flags in acute low back pain suggests the need for further investigation and possible specialist referral as part of the overall strategy. If there are no red flags present it is safe to reassure the patient and move ahead with the diagnosis process.

Red flag conditions	History or examination findings
Possible fracture	Major trauma Minor trauma in elderly, osteoporotic or those taking long term corticosteroids
Possible infection	Symptoms and signs of infection such as fever or chills Recent bacterial infection Risk factors for infection such as underlying disease process, immunosuppression or intravenous drug use
Possible tumour	Age >50 or <20 years History of cancer Constitutional symptoms such as weight loss Pain at multiple sites Pain worse at rest Pain worsening at night Failure to improve with treatment Pain persists for more than 4-6 weeks
Possible significant neurological deficit	Severe or progressive sensory alteration or weakness Bladder or bowel dysfunction Evidence of neurological deficit (in legs or perineum in the case of low back pain)

Table 1. Red flag conditions for back pain

A focus should be made on the patients age and the age related differential diagnoses prior to full characterisation of the symptoms in history taking. As with all medical diagnosis, it is important to find out key information including the site of pain, whether any pain referral or radiation exists, associated symptoms in particular neurological deficit and systemic features of illness potentially leading to back pain, when the onset of pain began, the course of the pain, quality of pain, the severity, aggravating and relieving factors and movements, previous history of back pain and back injuries and treatment approaches used and their various success. History should also include questioning of the mechanism of injury or the inciting event. This mechanism of injury allows the clinician to predict what potential injuries may have occurred with the force transmitted and facilitates developing a rehabilitation program and implementing preventive measures through technique or training alterations if applicable.

Lawrence et al. state that the patient's athletic background should be explored (Lawrence et al., 2006). This includes types of sports played, duration of involvement, the level of competition along with what stage of the season the athlete is at, upcoming competition and future goals. This is relevant as it may impact upon the management approaches to be used and their success if an athlete is unwilling to miss a period of training or competition or is going to be uncooperative with management recommendations. It is also important to get an idea of what multidisciplinary management team and coaching staff the athlete has surrounding them as co-management is typically necessary and often mandatory when dealing with the high level elite and professional athlete. These multidisciplinary resources should be embraced and a good working relationship developed as cooperation is often required to implement management programs in an athlete centre approach to care.

In low back pain research performed on the general population, guidelines recommend early identification of psychosocial factors that could prevent recovery from acute low back pain (Ramond et al., 2011). As discussed earlier in this chapter it is unclear whether the yellow flag model is applicable to the sporting population. The presence of yellow flags may highlight the need to address specific psychosocial factors as part of a multimodal management approach. Yellow flags are psychosocial indicators suggesting increased risk of progression to long-term distress, disability and pain. Yellow flags were designed for use in acute low back pain. In principle they can be applied more broadly to assess the likelihood of development of persistent problems from any acute pain presentation. Yellow flags can relate to the patient's attitudes and beliefs, emotions, behaviors, family, and workplace. The behavior of health professionals can also have a major influence. Key factors in low back pain are: the belief that pain is harmful or severely disabling; fear-avoidance behavior (avoiding activity because of fear of pain); low mood and social withdrawal; and expectation that passive treatment rather than active participation will help (New Zealand Low Back Pain Guide, 1997). Future research is required to investigate the relevance of these factors in athletic populations.

Following history taking, physical examination should be equally as thorough and incorporate standard observations and structural analysis, range of motion assessment, palpation and traditional orthopaedic and neurological testing procedures to inform possible investigations if required. The single-leg hyperextension test has been described and is a useful provocative test when differential diagnosis includes spondylolysis (Jackson et al., 1976).

5.1 Imaging of injuries

Much controversy exists surrounding the utility of plain film, computed tomography, magnetic resonance imaging, and bone scintigraphy in the evaluation of sports-related spine injuries (Hollenberg et al., 2003). Diagnostic imaging should be used in an evidence based and targeted fashion. The evidence to support the use of diagnostic imaging in non-specific, mechanical low back pain without red flags present is lacking and its use is often costly, time consuming and potentially harmful to the patient when radiation doses are considered. The topic of routine screening is also a dated process. In football code players, it is unclear whether they have a greater prevalence of radiographic lumbar spine abnormalities, including spondylolysis and spondylolisthesis, as age-matched controls (Jones et al., 1999).

In a large retrospective study of plain radiographs of the lumbar spine of 4243 athletic men and women with low back symptoms, 14% had a radiologic diagnosis of spondylolysis and 47% of these (or 7% of all athletes with back symptoms) had associated spondylolisthesis (Rossi & Dragoni, 2001). However, the diagnosis of spondylolysis and spondylolisthesis does not always equate to the symptoms present. The prevalence of spondylolysis in the general population has been estimated between 3% and 6% (Bono, 2006). Most commonly spondylolysis and spondylolisthesis occurs at L5 (85% to 95% of cases) and L4 (5% to 15%) (Standaert et al., 2000). Degenerative findings are known to be higher in athletes with low back demands on radiographic imaging (Sward et al., 1991). Again, their presence does not have to equate a source of symptoms in all cases.

When investigating spondylolysis, imaging should commence with plain radiographs, with anteroposterior, lateral and oblique views. Grading of the spondylolisthesis can be made on the lateral film using the Myerding system. Whilst plain films can be diagnostic, CT is superior in the diagnosis and is the imaging modality of choice for the diagnosis of spondylolysis (Teplick et al., 1986). SPECT is sensitive to metabolic bone changes and is positive in acute spondylolisthesis, however it can be normal in chronic pars defects (Lusins et al., 1994), helping to diagnose acute versus chronic injury and in attributing a source of symptoms. SPECT has been shown to have superior sensitivity to standard bone scans for detecting spondylolysis (Bellah et al., 1991). Magnetic resonance imaging can detect early changes in bone marrow oedema but not fracture, however marrow oedema is known to predate a frank pars defect (Gundry & Fritts., 1999). The use of magnetic resonance imaging in the evaluation of spondylosis has mixed opinions in the literature (Hollenburg et al., 2003), but given the lack of radiation its use is increasing particularly when repeated scanning is required for follow up of adolescent athletes.

Other injuries that require imaging to diagnose include disc herniations. Magnetic resonance imaging is the imaging of choice for the diagnosis of disc herniation, foraminal narrowing and other disc injuries. It can also demonstrate degenerative disc disease and facet arthropathy as causes of back pain (Hollenburg et al., 2003). However, the exact correlation between a degenerated disc and low back pain has been described as elusive as high rates of radiographic findings of degenerative discs are found in asymptomatic patients (Boden et al., 1990). Fatigue type sacral stress fractures are a potential cause of low back pain in athletically active premenopausal women (Johnson et al., 2001). Although plain films can be diagnostic, symptoms typically precede radiographic findings by weeks to months (Johnson et al., 2001). Additionally there is difficulty interpreting radiological findings in the sacral area. Bones scans are very sensitive for stress fractures, but non-specific: a normal scan virtually excludes the diagnosis. CT is sensitive and specific for most stress fractures (Hollenburg et al., 2003). More recently magnetic resonance imaging has been used for the diagnosis of stress fractures (Major et al., 2000) despite previously being thought of sensitive but not specific. In the very early stages magnetic resonance imaging can detect medullary oedema but is insensitive for detecting a fracture line.

6. Management

Success in dealing with athletes with back injuries likely requires efforts to address both the cause of the injury and the most appropriate rehabilitation therapy (McGill, 2002). In many cases, addressing the cause of low back pain involves the athlete changing technique but without exception, they have to change the way they train (McGill, 2002). However, evidence to support risk factors for the development of athletic low back pain is lacking. Evidence exists showing that coaching aimed at improving technique in cricket fast bowlers decreases the prevalence and progression of disc degeneration measured with magnetic resonance imaging (Elliot & Khangure, 2002). Whether this translates to improved clinical results or to other sports remains unknown. Despite personal opinions that exist in the literature on the benefits of rehabilitation, there is lack of clinical research recruiting subjects with low back pain from athletic populations into randomised controlled trials investigating rehabilitation protocols or other treatment approaches. Apart from one short-term small study (Hanrahan et al., 2005), the author is not aware of other randomised controlled trials for the treatment or rehabilitation of low back pain with subjects drawn from an athletic population. It is not possible to produce evidence based guidelines for the management of back pain in different sports until an adequate literature base is established.

Current published evidence based guidelines for low back pain management for acute pain in non-athletic populations generally advocate an approach to management that includes advice to: remain active, modify activity, remove only those activities that specifically aggravate and potential replace with other non aggravating activity (relative rest) and to stay at work (Koes et al., 2001; Arnau et al., 2006). Simple analgesic pharmacological agents and exercise and manual therapies are often also advised in a multimodal approach. For chronic conditions various exercise-based protocols are often recommended. It may be for an athlete that a too aggressive active approach to management and the tissue loading from incorrectly prescribed 'stabilization exercises' (Callaghan et al., 1998; Kavcic et al., 2004) may be aetiological or aggravating factors. In support of this assertion, it has been shown that there is no significant advantage of additional core-strengthening in reducing low back pain occurrence in athletes (Nadler et al., 2002). Future research should investigate different rehabilitation protocols in a range of athletes from different sports.

The core principles of published guidelines should be used in the management of athletes with low back pain until they are replaced with athlete specific research and guidelines. The published guidelines in many ways mirror many of McGill's suggestions on how to reduce the risk of low back injuries in athletes, which include (McGill, 2002):

- Avoiding end range of spine motion during exertion. Examples of this include golfers sparing the spine from full lateral bend and near full rotation by reducing the back swing and grooving abdominal patterns that lock the rib cage to the pelvis on follow through.
- Use techniques to reduce reaction moments, such as tackling athletes directing force vectors through the lumbar spine to minimize resulting compressive forces.

- Avoid prolonged sitting (or sitting at all) on the bench, as prolonged flexion through sitting exacerbates discogenic back problems together with ligament based syndromes and results in decreased lumbar flexibility after a warm up period (Greene et al., 2002).
- Do not train shortly after rising from bed if a large amount of lumbar motion is required.
- Have athletes capable of stabilizing their lumbar spine irrespective of their phase of ventilation.
- Have the athlete contract musculature to stabilize the spine to more effectively transmit forces, particularly when an athlete might experience an unexpected load, when using combinations of simultaneous moments and after speed and acceleration of body segments are required.
- Practice spine sparing movement patterns and stabilizing motor patterns.

As most low back pain in athletes is likely to result from repetitive micro-trauma and fatigue from the often monotonous and repetitive overuse situations in training, management must include modifications in training (Baranto et al., 2009). Discussions should be made with coaching staff to ensure a period of relative rest, activity modification and if relevant, technique alteration is made to prevent the cycle of recurrent exacerbations and chronic pain.

Given the natural history of low back pain in adolescence involves a significantly increased risk of adult low back pain, it might be counterproductive to postpone treatment of adolescent athletes until the problems become more severe and chronic (Hestbaek et al., 2006). Hestbaek et al. have suggested a change in focus from the adult to the young population in relation to research, prevention, and treatment of low back pain (Hestbaek et al., 2006). However, it remains to be seen whether a greater focus on prevention and treatment can eliminate the risk and consequences of future low back pain episodes and minimise future chronicity.

A growing body of literature exists suggesting that classification of patients with non-specific, mechanical low back pain into subgroups for the purpose of directing treatment decision-making is important to improve prognosis, quality of care and patient outcomes (Borkan et al., 1998; Beaton et al., 2001). Various approaches are used to classify patients including the McKenzie or Mechanical Diagnosis and Therapy (MDT) technique and the Delitto or Treatement Based Classification (TBC). An important clinical symptom observed during the MDT examination process is centralization. This is where spinal and referred pain is abolished in a in a distal-to-proximal direction in response to therapeutic movement and positioning strategies (McKenzie & May., 2003; Aina et al., 2004). With the TBC patients are classified into three stages based on condition severity, ranging from the acute to subacute and advanced rehabilitation stage (Delitto et al., 1995). Stage 1, where the goal is symptomatic relief identifies four basic treatment subgroups, i.e. manipulation, exercise, stabilization, and traction, using specific clinical signs and symptoms, has been extensively researched and supported in the literature (Fritz et al., 2003; Fritz et al., 2000; Fritz et al., 2007).

An increased body of literature is also developing to support clinical prediction rules, which are prognostic models aiming to identify patient characteristics and clinical signs and symptoms to assign patients to treatment approaches to predict patient outcomes. Although such models do not exist for athletic populations, their development is encouraged given the multitude of treatment modalities that currently exist for back pain. When looking at clinical prediction rules for low back pain, two separate models have been developed to identify patients who would respond best to manipulation (Flynn et al., 2002; Fritz et al., 2005). The original model used five criteria: no symptoms below the knee, recent onset of symptoms (<16 days), low fear avoidance belief questionnaire score for work, hypo-mobility of the lumbar spine, and hip internal rotation range of motion (>35 degrees for at least one hip) (Flynn et al., 2002). This was modified to two criteria that included no symptoms below the knee and recent onset of symptoms (<16 days), as a pragmatic alternative for identifying patients most likely to positively respond to manipulation (Fritz et al., 2005). The stabilization clinical prediction rule was developed to determine whether patients with low back pain are likely to favorably benefit from stabilization exercises (Hicks et al., 2005). It uses four classification criteria, which include: age <40, positive prone instability test, positive aberrant trunk movements, and average straight leg range of motion >91 degrees. Whether these clinical prediction rules can be applied to athletes or whether new rules are required to be developed for athletic populations remain to be seen.

When specifically looking at exercise based rehabilitation protocols McGill suggests several key principles that should be included when developing exercise programs (McGill, 2002):

- Muscle endurance, not strength is more important.
- Patients should be encouraged to maintain a neutral spine when under load and use abdominal contraction and bracing in a functional way.
- No single abdominal exercise challenges all of the abdominal musculature while sparing the back. Therefore more than one exercise is required and the 'big three' is recommended: curl ups, side bridge and leg and arm extensions in the birddog position.

Once the basics are developed, then higher challenges and advanced exercises can be incorporated. When specifically looking at athletes, McGill suggests a five stage paradigm based around an adequate foundation of stabilizing motion/motor patterns (McGill, 2002):

1. Identifying the essential motions and grooving appropriate motion/motor patterns.
2. Ensuring joint and whole body stabilizing patterns.
3. Develop muscle endurance around these patterns.
4. Enhance strength.
5. Establish power.

Non-operative management is the mainstay for athletes with low back injuries. If simple conservative approaches to management fail, therapeutic epidural spinal injections are often the next line of therapy recommended in a trial of therapy. However, there are conditions which will require early surgical opinion and management, whilst failed non-operative management of severe, chronic low back pain may also require surgical management. Typical conditions for surgical referral include spondylolisthesis, disc herniation and traumatic fracture. The natural history and risk of progression and the non-operative and operative treatment of spondylolysis has been extensively covered by Bono (Bono 2006) and other authors (Lennard TA & Crabtree M, 2005). Bono states that indications for early surgical management for spondylolysis are a neurological deficit related to spondylolisthesis, a progressive slip or a grade III or high grade slip at presentation. Other literature also exists discussing the management of disc degeneration and disc herniation (Lennard TA & Crabtree M, 2005; Bono, 2006; Lawrence et al., 2006).

7. Future research

Further high quality research into low back pain in athletic populations is required. A list of research projects identified in this chapter include:

- Conducting long term longitudinal studies assessing the true prevalence, severity and epidemiology of low back pain in junior and senior athletic populations, across different sports and different grades of competition. Ideally with homogeneity in study design and methodology to allow direct comparisons with data from non-athletic populations.
- Development of a validated sports specific functional based outcome measure for athletic populations for use in both research and in clinical settings.
- Identifying risk factors for the development of low back pain and injuries in junior and adult levels athletes
- Determine if these risk factors are reversed, that it results in reduced low back pain and injuries.
- Assess the role that psychosocial variables or yellow flags have in the development of low back injury and chronic pain during play and after a career has ended.
- Determine whether a current or previous history of low back injury renders athletes susceptible to developing other injuries and whether management approaches incorporating the low back can subsequently prevent injury.
- Identify deficits in athletic ability occurring secondary to low back pain and injury and whether rehabilitation protocols or other management approaches can reverse these changes.
- Conduct randomised controlled trials to determine optimal management approaches for the prevention and treatment of acute and chronic low back pain from subjects recruited from an athletic population
- Assess whether current clinical prediction rules for the management of low back pain can be used on athletic populations or whether new prediction rules recruiting subjects from a sporting background are required.

8. References

Aina A, May S, Clare H. The centralization phenomenon of spinal symptoms: a systematic review. Man Ther 2004;9:134– 43.

Arnau JM, Vallano A, Lopez A, Pellise F, Delgado MJ, Prat N. A critical review of guidelines for low back pain treatment. Eur Spine J 2006, 15:543-53.

Bahr R, Andersen SO, Loken S, Fossan B, Hansen T, Holme I. Low back pain among endurance athletes with and without specific back loading--a cross-sectional survey of cross-country skiers, rowers, orienteerers, and nonathletic controls. Spine 2004, 29:449-54.

Balagué F, Dudler J, Nordin M. Low-back pain in children. Lancet 2003, 361:1403-4.

Baquie P, Brukner P. Injuries presenting to an Australian sports medicine centre: a 12-month study. Clin J Sport Med. 1997;7:28-31.

Baranto A, Hellström M, Nyman R, Lundin O, Swärd L. Back pain and degenerative abnormalities in the spine of young elite divers: a 5-year follow-up magnetic resonance imaging study. Knee Surg Sports Traumatol Arthrosc. 2006;14(9):907-14.

Baranto A, Andersen TI, Sward L. Preventing low back pain. In: Bahr R, Engebretsen L (Eds). Sports Injury Prevention. Chapter 8, Blackwell Publishing, 2009.

Beaton DE, Bombardier C, Katz JN, Wright JG. A taxonomy for responsiveness. J Clin Epidemiol 2001;54:1204–17.

Bellah RD, Summerville DA, Treves ST, Micheli LJ. Low-back pain in adolescent athletes: detection of stress injury to the pars interarticularis with SPECT. Radiology. 1991;180:509-12.

Bennell K, Wajswelner H, Lew P, Schall-Riaucour A, Leslie S, Plant D, Cirone J. Isokinetic strength testing does not predict hamstring injury in Australian Rules footballers. Br J Sports Med. 1998;32(4):309-14.

Biedermann HJ, Shanks GL, Forrest WJ, Inglis J. Power spectrum analyses of electromyographic activity. Discriminators in the differential assessment of patients with chronic low-back pain. Spine. 1991;16(10):1179-84.

Bo Andersen L, Wedderkopp N, Leboeuf-Yde C. Association between back pain and physical fitness in adolescents. Spine 2006, 31(15):1740-4.

Boden SD, Davis DO, Dina TS, Patronas NJ, Wisel SW. Abnormal magnetic resonance scans of the lumbar spine in asymptomatic individuals. A prospective investigation. J Bone Joint Surg Am. 1990; 72:403-408.

Bono CM. Low-back pain in athletes. J Bone Joint Surg Am. 2004, 86-A:382-96.

Borkan JM, Koes B, Reis S, Cherkin DC. A report from the Second International Forum for Primary Care Research on Low Back Pain. Reexamining priorities. Spine 1998;23:1992–6.

Burton AK, Clarke RD, McClune TD, Tillotson KM. The natural history of low back pain in adolescents. Spine 1996, 21(20):2323-8.

Callaghan JP, Gunning JL, McGill SM. The relationship between lumbar spine load and muscle activity during extensor exercises. Phys Ther 1998, 78:8-18.

Carlson H, Thorstensson A, Nilsson J. Lumbar back muscle activity during locomotion: effects of voluntary modifications of normal trunk movements. Acta Physiol Scand 1988, 133:343-53.

Carragee E, Alamin T, Cheng I, Franklin T, Hurwitz E. Does minor trauma cause serious low back illness? Spine 2006, 31:2942-9.

Cholewicki J, McGill SM, Norman RW. Lumbar spine loads during the lifting of extremely heavy weights. Med Sci Sports Exerc. 1991;23:1179-86.

Cholewicki J, Greene HS, Polzhofer GK, Galloway MT, Shah RA, Radebold A. Neuromuscular function in athletes following recovery from a recent acute low back injury. J Orthop Sports Phys Ther 2002, 32:568-75.

Cholewicki J, Silfies SP, Shah RA, Greene HS, Reeves NP, Alvi K, Goldberg B. Delayed trunk muscle reflex responses increase the risk of low back injuries. Spine 2005, 30:2614-20.

Cupisti A, D'Alessandro C, Evangelisti I, Piazza M, Galetta F, Morelli E. Low back pain in competitive rhythmic gymnasts. J Sports Med Phys Fitness 2004, 44:49-53.

Cresswell A. Responses of intra-abdominal pressure and abdominal muscle activity during dynamic loading in man. Eur J App Phys. 1993;66:315-20.

Crisco JJ, III, Panjabi MM. The intersegmental and multisegmental muscles of the lumbar spine. A biomechanical model comparing lateral stabilizing potential. Spine. 1991;16(7):793–9.

Demoulin C, Distrée V, Tomasella M, Crielaard JM, Vanderthommen M. Lumbar functional instability: a critical appraisal of the literature. Ann Readapt Med Phys 2007, 50(8):677-84.

Dimar JR 2nd, Glassman SD, Carreon LY. Juvenile degenerative disc disease: a report of 76 cases identified by magnetic resonance imaging. Spine J 2007, 7(3):332-7.

Druss BG, Marcus SC, Olfson M, Pincus HA. The most expensive medical conditions in America. Health Aff (Millwood) 2002;21:105–11.

Edgerton V, Wolf S, Levendowski D, Roy RR. Theoretical basis for patterning EMG amplitudes to assess muscle dysfunction. Med Sci Sports Exerc. 1996;28(6):744-751.

Elliott B, Khangure M. Disk degeneration and fast bowling in cricket: an intervention study. Med Sci Sports Exerc. 2002;34(11):1714-8.

Eriksson K, Nemeth G, Eriksson E. Low back pain in elite cross-country skiers. A retrospective epidemiological study. Scand J Med Sci Sports 1996, 6:31-5.

Ford D, Bagnall KM, McFadden KD, Greenhill B, Raso J. Analysis of vertebral muscle obtained during surgery for correction of a lumbar disc disorder. Acta Anat (Basel). 1983;116(2):152-7.

Flynn T, Fritz J, Whitman J, Wainner R, Magel J, Rendeiro D, et al. A clinical prediction rule for classifying patients with low back pain who demonstrate short-term improvement with spinal manipulation. Spine 2002;27:2835–43.

Fritz JM, Cleland JA, Childs JD. Subgrouping patients with low back pain: evolution of a classification approach to physical therapy. J Orthop Sports Phys Ther 2007;37:290–302.

Fritz JM, Delitto A, Erhard RE. Comparison of classification- based physical therapy with therapy based on clinical practice guidelines for patients with acute low back pain: a randomized clinical trial. Spine 2003;28:1363–71.

Fritz JM, George S. The use of a classification approach to identify subgroups of patients with acute low back pain. Interrater reliability and short-term treatment outcomes. Spine 2000;25:106–14.

Fritz JM, Childs JD, Flynn TW. Pragmatic application of a clinical prediction rule in primary care to identify patients with low back pain with a good prognosis following a brief spinal manipulation intervention. BMC Fam Pract 2005;6:29.

Gatchel RJ, Polatin PB, Mayer TG. The dominant role of psychosocial risk factors in the development of chronic low back pain disability. Spine 1995, 20:2702-9.

Gatt CJ Jr, Boland AL. Rowing injuries. Postgrad Adv Sports Med. 1989;III:1-17.

Gatt CJ Jr, Hosea TM, Palumbo RC, Zawadsky JP. Impact loading of the lumbar spine during football blocking. Am J Sports Med. 1997;25:317-21.

Gibbs NJ, Cross TM, Cameron M, Houang MT. The accuracy of MRI in predicting recovery and recurrence of acute grade one hamstring muscle strains within the same season in Australian Rules football players. J Sci Med Sport. 2004;7(2):248-58.

Greene HS, Cholewicki J, Galloway MT, Nguyen CV, Radebold A. A history of low back injury is a risk factor for recurrent back injuries in varsity athletes. Am J Sports Med 2001, 29:795-800.

Gregoire L, Veeger HE, Huijing PA, van Ingen Schenau GJ. Role of mono- and biarticular muscles in explosive movements. Int J Sports Med. 1984;5(6):301-5.

Gundry CR, Fritts HM. MR imaging of the spine in sports injuries. Magn Reson Imging Clin N Am. 1999; 7:85-103.

Hanrahan S, Van Lunen BL, Tamburello M, Walker ML. The short-term effects of joint mobilizations on acute mechanical low back dysfunction in collegiate athletes. J Athl Train 2005, 40:88-93.

Harreby M, Kjer J, Hesselsøe G, Neergaard K. Epidemiological aspects and risk factors for low back pain in 38-year-old men and women: a 25-year prospective cohort-study of 640 Danish school children. Eur Spine J 1996, 5(5):312-8.

Hestbaek L, Leboeuf-Yde C, Engberg M, Lauritzen T, Bruun NH, Manniche C. The course of low back pain in a general population. Results from a 5-year prospective study. J Manipulative Physiol Ther 2003, 26:213-9.

Hestbaek L, Leboeuf-Yde C, Kyvik KO, Manniche C. The course of low back pain from adolescence to adulthood: eight-year follow-up of 9600 twins. Spine 2006, 31(4):468-72.

Hicks GE, Fritz JM, Delitto A, McGill SM. Preliminary development of a clinical prediction rule for determining which patients with low back pain will respond to a stabilization exercise program. Arch Phys Med Rehabil 2005;86:1753–62.

Hides JA, Richardson CA, Jull GA. Multifidus muscle recovery is not automatic after resolution of acute, first-episode low back pain. Spine. 1996;21(23):2763-9.

Hill JJ, Keating JL. Risk factors for the first episode of low back pain in children are infrequently validated across samples and conditions: a systematic review. J Physiother. 2010;56(4):237-44.

Hodges P, Richardson C. Inefficient muscular stabilization of the lumbar spine associated with low back pain: a motor control evaluation of transversus abdominus. Spine. 1996;21(22):2640-50.

Hodges PW, Richardson CA. Delayed postural contraction of transversus abdominis in low back pain associated with movement of the lower limb. J Spinal Disord. 1998;11(1):46-56.

Hodges PW, Moseley GL, Gabrielsson AH, Gandevia SCl. Acute experimental pain changes postural recruitment of the trunk muscles in pain free humans. Exp Brain Res. 2003;151(2):262-271.

Hollenberg GM, Beitia AO, Tan RK, Weinberg EP, Adams MJ. Imaging of the spine in sports medicine. Curr Sports Med Rep. 2003;2(1):33-40.

Hosea TM, Gatt CJ, McCarthy KE, Langrana NA, Zawadsky JP. Analytical computation of rapid dynamic loading of the lumbar spine. Trans Orthop Res Soc. 1989; 14:358.

Hungerford B, Gilleard W, Hodges P. Evidence of altered lumbo-pelvic muscle recruitment in the presence of sacroiliac joint pain. Spine. 2003; 28(14): 1593-600.

Hutchinson MR. Low back pain in elite rhythmic gymnasts. Med Sci Sports Exerc 1999, 31:1686-8.

Iwai K, Nakazato K, Irie K, Fujimoto H, Nakajima H. Trunk muscle strength and disability level of low back pain in collegiate wrestlers. Med Sci Sports Exerc 2004, 36:1296-300.

Iwamoto J, Abe H, Tsukimura Y, Wakano K. Relationship between radiographic abnormalities of lumbar spine and incidence of low back pain in high school and college football players: a prospective study. Am J Sports Med 2004, 32:781-6.

Iwamoto J, Abe H, Tsukimura Y, Wakano K. Relationship between radiographic abnormalities of lumbar spine and incidence of low back pain in high school rugby players: a prospective study. Scand J Med Sci Sports 2005, 15:163-8.

Jacob T, Baras M, Zeev A, Epstein L. Physical activities and low back pain: a community-based study. Med Sci Sports Exerc 2004, 36:9-15.

Jacobs R, Bobbert MF, van Ingen Schenau GJ. Mechanical output from individual muscles during explosive leg extensions: the role of biarticular muscles. J Biomech. 1996;29(4):513-23.

Jackson DW, Wiltse LL, Cirincoine RJ. Spndylolysis in the female gymnast. Clin Orthop. 1976;117:68-73.

Janda V. Evaluation of muscular imbalance. In: Liebenson C. Rehabilitation of the spine. Baltimore: Lippincott Williams & Wilkins, 1996; ch 6, pp 97-112.

Johnson AW, Weiss CB Jr, Stento K, Wheeler DL. Stress fractures of the sacrum: an atypical cause of low back pain in the female athlete. Am J Sports Med. 2001; 29:498-508

Jones DM, Tearse DS, el-Khoury GY, Kathol MH, Brandser EA. Radiographic abnormalities of the lumbar spine in college football players. A comparative analysis. Am J Sports Med 1999, 27(3):335-8.

Jorgensen K, Nicholaisen T, Kato M. Muscle fiber distribution, capillary density, and enzymatic activities in the lumbar paravertebral muscles of young men. Significance for isometric endurance. Spine. 1993;18(11):1439-50.

Kaigle A, Holm S, Hansson T. Experimental instability in the lumbar spine. Spine. 1995;20(4):421-430.

Kankaanpää M, Taimela S, Laaksonen D, Hänninen O, Airaksinen O. Back and hip extensor fatigability in chronic low back pain patients and controls. Arch Phys Med Rehabil. 1998;79(4):412–7.

Kavcic N, Grenier S, McGill SM. Quantifying tissue loads and spine stability while performing commonly prescribed low back stabilization exercises. Spine 2004, 29:2319-29.

Keyserling WM. Workplace risk factors and occupational musculoskeletal disorders, Part 1: A review of biomechanical and psychophysical research on risk factors associated with low-back pain. AIHAJ. 2000;61(1):39-50.

Koes BW, van Tulder MW, Ostelo R, Kim Burton A, Waddell G. Clinical guidelines for the management of low back pain in primary care: an international comparison. Spine 2001, 26:2504-13.

Kornberg C, Lew P. The effect of stretching neural structures on grade one hamstring injuries. Journal Orthop Sports Phys Ther. 1989;10:481-487.

Kujala UM, Salminen JJ, Taimela S, Oksanen A, Jaakkola L. Subject characteristics and low back pain in young athletes and nonathletes. Med Sci Sports Exerc. 1992, 24(6):627-32.

Kujala UM, Sarna S, Kaprio J, Koskenvuo M. Hospital care in later life among former world-class Finnish athletes. JAMA 1996, 276:216-20.

Kujala UM, Taimela S, Erkintalo M, Salminen KK, Kaprio J J. Low-back pain in adolescent athletes. Med Sci Sports Exerc 1996, 28(2):165-70.

Kujala UM, Taimela S, Oksanen A, Salminen JJ. Lumbar mobility and low back pain during adolescence. A longitudinal three-year follow-up study in athletes and controls. Am J Sports Med 1997, 25(3):363-8.

Lamoth CJ, Meijer OG, Daffertshofer A, Wuisman PI, Beek PJ. Effects of chronic low back pain on trunk coordination and back muscle activity during walking: changes in motor control. Eur Spine J 2005, 15:23-40.

Leboeuf-Yde C, Lauritsen JM. The prevalence of low back pain in the literature. A structured review of 26 Nordic studies from 1954 to 1993. Spine 1995, 20:2112-8.

Leinonen V, Kankaanpää M, Airaksinen O, Hänninen O. Back and hip extensor activities during trunk flexion/extension: effects of low back pain and rehabilitation. Arch Phys Med Rehabil. 2000;81(1):32–7.

Lennard TA, Crabtree M. Spine in Sports. Mosby, 2005.

LeResche L, Mancl LA, Drangsholt MT, Saunders K, Korff MV. Relationship of pain and symptoms to pubertal development in adolescents. Pain 2005, 118(1-2):201-9.

Lundin O, Hellstrom M, Nilsson I, Sward L. Back pain and radiological changes in the thoraco-lumbar spin of athletes. A long term follow up. Scand J Med Sci Sports 2001, 11:103-9.

Lusins JO, Elting JJ, Cicoria AD, Goldsmith SJ. SPECT evaluation of lumbar spndylolysis and spondylolisthesis. Spine. 1994; 5:608-612.

Macintosh JE, Valencia F, Bogduk N. The morphology of the human lumbar multifidus. Clin Biomech. 1986;1:196-204.

Mahlamaki S, Soimakallio S, Michelsson JE. Radiological findings in the lumbar spine of 39 young cross-country skiers with low back pain. Int J Sports Med 1988, 9:196-7.

Major NM, Helms CA. Sacral stress fractures in long distance runners. Am J Roentgenol. 2000; 174:727-729.

Manek NJ, MacGregor AJ. Epidemiology of back disorders: prevalence, risk factors, and prognosis. Curr Opin Rheumatol. 2005;17(2):134-40.

McGill S. Low back disorders: Evidence based prevention and rehabilitation. Human Kinetics. United States of America. 2002.

McKenzie R, May S. The Lumbar spine: mechanical diagnosis and therapy. 2nd ed. Waikanae: Spinal Publication Ltd; 2003. 6 Delitto A, Erhard RE, Bowling RW. A treatment-based classification approach to low back syndrome: identifying and staging patients for conservative treatment. Phys Ther 1995;75:470–85; invited commentary 485–8.

McGill SM. Kinetic potential of the lumbar trunk musculature about three orthogonal orthopaedic axes in extreme postures. Spine. 1991;16(7):809-15.

McHardy A, Pollard H. Low back pain in golfers: a review. J Chiropr Med 2005, 4:135-43.

McManus A, Stevenson M, Finch CF, Elliot B, Hamer P, Lower A, Bulsara M. Incidence and risk factors for injury in non-elite Australian Football. J Sci Med Sport 2004, 7:384-91.

Meir RA, McDonald KN, Russell R. Injury consequences from participation in professional rugby league: a preliminary investigation. Br J Sports Med 1997, 31:132-4.

Micheli LJ, Wood R. Back pain in young athletes. Significant differences from adults in causes and patterns. Arch Pediatr Adolesc Med 1995, 149(1):15-8.

Mundt DJ, Kelsey JL, Golden AL, Panjabi MM, Pastides H, Berg AT, Sklar J, Hosea T. An epidemiologic study of sports and weight lifting as possible risk factors for herniated lumbar and cervical discs. The Northeast Collaborative Group on Low Back Pain. Am J Sports Med. 1993;21(6):854-60.

Nadler SF, Malanga GA, Feinberg JH, Prybicien M, Stitik TP, DePrince M. Relationship between hip muscle imbalance and occurrence of low back pain in collegiate athletes: a prospective study. Am J Phys Med Rehabil. 2001;80(8):572–7.

Nadler SF, Malanga GA, Bartoli LA, Feinberg JH, Prybicien M, Deprince M. Hip muscle imbalance and low back pain in athletes: influence of core strengthening. Med Sci Sports Exerc 2002, 34:9-16.

Nadler SF, Moley P, Malanga GA, Rubbani M, Prybicien M, Feinberg JH. Functional deficits in athletes with a history of low back pain: a pilot study. Arch Phys Med Rehabil 2002, 83:1753-8.

New Zealand Low back Pain Guide. Accident Rehabilitation and Compensation Insurance Corporation of New Zealand and the National Health Committee. Wellington 1997.

Ong A, Anderson J, Roche J. A pilot study of the prevalence of lumbar disc degeneration in elite athletes with lower back pain at the Sydney 2000 Olympic Games. Br J Sports Med. 2003;37(3):263-6.

Ogon M, Riedl-Huter C, Sterzinger W, Krismer M, Spratt KF, Wimmer C. Radiologic abnormalities and low back pain in elite skiers. Clin Orthop Relat Res 2001, 390:151-62.

O'Kane JW, Teitz CC, Lind BK. Effect of preexisting back pain on the incidence and severity of back pain in intercollegiate rowers. Am J Sports Med 2003, 31:80-2.

Orchard J. The use of local anaesthetic injections in professional football. Br J Sports Med 2001, 35:212-3.

Orchard JW. Intrinsic and extrinsic risk factors for muscle strains in Australian football. Am J Sports Med. 2001;29(3):300-303.

Orchard J. Missed time through injury and injury management at an NRL club. Sport Health 2004, 22:11-9.

Orchard JW. Is it safe to use local anaesthetic painkilling injections in professional football? Sports Med 2004, 34:209-19.

Orchard JW, Farhart P, Leopold C. Lumbar spine region pathology and hamstring and calf injuries in athletes: is there a connection? Br J Sports Med. 2004;38(4):502-4.

Orchard J, Seward H. Epidemiology of injuries in the Australian Football League, seasons 1997-2000. Br J Sports Med 2002, 36:39-44.

Orchard J, Wood T, Seward H, Broad A. Comparison of injuries in elite senior and junior Australian football. J Sci Med Sport 1998, 1(2):83-8.

Panjabi MM. The stabilizing system of the spine. Part I. Function, dysfunction, adaptation, and enhancement. J Spinal Disord. 1992a;5(4):383-9.

Panjabi MM. The stabilizing system of the spine. Part II. Neutral zone and instability hypothesis. J Spinal Disord. 1992b;5(4):390-6.

Panjabi MM. Clinical spinal instability and low back pain. J Electromyogr Kinesiol. 2003;13(4):371-9.

Renkawitz T, Boluki D, Grifka J. The association of low back pain, neuromuscular imbalance, and trunk extension strength in athletes. Spine J. 2006;6(6):673-83.

O'Sullivan P, Twomey L, Allison G. Dysfunction of the neuro-muscular system in the presence of low back pain - implications for physical therapy management. J Man Manip Ther. 1997a;5(1):20-26.

O'Sullivan P, Twomey L, Allison G, Sinclair J, Miller K. Altered patterns of abdominal muscle activation in patients with chronic back pain. Aust J Physiother. 1997b;43(2):91-98.

Ramond A, Bouton C, Richard I, Roquelaure Y, Baufreton C, Legrand E, Huez JF. Psychosocial risk factors for chronic low back pain in primary care--a systematic review. Fam Pract. 2011;28(1):12-21.

Ranson CA, Burnett AF, Kerslake RW. Injuries to the lower back in elite fast bowlers: acute stress changes on MRI predict stress fracture. J Bone Joint Surg Br. 2010;92(12):1664-8.

Rossi P, Dragoni S. The prevalence of spondylolysis and spondylolisthesis in symptomatic elite athletes: radiographic findings. Radiography. 2001; 7:37-42.

Roy SH, De Luca CJ, Snyder-Mackler L, Emley MS, Crenshaw RL, Lyons JP. Fatigue, recovery, and low back pain in varsity rowers. Med Sci Sports Exerc. 1990;22(4):463-9.

Salminen JJ, Erkintalo M, Laine M, Pentti J. Low back pain in the young. A prospective three-year follow-up study of subjects with and without low back pain. Spine 1995, 20(19):2101-7.

Salminen JJ, Erkintalo MO, Pentti J, Oksanen A, Kormano MJ. Recurrent low back pain and early disc degeneration in the young. Spine 1999, 24(13):1316-21.

Simonsen EB, Thomsen L, Klausen K. Activity of mono- and biarticular leg muscles during sprint running. Eur J Appl Physiol Occ Physiol. 1985;54(5):524-32.d Rehabil Clin N Am. 2000;11:785-803.

Standaert CJ, Herring SA, Halpern B, King O. Spondylolysis. Phys M, Sward L, Hellstrom M, Jacobsson B, Peterson L. Back pain and radiologic changes in the thoraco-lumbar spine of athletes. Spine 1990, 15:124-9.

Teplick JG, Laffey PA, Berman A, Haskin ME. Diagnosis and evaluation of spondylolisthesis and/or spondylolysis on axial CT. Am J Neuroradiol. 1986; 7:479-491.

Teitz CC, O'Kane JW, Lind BK. Back pain in former intercollegiate rowers. A long-term follow-up study. Am J Sports Med 2003, 31:590-5.

Tewes DP, Fischer DA, Quick DC, Zamberletti F, Powell J. Lumbar transverse process fractures in professional football players. Am J Sports Med. 1995;23(4):507-9.

Toueg CW, Mac-Thiong JM, Grimard G, Parent S, Poitras B, Labelle H. Prevalence of spondylolisthesis in a population of gymnasts. Stud Health Technol Inform. 2010;158:132-7.

Turl SE, George KP. Adverse neural tension: a factor in repetitive hamstring strain?. J Orthop Sports Phys Ther. 1998;27(1):16–21.

van Tulder M, Koes B, Bombardier C. Low back pain. Best Pract Res Clin Rheumatol 2002, 16:761-75.

Verrall GM, Slavotinek JP, Barnes PG, Fon GT, Spriggins AJ. Clinical risk factors for hamstring muscle strain injury: a prospective study with correlation of injury by magnetic resonance imaging. Br J Sports Med. 2001;35(6):435-439.

Verrall GM, Slavotinek JP, Barnes PG, Fon GT. Diagnostic and prognostic value of clinical findings in 83 athletes with posterior thigh injury: comparison of clinical findings with magnetic resonance imaging documentation of hamstring muscle strain. Am J Sports Med. 2003;31(6):969-73.

Videman T, Sarna S, Battie MC, Koskinen S, Gill K, Paananen H, Gibbons L. The long-term effects of physical loading and exercise lifestyles on back-related symptoms, disability and spinal pathology among men. Spine 1995, 20:699-709.

Videman T, Saarela J, Kaprio J, Näkki A, Levälahti E, Gill K, Peltonen L, Battié MC. Associations of 25 structural, degradative, and inflammatory candidate genes with lumbar disc desiccation, bulging, and height narrowing. Arthritis Rheum. 2009;60(2):470-81.

Vogt L, Pfeifer K, Banzer W. Neuromuscular control of walking with chronic low-back pain. Man Ther. 2003;8(1):21-28.

Waicus KM, Smith BW. Back injuries in the pediatric athlete. Curr Sports Med Rep. 2002, 1(1):52-8.

Walden M, Hagglund M, Ekstrand J. UEFA Champions League study: a prospective study of injuries in professional football during the 2001-2002 season. Br J Sports Med 2005, 39:542-6.

Walker BF. The prevalence of low back pain: a systematic review of the literature from 1966 to 1998. J Spinal Disord 2000, 13:205-17.

Walker BF, Muller R, Grant WD. Low back pain in Australian adults: prevalence and associated disability. J Manipulative Physiol Ther 2004, 27:238-44.

Watson KD, Papageorgiou AC, Jones GT, Taylor S, Symmons DP, Silman AJ, Macfarlane GJ. Low back pain in schoolchildren: occurrence and characteristics. Pain 2002, 97:87-92.

Werneke M, Hart D, Oliver D, McGill T, Grigsby D, Ward J, Weinberg J, Oswald W, Cutrone G. Prevalence of classification methods for patients with lumbar impairments using the McKenzie syndromes, pain pattern, manipulation, and stabilization clinical prediction rules. J Man Manip Ther. 2010; 18(4):187-204.Wilke H, Wolfe S, Claes L, Arand M, Wiesend A. Stability increase of the lumbar spine with different muscle groups. Spine. 1995;20(2):192-98.

Wojtys EM, Ashton-Miller JA, Huston LJ, Moga PJ. The association between athletic training time and the sagittal curvature of the immature spine. Am J Sports Med 2000, 28(4):490-8.

Woods C, Hawkins RD, Maltby S, Hulse M, Thomas A, Hodson A. The Football Association Medical Research Programme: an audit of injuries in professional football--analysis of hamstring injuries. Br J Sports Med. 2004;38:36-41.

4

Relationship of Duration and Intensity of Pain with Depression and Functional Disability Among Patients with Low-Back Pain

Michael O. Egwu[1] and Afolabi O. Olakunle[2]
[1]*Department of Medical Rehabilitation, Obafemi Awolowo University, Ile-Ife, Nigeria and Consultant Physiotherapist Department of Physiotherapy, Obafemi Awolowo University Teaching Hospitals Complex, Ile-Ife,*
[2]*Department of Physiotherapy, Obafemi Awolowo University Teaching Hospitals, Nigeria*

1. Introduction

Low Back Pain (LBP) is a common musculoskeletal disorder causing huge humanitarian and economical costs (Andersson, 1999). It is often classified, according to duration of pain, as acute (short term), sub-acute (intermediate) and chronic (long-term) and is typically referred to as being specific or non-specific (Andersson, 1999; Merskey and Bogduk, 1994). Specific LBP refers to symptoms caused by 'red flags' such as spinal fractures, cancers, infections, and cauda equina syndrome. However, approximately 90% of cases of back pain have no identifiable cause and are designated as non-specific (Deyo and Weinstein, 2001).

Non-specific LBP is described as a "mechanical" back pain of musculoskeletal origin in which symptoms vary with physical activity. Previouse studies have linked it's origin to various sources as follows: Matthews and Yates (1962) had demonstrated, with the help of epidurography, the presence of disc hernia which was resolved following mobilization; Irritation of spinal nerves causes spinal segmental sensitization, which limits the dynamic range of spinal segment mobility (Naguszewski et al, 2001; Cassius et al, 2002); unguided movement at the spine may strain the interspinous ligament to irritate the spinal segment (Lamb 1979; Cassius et al, 2002). Similarly, disc injury or gradually progressive micro trauma ends up in motion segment fusion which facilitates the deposition of collagen, hypo-mobility and pain (Lamb, 1979; Gose et al, 1998). Also, degenerative changes place the sclerotome, autonomic, motor and sensory systems in a hyper-excitable state, increases blood vessel tone, thus facilitating the release of endogenous algesic chemicals that irritate nociceptors (Lamb, 1979; Shacklock, 1995; Egwu et al, 2003). In addition, degeneration of the disc leads to a loss in disc height, thus reducing interpedicular distance, neural foraminal vertical height which may become stenotic (Matthews and Yates, 1962; Naguszewski et al, 2001). Facet changes and end-plate degeneration lead to osteophytes and leaping, which may encroach on the neural foramina anteriorly and/or posterior (Naguszewski et al, 2001). All of these eventuate into irritant focus, dysfunction and distorted neuro-dynamics with ectopic discharge that are the problems challenging the back pain patient. (Naguszewski et al, 2001; Amir et al, 1997; Amir et al, 1999; Devor, 1999). Moreso, physiological evidence shows that ectopic discharge of noxious impulses from nerve irritation sustains pain by triggering or enhancing sinusoidal voltage oscillation in dorsal root ganglion membrane potential (Amir et al, 1997; Amir et al,1999; Devor, 1999).

However, current reports suggest that the varieties of response to a painful experience are shaped by culture, literacy level and socio-economic status and are associated with the feelings of suffering, distress, functional disability, depression and so on (Merskey and Bogduk, 1994; Andersson, 1999). For instance Green et al (2003) observed that Caucasians report their pain promptly while African Americans with chronic pain report pain late and have more pain, depression, post traumatic stress disorders and impairment in their physical, emotional and social health. The above findings suggest that mood and other psycho-social states such as functional disability and depression may be crucial factors in determining who complains of LBP and their psycho-social response to it.

Depression (Dn) is a psychosocial condition characterized by difficulty in sleeping and concentration, decreased appetite and libido for at least 14 days. Other symptoms of depression are loss of interest and enjoyment, reduced energy, being easily fatigued, diminished activity, marked tiredness on slight effort, reduced concentration and attention on a task, reduced confidence and self-esteem, feeling of guilt and unworthiness, bleak and pessimistic views of the future and ideas or acts of self-destruction or suicide (WHO, 2001; Worz, 2003). On the other hand, Functional Disability (FD) is impairment in performing age-appropriate physical, mental, and social activities in daily life. It could be caused by pain, physical, cognitive and other mental impairments (Anthony and Schanberg, 2003). Non specific LBP is known to have a relationship with Dn and FD; however, the influence of duration and intensity of LBP on levels of Dn and FD is not clearly understood (Dworkin and Gitlin, 1991; Croft et al., 1995; Fishbain et al, 1997).

In addition, the term 'depression' is a continuum that includes lower mood states lacking clinical significance. However, at the other extreme of the continuum includes major (clinical) depressive disorders requiring clinical attention. Major Depression (MDn) is a mental disorder characterized by an all encompassing low mood, accompanied by low self – esteem and by loss of interest or pleasure in normally enjoyable activities (American Psychiatric Association, 1994). It is known that patients with MDn often do not comply with prescribed treatment regime, and if not detected in time worsen the person's physical health and slow down recovery from other ailments (American Psychiatric Association, 1994; Worz, 2003).

The prevalence rate of MDn has been reported in developed countries (Sullivan et al., 1992; Banks and Kerns, 1996; Hope and Foreshaw, 1999; Caragee, 2001), however, information concerning the prevalence of MDn among Nigerian patients with Chronic LBP (CLBP) is scant. It is important to know the rate occurrence of MDn among Nigerian patients as it will help clinicians to look out for red flags indicating the presence of MDn to facilitate decision on which LBP patient will need psychoanalysis and therapy to enhance compliance and/or efficacy of chemotherapy or physiotherapy for LBP. This study therefore examined the relationship of CLBP duration and intensity with, Dn and FD and also estimated the prevalence rate of MDn among Nigerian patients with CLBP.

2. Method

Subjects: one hundred patients (41 male, 59 female) with chronic non specific LBP (aged range 20 to 85 years, mean age 54±12.84years) participated in this study. These were patients with LBP of not less than 3 months duration seen in the Physiotherapy Departments of Obafemi Awolowo University Teaching Hospitals Complex (OAUTHC), Ile-Ife; Ladoke Akintola University Teaching Hospital,Osogbo; Osun State Hospital,Asubiaro and Ilesa units and National Orthopaedic Hospital,Igbobi, Lagos all in south west Nigeria between February 5 and September 19, 2010. Prior to this, ethical clearance was obtained from the Ethics and Research Committee of OAUTHC, permission was sort and obtained from the head of Physiotherapy Department of each participating hospital and informed consent was obtained from the patient after explaining the research procedure.

In oder to ensure particpation of indigenouse Yoruba speaking patients, Beck Depression Inventory II and Oswestry Disability Index 2.0 were translated into Yoruba language at the Department of Linguistics and African language, Obafemi Awolowo University, Ile-Ife, Nigeria. Similarly, to ensure validity of the translated questionnaires, ten patients with non-specific low back pain low back pain who were literate in both English and Yoruba language were requested to respond to the English version of the questionnaires and after ten minutes, they were also requested to respond to the translated one without prior knowledge that they would be filling the Yoruba translation. The translated questionnaires were found to be valid as all of them chose the same options in each of the questionnaires.

Inclution criteria:

1. Clinical diagnosis of LBP by a physician,
2. Pain duration not less than three months
3. Absence of any other ailment such as headache, infection, fever etc that have pain as one of its symptoms,
4. No history of mental illness,
5. No physical disability,
6. No history of recent life stressing events such as bereavement, huge financial loss or job loss,
7. literacy in either Yoruba or English language

Procedure: On arriving into the consulting room, patient's height was measured using a validated height metre. The subjects stood erect, barefoot on a flat surface, with the occiput, upper back, buttocks and heels, touching the height metre. In line with the view of Steele & Spurgeon (1983), the upper margin of the external auditory canal opening were in the Frankfurt horizontal plane, the point of greatest height to the nearest 0.1cm was then marked off on the height metre. Weight was measured to the nearest 0.1Kg with a weighing scale (Hanson Company, Ireland) and Body Mass Index (BMI) was later calculated by dividing the respondent's weight by the square of his or her height (Egwu et al, 2007). All measurements were taken by the same examiner with subjects on minimum clothing. Semantic differential scale (Olaogun et al, 2004) which has both English and Yoruba versions was giving to the subject to respond to alongside the above questionnaires. The subjects were then requested to choose and respond to either the English or the Yoruba version of the questionnaires and rate their pain accordingly.

3. Data analysis

Descriptive statistics of percentage, mean and standard deviation were used to summarize the subjects' age, height, weight, BMI, pain duration, pain intensity, Dn and FD scores. Spearman rank order correlation coefficient and Chi-square test were used to assess relationships and differences respectively, among the variables. Data were analyzed using Statistical Package for Social Sciences (SPSS) software version16. Significance was fixed at an alpha level of 0.05.

4. Results

The minimum, maximum, range, mean and standard deviation of the physical characteristics (age, weight, height, BMI), duration and intensity of pain, Dn and FD scores of the patients are shown in table 1. It can be seen from this table that on the average, subjects were 54 years old weighing 74 kilogram with BMI of 28 Kg/m². their pain had lasted for an average of 26 months with an intensity of 6/10 in the semantic differencial scale, their Dn score was 12 and FD score was 33. An analysis of the pain intensity distribution (not shown) reveals that 12% (N=12) had mild (1-3) pain, 58% (N=58) had moderate pain (4-6), while 30% (N=30) had severe pain (7-10). Also, their level of Dn was as follows: minimum (N=63, 63%), mild (N=21, 21%), moderate (N=12, 12%), and severe

Variables	Minimum	Maximum	Range	Mean ± SD
Age(Years)	20.00	85.00	65.00	54.00 ± 12.84
Weight(Kg)	35.00	150.00	115.00	73.89 ±17.73
Height(m)	0.96	1.90	0.94	1.64 ± 0.11
BMI (Kg/m²)	15.77	50.70	34.94	27.45 ±5.99
Pain duration (months)	6.00	180.00	174.00	26.06 ±32.37
Pain intensity	2.00	9.00	7.00	5.63 ± 1.84
Depression score	0.00	37.00	37.00	12.20± 8.33
Disability score	0.00	80.00	80.00	33.40 ± 18.10

BMI - Body mass index, Kg – Kilogram, m – metre, m² - metre squared

SD - Standard deviation

Table 1. General characteristics of the respondents and their psycho-physiological variables

	Functional Disability					
Level of depression	Minimal (N=28)	Mild (N=34)	Moderate (N=32)	Severe (N=6)	χ^2	p
Mild (N=63)	27	1	0	0	30.25	0.01*
Moderate (N=21)	17	13	4	0		
Severe (N=12)	17	6	6	3		
Crippled (N=4)	2	1	2	1		

*Significant at 0.05 alpha level

Table 2. Chi-square test of association between levels of depression and functional disability in patients with chronic low back pain.

(N=04, 04%). 28% (N=28) had minimal FD, 34%(N=34) reported moderate FD, others had 32% (N=32) and 6% (N=6) severe and cripling FD respectively (table 2). Level of Dn significantly ($P<0.05$) correlated to FD and BMI while pain intensity correlated significantly ($P<0.01$) with both Level of Dn and FD (tables 2-4, fig. A & B). Age, gender and pain duration did not significantly relate to level of Dn and FD.

Variables	Relationship	ρ	p
Pain Intensity	Pain Duration	-0.147	0.145*
	Depression	-0.049	0.628*
	Functional Disability	-0.079	0.443*
Pain Intensity			
	Pain Duration	-0.147	0.145*
	Depression	0.325	0.001**
	Functional Disability	0.348	0.001**
Depression			
	Functional Disability	0.406	0.001**
	Gender	0.114	0.257*
	Age	0.043	0.668*

*Not significant.

** Correlation is significant at the 0.01 level (2-tailed).

Table 3. Spearman Rank Correlation Coefficients showing relationship among the studied psycho-physiological variables.

	Levels of Depression					
BMI Rating	Minimal	Mild	Moderate	Severe	χ^2	p
Underweight (>18.5)	1	3	0	0	18.84	0.03*
Normal (18.5-24.9)	19	2	2	1		
Overweight (25.0- 29.9)	34	8	5	3		
Obese (30.0-39.9)	9	8	5	0		

*Significant at 0.05 alpha level.

Table 4. Chi-square test of association between levels of Depression and Body Mass Index (BMI) of subjects (N=100).

5. Discussion

Standard internationally accepted definition of chronic pain (CP) is not available (Harshall and Ospina 2003), however IASP defines CP as pain without apparent biological value that has persisted beyond the normal tissue healing time of 1-6months and recommended 3 months as a good cut off point between acute and CP (Merskey and Bogduk, 1994). Also, it is known that most LBP patients who attend physiotherapy clinics are chronic episodic back pain sufferers experiencing a flare–up and who have been on and off chemotherapy and/or physiotherapy (Egwu and Nwuga, 2008). Therefore, no attempt was made in this study to control subject's therapy.

In this study, the relationship of duration and intensity of LBP with Dn and FD was investigated among patients whose pain had lasted for at least 3months. The result reveals that level of Dn significantly ($P<0.05$) correlated to FD while pain intensity correlated significantly ($P<0.01$) with both Level of Dn and FD.

In oder to understand and explain why rise in pain intensity and not the duration of pain affect Dn and function, the ambient mood state, culture and belief system of this cohort need to be clearly understood. Pain is known to have affective, cognitive, emotional and sensory components and it has been observed that majority of Nigerian patients with LBP are in the low-medium socio-economic status and report for treatment late (Merskey and Bogduk, 1994; Egwu and Nwuga, 2008). The reason why patients report late for treatment is believed to be due to the fact that in Africa, pain is culturally interpreted as a harmless experience that accompany ageing (Onyejeme et al, 2002; Egwu and Nwuga, 2008)). Consequently, complaining of pain is seen as a sign of weekness and facial expression of pain is subdued as much as possible (fig. A) until the individual's tolerance limit is exceeded. This behaviour had been explained by Zola (1973) who observed that people seeking help for a symptom

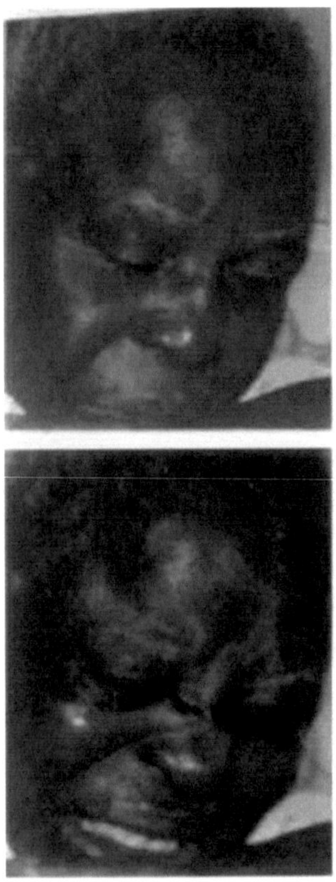

A = pain, B = increased pain intensity

Fig. 1. Pain is endured culturally without complain and without facial expression suggesting its presence (fig.A), until the individual's endurance limit is exceeded (fig.B) before patient finds need to seek help to stop this suffering from rising pain intensity leading to depression and functional disability.

for the first time do so because they are unable to tolerate it any longer. Also, perception of the nature and meaning of incoming sensory information, how the body responds physiologically and what actions are taken, as well as anticipation of what the future holds, are inextricably intertwined (Fordyce, 1995). Further, emotional states influence whether and how an aversive stimulus like pain is perceived. Emotional states also influence physiological processes such as heart rate, blood pressure and muscle tension, which then feed back to colour the perception of what is happening, the meanings assigned to it, the consequences inferred to follow, and the actions taken in response.

Therefore, the observed corelation between pain intensity, Dn and FD indicate that rise in pain intensity (fig.B) [reflecting the rising ectopic discharge of noxious impulses from nerve irritation enhancing the sinusoidal voltage oscillation in the dorsal root ganglion mambrane potentials] is one exercabating symptom that become intolerable among patient with CLBP driving them from the lower to the higher extremes of the Dn continuum. In addition, pain is known to focus emotions during difficult life situations, and when we assess ourselves in a situation and don't like where we are, where we have been, or where we are going and we can take no action to close the gap, we consider ourselves as suffering (Budd 1992; Worz 2003). Consequently part of the factors that determine tolerance limit is the feeling of suffering and/or perceived threat to life both of which affect level of Dn and FD. Thus, the level of Dn (minimum - severe) and FD (minimum – cripling) relates to the level of suffering percieved due to the worsening impact of poverty, high number of life stressing events and rising intensity of pain on work, motor activity and social role perfomance until some of the patients become severely(endogenously) depressed and/or crippled (unable to walk properly).

A 4% rate occurance of severe (major) Dn was observed in this study, this is very low compared to reports (16% - 37%) from advanced countries (Sullivan et al., 1992; Banks and Kerns, 1996; Hope and Foreshaw, 1999; Caragee, 2001; Cairns et al, 2003; Currie and Wang, 2003). However, it falls within the prevalence range (1.5% - 57%) according to the diagnostic and statistical manual of American Psychiatric Association (1980). This wide variation in the estimates of MDn is said to be dependent upon the setting, population and diagnostic instrument used.

The reason for this low prevalence of MDn in south-west Nigeria may therefore be related to the high tolerance for pain by an average African and the peculiar Yoruba culture of denial (American Pain Society, 2005; Green et al, 2003). It has been reported that Caucasians report their pain promptly and take more opioids while African Americans with chronic pain report pain late and take less quantity of opioids, have more pain, depression, post traumatic stress disorders and impairment in their physical, emotional and social health (Green et al, 2003, Meldrum, 2003). Thus, based on the theory of stimulus and habituation, they have less likelihood of rating depression as severe (Green et al, 2003; Egwu and Nwuga 2008). Besides, Yoruba culture mixed with christien religion don't admit or orally express negative emotions. For instance, somebody who is weak or in pain will rather say 'I am strong', while somebody who is penniless will say 'I have too much money'. Consequently, some patients whose Dn may have been of clinical level may have played it down thus explaining the low prevalence of MDn observed in this study.

Group health cooperative centre for health studies (2006) pointed out that there is significant differences among socio-cultural groups in the link between obesity and Dn. They noted that in groups were obesity is more common (low-medium socio-economic status non Caucasians) there is less Dn among obese people because they are not stigmatized. This report is consistent with our current finding that despite a significant relationship between BMI and Dn, non of the 22 obese respondent was severely depressed and it is in tandem with the perception in poor countries that being fat is a sign of wealth (Onyejeme et al, 2002; Haslam and James, 2005).

6. Conclusion

Pain intensity (not duration) correlate significantly with both level of Dn and FD without age and gender bias. Level of Dn also significantly correlate to FD and BMI with a 4% rate occurence of MDn underscoring the importance of the bio-psycho-social approach to CLBP therapy.

7. Acknowledgement

The authors wish to thank Mrs Adeola Faleye and Mrs Boboye who translated the questionnaires into Yoruba; Heads and clinicians of Physiotherapy departments in OAUTHC,Ile-Ife, LAUTECH,Osogbo, Osun State Hospital,Asubiaro and Ilesa and National Orthopaedic Hospital,Igbobi, Lagos for for their support during this work and all those who responded to the questionnaires for without them this work will not have been possible.

8. References

American Pain Society (2005). *Racial and ethnic identifiers in pain management: the importance to Research, clinical practice, and public health policy*, Williams D A (editor), APS Bulletin 2005; 15: 2, 1-4. Available at http://www.ampainsoc.org/pub/bulletin/spr05/sig1.htm, Accessed 5th May, 2006.

Amarican Psychiatric Asociation (1994) Staff, Diagnostic and Statistical Manual of mental disorders:DSM-IV, 4th ed., Amarican Psychiatric Asociation, Woshinton D.C.

Amir R, Devor M. Spike-evoked suppression and burst patterning in dorsal root ganglion neurons. Journal of physiology 501:183-196, 1997.

Amir R, Michaelis M, Devor M. Membrane potential oscillation in dorsal root ganglion neurons: Role in normal electrogenesis and neuropathic pain. The journal of Neuroscience 19:8589-8596, 1999.

Anderson BOA. (1999) The epidemiological features of chronic low back pain.Lancet 354, 581-585.

Anthony KK, & Schanberg LE (2003) Pain in children with arthritis: a review of the literature. Science 285, 409-412.

Banks, S.M. and Kerns, R.D (1996) Explaining high rates of depression in chronic pain: a diathesis-stress framework. Psychology Bulleting, 119:95–110.

Budd MA. Human suffering: the road to illness or gate-way to learning? Paper presented at Lee Travis institute for biopsychosocial research and U.S. Public Health Service, Boston Massachusset 1992; 1-17

Cairns M.C, Foster N.E, Wright C.C, Pennignton D. (2003) Level of Distress in a Recurrent Low Back Pain Population Referred for Physiotherapy. Spine 28, 953-959.

Cassius DA, Fisher A., Dubo H, Imamura M. Spinal segmental sensitization as a representation of all pain. Diagnosis by a new examination technique. Proceedings of the 10th World Congress on Pain. IASP Press, Seatle. 2002; P. 342.

Caragee E.J. (2001) Psychological and Functional Profiles in Selected Subjects With Low Back Pain. Spine1,198-204.

Croft PR, Papageorgiou AC, Ferry S, Thomas E, Jayson MI & Silman AJ (1996) Psychological distress and loe back pain. Evidence from a prospective study in the general population. Spine 20, 2731-7.

Currie SR & Wang J (2004) Chronic back pain and major depression in the general Canadian population. Pain 107, 54-60.

Deyo RA & Weinstein JN (2001) low back pain. New England Journal of Medicine. 344, 363-370.

Devor M. central changes mediating neuropathic pain. In : pain research and clinical management, vol. 3, proceedings of the Vth world congress on pain(Dubner R, Ghebhart GF, Bond MR, eds) pp114-128. Amsterdam: Elsevier, 1999.

Dworkin SF & Gitlin MJ (1991) Clinical aspect of depression in chronic pain patients. Clinical Journal of Pain 7, 79-94.

Egwu MO, Alabi MM, Nwuga VCB. Effect of Vertical Oscillatory Pressure on neck Pain and some cardiovascular variables. Physiotherapy 89:666-674, 2003.

Egwu MO, Nwuga VCB (2008) Relationship between low back pain and life- stressing events among Nigerian and Caucasian patients. Physiotherapy; 94:133-140.

Egwu MO, Adewale AO, Olaogun MOB. The Effect of Vertical Oscillatory Pressure on youths and Elderly adult low back Pain intesnsity and lumbosacral mobility. Journal of Japanese Physical Therapy Association 10:17-26, 2007.

Fishbain DA, Cutler R, Rosomoff RS (1997) Chronic pain associated depression: antecedent or cobsequence of chronic pain? A review. Clinical journal of pain 13, 116-137.

Fordyce WE. Back pain in work place: Management of disability in non-specific conditions. IASP press, Seattle, 1995; 35 – 70.

Green C, Baker J, Ndao-Brumblay, Nagrant A,Washinton J. Disparities between African Americans and Caucasians in pain and its effects.2003. Available at http://www.med.umich.edu/opm/newspage/2003/racialpain.htm, accessed 5th May, 2006.

Gose EE, NaguszewskiWK, Naguszewski RK. Vertebral axial decompression therapy for pain associated with herniated or degenerated disc or facet syndrome: An outcome study. Neurological Research 20:186-190.

Group Health Coopertive Centre for Health Studies (2006) Obesity and depression: link is strong among Caucaians and those with more education Xagena Medicine. Accessed march 31, 2010.10:31a.m.

Gunn CC, Milbrandt T. Early subtle signs in low back sprain. Spine 3:3, 1978.

Haslam DW & James WP (2005) 'Obesity'. Lancet 366, 1197-1209.

Harshall C & Ospina M (2003) How prevalent is chronic pain? Pain clinical updates 11, 1-4.

Hope P & Forshaw MJ (1999) Assessment of psychological distress is important in patients presenting with low back pain. Physiotherapy 85, 563-570.

Lamb DW. The neurology of spinal pain. Physical therapy 59:971-973, 1979.

Matthews JA, YatesDAH. Reduction of lumbar disc prolapse by manipulation. British Medical Journal 3:696-697, 1962.

Meldrum ML (2003). Opioids and Pain Relief: a historical perspective. Progress in Pain Research. Seattle, IASP Press 25.

Merskey H, Bogduk N (1994) Classification of chronic pain. IASP press, Seattle; 189 - 200.

Naguszewski WK, Naguszewski RK, Crose EE. Dermatomal Somatosensory evoked potential demonstration of nerve root decompression after VAX-D Therapy. Neurological Researcch 2001; 23, 706-714, 2001

Olaogun MOB, Adedoyin RA, Ikem IC and Anifaloba RO (2004) Reliability of rating low back pain with a visual analogue scale and a semantic differential scale. Physiotherapy Theory and Practice 20, 135-142.

Onyejeme BO, Onyeneke EC, Erigymremu GE (2002) The effect of social and cultural variables on the treatment of pain in Eastern Part of Nigeria - Epidemiological Study', 10th World Congress on Pain, August 17 - 22, Sandiego, California,79.

Steele MF, Spurgeon JB (1983) Body size, Body form and Nutritional intake of Blacks Ages 9 Living in Rural and Urban Regions of Eastern North Carolina. Growth 47, 207-216.

Sullivan, M.J.L., Reesor, K., Mikail, S.F., Fisher, R. (1992) The treatment of depression in chronic low back pain: review and recommendations. Pain 50, 5–13.

World Health Organization (2001) Mental health: New understanding, new hope. Geneva, Switzerland.

Worz R (2003). Pain in Depression - Depression in Pain. Pain clinical updates, Seattle, IASP Press 11.

Zola IK (1973) Pathways to the doctor- from person to patient. *Social science medicine*, 677-689

5

Psychosocial Risk Factors in the Development of LBP

Simone Ho
The Chinese University of Hong Kong,
Hong Kong

1. Introduction

Low back pain (LBP) is the most prevalent and expensive musculoskeletal problem worldwide. As many as eight in every ten adults experience LBP at some point in their lifetime (Dionne et al., 2001). An episode of LBP usually lasts a few days in most people, unfortunately in some individuals, the pain deteriorates and persists for an extended period of time; to a point where it considerably limits daily activities including work and leisure (Truchon & Fillion, 2000). Substantial economic burden is incurred due to sickness absence, with loss of productivity and healthcare costs relating to treatment for these chronic or recurrent LBP patients. It was estimated that the total healthcare expenses incurred by LBP patients costs $91 billion in the US (Luo et al., 2004).

LBP is a complex multifactorial phenomenon. The complexity of LBP may be viewed as the multiple biomechanical, psychosocial and individual factors which are closely-interacting with each other (National Institute of Occupational Safety and Health, 1997). Although the etiologic mechanisms for LBP are still poorly understood, many reviews and studies have concluded that the ergonomic work factors are most common LBP risk factors. They include lifting, forceful movement, whole body vibration, and awkward postures (Cohen et al., 2010, Vandergrift et al., 2012; Punnett et al., 2005). Increasing evidence exists that suggests psychosocial phenomena are also linked to low back problems, although to date the evidence for these is less conclusive. Understanding the importance of the psychosocial pathway in the development of LBP lies not only in the advancement of knowledge in the phenomenon, but also in designing preventive interventions. Five plausible explanations have been suggested to account for associations between work-related psychosocial factors and musculoskeletal symptoms (National Institute of Occupational Safety and Health, 1997).

2. Psychosocial factor

A psychosocial factor "may be defined as a measurement that potentially relates psychological phenomena to the social environment and to the pathophysiological changes" (Hemingway & Marmot, 1999). The concept of psychosocial factors includes a vast array of conditions that fall within three separate domains (National Institute of Occupational Safety and Health, 1997): (1) factors associated with the job and work environment, (2) factors associated with outside of work environment (3) characteristics of the individual.

Included in the domain of job and work environment include various aspects of job content, for example, perceived workload, monotonous work, low job control, low job satisfaction, limited social support (Truchon, 2001). Outside of work environment parameters typically include factors associated with demands and responsibilities in family situation and leisure time, and social/familial relationship and support. Finally, individual factors are generally of two types corresponding to: (1) sociodemographic factors, for example, age, social class, culture, educational status; life style (smoking) and (2) psychological factors, for example, affective variables (anxiety and depression), personality traits, cognitive variables (fear avoidance and life satisfaction) and coping strategies (catastrophizing) (Pincus et al., 2002a; Truchon, 2001).

It is important to note, however, that the linkages between psychosocial factors and LBP are complex and influenced by a multitude of conditions. Psychosocial factors may exert effect alone or combine in clusters, and may act at different stages of LBP. In particular, both personal and situational characteristics may lead to differences in the way individuals exposed to the same situation may perceive and/or react to the situation (Burton & Erg, 1997; National Institute of Occupational Safety and Health, 1997).

2.1 Psychosocial work characteristics

Psychosocial factors related to job and work environment are characterized by high perceived workload, monotonous work, low job control, low job satisfaction, and limited social support. A number of reviews have shown varying levels of associations between measures of work-related psychosocial factors and self-reported back pain (Bongers et al., 1993; Davis & Heaney, 2000; Hartvigsen et al., 2004; Hoogendoorn et al., 2000; National Institute of Occupational Safety and Health, 1997; Ramond et al., 2011).

Bongers et al. (1993) reviewed 46 articles published between 1973 and 1992, in which some evidence for an association between low back disorders and monotonous work was found. Evidence shows that there is a contradictory relationship between low back disorders and work demands while the evidence for a relationship between poor social support and low back disorders is mixed. In a similar vein, among the 13 studies published between 1973 and 1994 reviewed by NIOSH (National Institute of Occupational Safety and Health, 1997) found mixed evidence for an association between monotonous work and back disorders and contradicting evidence for an association between back disorders and job dissatisfaction. Weak evidence is found in the association between social support and back disorders while the relationship between low job control and back disorders has limited evidence. However, significant association was found between back disorders and perceptions of intensified workload (National Institute of Occupational Safety and Health, 1997). It is important to note that potential covariates were controlled in most studies that have been reviewed by NIOSH. Davis & Heaney (2000) reviewed 66 articles that were published before 1999. They concluded that there seems to be a consistent relationship between low job satisfaction and job stress with the development of LBP in those better quality studies. Likewise, Hoogendoorn et al. (2000), after reviewing 11 cohort and two case-control studies, concluded that there was strong evidence for low job satisfaction and low social support in the workplace as risk factor for back pain.

On the contrary, Hartivigsen et al. (2004), after critically assessing 40 epidemiological literature published between 1990 and 2002, found moderate to strong evidence for no association between LBP and consequences of LBP and perception of work, organizational aspects of work, social support and stress at work. The major strengths in this review are the inclusion of prospective cohort studies and that both the level and the strength of evidence are reported. Similarly, a recent study by Clays et al. (2007), in a longitudinal study of 2556 middle-aged workers, found a non-significant association between LBP and low decision latitude, high job insecurity, feeling stressed at work after adjusted individual and physical risks. The major flaw in this study is a large drop out rate, which may probably lead to selection bias.

Despite the findings were inconsistent, some authors (Bongers et al., 1993; National Institute of Occupational Safety and Health, 1997) suggest that intensified workload, monotonous work, low job satisfaction, low social support, low job control, and job stress may be associated with LBP or low back disorders. However, the possible effect of gender has not been evaluated in most of these epidemiological studies. There are only few reviews and studies that have analyzed female group and male group separately or have investigated the psychosocial risk factors among women only.

In a cross-sectional study by Josephson et al. (1998) of 269 female nursing personnel, found insufficient social support had the highest risk for care seeking behavior for low back pain, as assessed by physical examination and blinded interviews after confounder adjustments for age and smoking. Barnekow-Bergkvist (1998) defined, in a 18-year study of 425 Swedish students, an outcome as self-report of low back symptoms. The students had psychosocial assessment which included psychosocial stress at work, sociodemographic factors and stress, and individual attributes. Among the women, low back problems were related to monotonous work. The strength of this study is that it analyzed the influence of physical workload, psychosocial stress, and sociodemographic and individual factors together. However, those who worked less than 16 hours/week were excluded (women > men), the results of the analyses of associations between work-related risk factors and low back problem of women may be underestimated.

In contrast, Vingard et al. (2000), in a study of 1193 working women and 925 working men, found monotonous work and low job satisfaction had very limited influence in women when compared to men, after adjustment for lifestyle, and physical loads in leisure time and sport activities. They explained that possibly women are more satisfied with their work situation or have lower expectations than men. In a similar vein, Hofftman et al. (2004) reviewed 14 studies regarding gender differences in the effect of risk factors on back complaints and found that psychosocial work factors are shown to be important, but to a lesser extent in women when compared to men.

In a recent systematic review, Ramond et al. (2011) examined 23 prospective studies to review the evidence for psychosocial risk factors and LBP outcome in primary care, in which 16 psychosocial factors were included in the analysis. The review found that social support was not associated with LBP whilst job satisfaction was mostly not associated with LBP, and LBP requiring compensation was shown to be a predictive factor for negative LBP outcome. A longitudinal study by Vandergrift et al. (2012) investigated the relationship between physical and psychosocial risk factors for LBP among 1181 workers of automobile manufacturing company. An association was found between the psychosocial risk factors of low job control and high job demand and the development of LBP only in workers with high physical exposures.

2.2 Psychosocial role (outside of work)

The main focus in the relationship between psychosocial factors and LBP has chiefly been work related in most studies. However, there is increasing awareness that psychosocial factors which are unrelated to work, may also play an important role in the development of LBP. These psychosocial factors are characterized by family or social or emotional support, leisure time activity or social contact and participation, spousal relationship and housework satisfaction. Only a few prospective studies have addressed individual and outside of work environmental psychosocial factors.

Yip et al. (2004), studying 417 middle-aged women in a case control study, found an association between psychosocial stress related to housework and LBP after adjusted for working status and source of recruitment. However, no association was found between self reported poor relationship with cohabitants, housework satisfaction, living alone and the risk of LBP. Likewise, a cross sectional longitudinal study by Barnekow-Bergkvist (1998) described earlier found no association between physical activity at leisure time and low back symptoms.

Simiarly, Hoogendoorn et al.'s review (2000), as described above, also included assessment of psychosocial factors in private life such as family support, presence of a close friend or neighbour, social contact, social participation, emotional support and concluded that insufficient evidence was found for an effect of psychosocial factors in private life because the data were very limited.

However, Brulin et al. (1998) conducted a cross sectional study of 361 women in a Swedish home care service and found having children at home decreased the risk of low back complaints (OR 0.5), even after age adjusted. It was argued that having children in the family can protect against social isolation, and that social isolation was found to increase the risk of low back pain (Frymoyer & Cats-Baril, 1987). The strength of this study lies in its combined focus on sociodemographic, physical and psychosocial factors at work and physical activity during leisure time. This study is based on a single occupation therefore the sample may not be representative of the general population. Likewise, Thorbjornsson et al. (1998), in a longitudinal study over 24 years, found that social relationship satisfaction has a long-term effect on LBP among 252 women and 232men.

2.3 Individual characteristics

Since the complexity of LBP may be represented by the interactive effect of psychosocial, biomechanical and individual factors (Chany et al., 2006), a better understanding in individual factors may elucidate the complex reactions. These individual factors are characterized by history of previous low back pain or disorder, age, gender, socioeconomic characteristics, smoking habit, psychological or emotional distress, personality trait, cognitive appraisals and coping strategies.

2.3.1 History of previous low back pain/disorder

A previous pain/injury history to the lower back is consistently a strong risk factor for future reports of LBP in the work environment (Dempsey et al., 1997; National Institute of Occupational Safety and Health, 1997). In women during/after pregnancy, there is strong evidence for the risk of developing LBP in those who had previous LBP and previous lumbopelvic pain during or after pregnancy (Wu et al., 2004).

2.3.2 Age

Age has been suggested to play a role in the development of LBP in a review by Dempsey et al. (1997). The review found that the occurrence of LBP increases with increasing age up to about 50 to 60 years of age in several community-based studies, after which there seems to be a decline. Several biological plausibility of the role of age in LBP exist, in that accumulated work-related spinal damage including microtruama, natural degeneration of the spine, and decreased spinal load bearing capacity. With regards to LBP during pregnancy, Wu et al.'s review (2004) found the evidence for maternal age was conflicting. In the review, nine studies suggested a higher risk in younger women, two studies suggested a higher risk for older women, and 12 had no effect.

2.3.3 Gender

The risk of LBP is higher among women as consistently shown in community-based surveys, with odds ratios varying between 1.30 to 1.57 (Houtman et al., 1994; Skovron et al., 1994). The risk of back pain increases by twofold for women with back pain history and increases for women who have been pregnant before (Ostgaard & Andersson, 1991). Women are expected to suffer from more lower back pain because of their wider pelvis, the stress of hormonal changes, and childbirth (Meisler, 2003). Retrospective studies showed that 10-25% of women with chronic LBP report the first symptom of back pain during pregnancy (Biering-Sorensen, 1983; Svensson et al., 1990). During pregnancy, 50%-80% of women experience some degree of pregnancy-related low back pain (PLBP) and pelvic girdle pain (PPGP). Women who have previously had pelvic pain during pregnancy experience a relapse during 85% of a subsequent pregnancy (Mens et al., 1996). The pain symptoms often impact on daily activities, sleep and sometimes lead to work absenteeism and even chronic disability. A study shows lower quality of life during pregnancy among women with back problems (Olsson & Nilsson-Wikmar, 2004). Among these women who are affected by LBP, pain sometimes becomes chronic or recurrent (Larsen et al., 1999; Ostgaard et al., 1997).

2.3.4 Socioeconomic characteristics

It was found that low back symptoms were consistently more common among the women in the highest socioeconomic class (professionals, managers, and salaried employees) (Barnekow-Bergkvist et al., 1998). Likewise, Papageorogiou et al. (1997) found significant associations between LBP and higher social class and perceived inadequacy of income in a prospective population-based cohort study of 1412 working adults, the association was more marked in women. It was argued that higher stress levels among women in high and middle socioeconomic classes because of the combination of work-related stress and stress related to responsibilities for the family (Lundberg, 1999).

2.3.5 Smoking

Smoking is suggested to be a risk factor for low back disorders. NIOSH's review (National Institute of Occupational Safety and Health, 1997) found that the evidence is conflicting as smoking history has a positive relationship with low back pain, sciatica or intervertebral herniated disc in some studies whereas in others the relationship was negative. Josephson et al. (1998) described earlier found no association between smoking and LBP among female nursing personnel. Whereas McaGregor et al. (2004), investigation of 1064 women in a case control study, found smoking was associated with LBP.

Several explanations for the association have been proposed (Dempsey et al., 1997; National Institute of Occupational Safety and Health, 1997). It is speculated that back pain is caused by coughing associated with smoking which increases intradiscal pressure, leading to disc bulging and herniation. Another explanation postulated is nicotine's effect in diminishing blood flow to vertebral body and thus impacting discal metabolism and reducing mineral content of bone causing microfractures. However, it has been pointed out that a number of confounding risk factors have been linked with smoking, including lower economic class, education level, occupational exposure to heavy work, and psychosocial and life style factors (Dempsey et al., 1997).

2.3.6 Psychological/emotional distress

Given that psychological or emotional distress such as anxiety and depression may arise from work environment, they may also result from non-work environment. Strong evidence suggests that high comorbidity between psychological distress and pain, in particular among chronic pain patients (Gatchel & Gardea, 1999). The comorbid presentation of pain and depression is observed in as many as 50% of patients who suffer from chronic pain conditions (Gallagher, 2003). Similarly, in a recent review by Ramond et al, (2011), as described earlier, the association between negative LBP outcome and depression and psychological distress were found.

A sex-specific effect of anxiety on pain report is apparent. There is a significant association between anxiety and pain report in men but not in women. However, among women, depression, catastrophizing, anxiety sensitivity, stress, low energy and pain reports were significantly associated (Korovessis, 2010; Robinson et al., 2005).

Variability in psychological distress has been linked to LBP in women. Quint et al. (1998) in a case control study found that women had higher levels of psychological distress than men in a group of hospitalized patients with LBP. Clays et al. (2007), in the study of Belstress workers described above, found feeling depressed increased the relative risk for LBP in 30% women. However, Robinson et al. (2005), in a cross sectional study of 53 chronic LBP patients, found significant relationships between anxiety and the induced pain for men but not for women.

Some authors have reported other psychological variables to be related to LBP such as stressful life events and deficit in emotional awareness. Yip et al. (2004), as described above, found an association between stressful life event in the past 12 months and LBP with adjustments. However, Skillgate et al. (2007) failed to support this relationship and found no association between LBP and two or more life events or critical life changes experienced during the preceding 5 years. When women and men were analyzed separately, no systemic differences were observed regarding the estimated ORs of LBP.

Mehling et al. (2005), in a cross sectional study of 1180 transit operators, found alexithymia (deficit in emotional awareness) was associated with higher odds of LBP after controlled for demographic, behavioral and physical and psychosocial factors (OR=2.0, 95% CI 1.31-3.0). The association was stronger in women (OR=4.35) than in men (OR=1.83) with the factor 'difficult to identifying feelings' showing the strongest association. However, the authors pointed out that the study is limited by not controlling for depression or somatization, both factors are associated with alexithymia and LBP.

2.3.7 Personality trait

While the presence of psychosocial factors may be characteristic of the job and work environment, the biomechanical response of the individual to these psychosocial stressors may be dependent on the individual's perceptions about stress (Chany et al., 2006). Personality is one of the factors that may provide more clues to individual responses.

A study found that personality traits have association with muscle recruitment patterns, which may lead to variations in spinal loading as the individual is under psychosocial stress (Marras et al., 2000). The hypothesis posits that job-matched personality traits allow the individual to generate the appropriate biomechanical response with reduction of spinal loading (Chany et al., 2006). Contrarily, job-mismatched personality traits may provoke psychosocial stress which increases the trunk muscle activities, in turn, causes an increase in the spinal loading.

Chany et al. (2006), in a laboratory-based study of 12 experienced and 12 novice materials handlers (3 females and 21 males), ages ranged from 19 to 33 years, investigated the long term effect of repetitive lifting on the spinal loading of workers with different personality types. They found that intuitors personality had higher shear spinal loading compared with sensor type, and that perceiver personality had higher compressive and shear forces compared with judgers personality trait.

They suggested explanations for the trends, in that a personality of intuitors prefers to learn new skills, repetitive lifting task seems less matched to the intuitors' preferences, which was the primary influence on how the intuitors coactivated their muscles to high levels resulting in high spinal loads. Perceivers enjoy variations in circumstances, whereas judgers prefer scheduled work plans, the task appears to be a better match for the judgers (Chany et al., 2006). They concluded that inherent personality characteristics may play a role in one's motor control strategies when performing a repetitive lifting task and that the perceived stress (of repetitive lifting) manifests itself by increases in muscle coactivity which results in higher spinal loading. The study was designed to assess the spinal loading during lifting, with the assumption that increased spinal loading is associated with LBP, therefore, these results cannot be used to determine in other work situations.

2.3.8 Cognitive appraisals and coping strategies

Cognitive variables are among the best predictors of LBP-related chronic disability (Truchon & Fillion, 2000), but not extensively documented at the early stage of LBP (Truchon, 2001).

Truchon's review found that there may be a link that exists between a negative cognitive appraisal, negative affective state (depression, anxiety or anger) that it generates, and passive coping strategies (avoid the threat). Negative cognitive appraisals include pain catastrophizing, blaming oneself or others, anticipation of negative consequences associated with the disease, inaccurate interpretations of the significance of the physical symptoms and the effectiveness of the medical treatments.

Studies have focused on how pain catastrophizing has influenced LBP (Grant et al., 2002; Robinson et al., 2005). Pain catastrophizing is defined as an 'overappraisal' of the negative aspects/consequences of an experience (Raak et al., 2002) wherein an individual has a tendency to focus on and exaggerate the threat value of painful stimuli and negatively evaluate one's ability to deal with pain. Grant et al. (2002) studied the associations among pain appraisals, coping strategies, personal characteristics, perceived spousal responses and daily changes in mood and pain in 88 chronic LBP women. They found that catastrophizing appraisals and praying and hoping coping strategies were associated with an increase in negative mood or pain. For chronic back pain patients, catastrophizing was associated with increased pain intensity, anxiety and depression levels, after controlling for pain (Grant et al., 2002).

Likewise, Robinson et al. (2005), in a cross sectional study of 53 chronic LBP patients, found significant relationships between pain catastrophizing measures and the induced pain for men but not for women. In a similar vein, there seems to be a link between pain catastrophizing and pain reporting. Crombez et al. (2002) found pain catastrophizing (as measured by the Pain Catastrophizing Scale) was related to overpredictions of pain in 37 patients with LBP during performance tests of toe touch in standing position and straight leg raise in supine position.

Other coping strategy that exists may also have an effect on LBP. Busch (2005) interviewed 22 chronic LBP patients (15 women and 7 men) of working age in a rehabilitation clinic in Sweden and found that the majority of participants used disregarding strategy in response to chronic LBP. This disregarding strategy process developed from a psychological defense to a conscious coping strategy with changing pain-related behaviors. The change in pain-related behaviours of an increased sense of responsibility for pain and pain management helped rehabilitation of chronic LBP.

3. Conclusion

In an attempt to uncover a representative sample of publications that have investigated the psychosocial risk factors for developing LBP, there are some conclusions that may be drawn, although this is not a fully exhaustive review of publications. Most of the research has focused on LBP and psychosocial risk factors associated with work environment and much less in the area of factors associated with outside of work environment and individual characteristics. Women seem to have different psychosocial risk factors from men for the development of LBP.

The current state of knowledge suggests that psychosocial factors at work play an important role in the development of LBP, although the underlying mechanisms are not fully understood (Bongers et al., 1993; Hoogendoorn et al., 2000; National Institute of Occupational Safety and Health, 1997; Pincus et al., 2002b). It is unclear as to which psychosocial factors are definitively related to LBP because of the inconsistency of the results. It seems that individual's reactions to psychosocial work characteristics, for example job dissatisfaction and job stress, are more consistently related to LBP than are the psychosocial work characteristics themselves, for example work demand, low job control, low social support (Davis & Heaney, 2000). However, low job control and high work demand are associated with LBP in people with increased physical exposures (Vandergrift et al., 2012).

The inconsistent results in the studies may be attributed by some methodological problems in the majority of studies. Earlier studies involved working population of specific occupation and were not population based therefore limits the generalizability of the results. In an extensive review by Davis et al. (2000), two major methodological limitations were revealed in the critique of 66 articles pertaining to psychosocial work factors and LBP. First, very few studies had adequately controlled the potential confounding effect of biomechanical factors. Second, there is a paucity of high quality measures of both psychosocial work characteristics and biomechanical demands. Another major limitation includes insufficient prospective study designs rendering weak causal inferences. Thus, it is argued that psychological distress may simply be a consequence of chronic LBP without etiologic role in the development of low back disorder, or it may play a role in the etiology of LBP. Furthermore, there is random error in the operationalization of LBP and psychosocial variables due to the use of non-standardized questionnaires.

It is therefore important to consider the multitude of psychosocial factors and physiologic factors that are intertwined and ultimately produce the back pain experience. Understanding the complex and the interactive processes involved in LBP will not only help to predict those who develop LBP and chronic disability, as well as to develop more effective treatments for LBP patients, but also to design better epidemiological and intervention studies by the inclusion of potential psychosocial covariates.

4. References

Barnekow-Bergkvist, M., G. Hedberg, U. Janlert, & E. Jansson. (1998). Determinants of self-reported neck-shoulder and low back symptoms in a general population. *Spine,* 23, 2, pp. 235-243.

Biering-Sorensen, F. (1983). A prospective study of low back pain in a general population. I. Occurrence, recurrence and aetiology. *Scandivian Journal of Rehabilitation Medicine,* 15, pp. 71-79.

Bongers, P. M., C. R. de Winter, M. A. Kompier, & V. H. Hildebrandt. (1993). Psychosocial factors at work and musculoskeletal disease. *Scandivian Journal of Work Environmental Health,* 19, 5, pp. 297-312.

Brulin, C., B. Gerdle, B. Granlund, J. Hoog, A. Knutson, & G. Sundelin. (1998). Physical and psychosoical work-related risk factors associated with musculoskeletal symptoms among home care personnel. *Scandinavian Journal of Caring Science,* 12, pp. 104-110.

Burton, A. K. & E. Erg. (1997). Back injury and work loss. Biomechanical and psychosocial influences. *Spine,* 22, 21, pp. 2575-2580.

Busch, H. (2005). Appraisal and coping processes among chronic low back pain patients. *Scandinavian Journal of Caring Science,* 19, 4, pp. 396-402.

Chany, A. M., J. Parakkat, G. Yang, D. L. Burr, & W. S. Marras. (2006). Changes in spine loading patterns throughout the workday as a function of experience, lift frequency, and personality. *The Spine Journal,* 6, 3, pp. 296-305.

Clays, E., Bacquer D. De, F. Leynen, M. Kornitzer, F. Kittel, & Backer G. De. (2007). The impact of psychosocial factors on low back pain: longitudinal results from the Belstress study. *Spine,* 32, 2, pp. 262-268.

Cohen, S. P., Gallagher, R. M., Davis, S. A., Griffith, S. R. & Carragee, E. J. (2011). Spine-area pain in military personnel: a review of epidemiology, etiology, diagnosis, and treatment. *The Spine Journal,* (In press).

Crombez, G., C. Eccleston, J. W. Vlaeyen, D. Vansteenwegen, R. Lysens, & P. Eelen. (2002). Exposure to physical movements in low back pain patients: restricted effects of generalization. *Health Psychology,* 21, 6, pp. 573-578.

Davis, K. G. & C. A. Heaney. (2000). The relationship between psychosocial work characteristics and low back pain: underlying methodological issues. *Clinical Biomechanics,* 15, 6, pp. 389-406.

Dempsey, P. G., A. Burdorf, & B. S. Webster. (1997). The influence of personal variables on work-related low-back disorders and implications for future research. *Journal of Occupational and Environmental Medicine,* 39, 8, pp. 748-759.

Dionne, C. E., Korff M. Von, T. D. Koepsell, R. A. Deyo, W. E. Barlow, & H. Checkoway. (2001). Formal education and back pain: a review. *Journal of Epidemiology and Community Health,* 55, 7, pp. 455-468.

Frymoyer, J. W. & W. Cats-Baril. (1987). Predictors of low back pain disability. *Clinical Orthopaedics and Related Research,* 221, pp. 89-98.

Gallagher, R. M. 2003. The pain depression conundrum: bridging the body and mind. 16 September 2011. Available from: http://www.medscape.com/viewprogram/2030.

Gatchel, R. J. & M. A. Gardea. (1999). Psychosocial issues: their importance in predicting disability, response to treatment, and search for compensation. *Neurologic Clinics,* 17, 1, pp. 149-166.

Grant, L., B. Long, & J. D. Willms. (2002). Women's adaptation to chronic back pain: Daily appraisals and coping strategies, personal characteristics and percieved spousal responses. *Journal of Health Psychology,* 7, 5, pp. 545-563.

Hartvigsen, J., S. Lings, C. Leboeuf-Yde, & L. Bakketeig. (2004). Psychosocial factors at work in relation to low back pain and consequences of low back pain; a systematic, critical review of prospective cohort studies. *Occupational and Environmental Medicine,* 61, 1, pp. 2.

Hemingway, H. & M. Marmot. (1999). Evidence based cardiology: psychosocial factors in the aetiology and prognosis of coronary heart disease. Systematic review of prospective cohort studies. *Bristish Medical Journal,* 318, 7196, pp. 1460-1467.

Hooftman, W., M. van Poppel, & A. van der Beek. (2004). Gender differences in the relations between work-related physical and psychosocial risk factors and musculoskeletal complaints. *Scandivian Journal of Work Environmental Health,* 30, pp. 261-278.

Hoogendoorn, W. E., M. N. van Poppel, P. M. Bongers, B. W. Koes, & L. M. Bouter. (2000). Systematic review of psychosocial factors at work and private life as risk factors for back pain. *Spine,* 25, 16, pp. 2114-2125.

Houtman, I. L., P. M. Bongers, & P. G. Smulders. (1994). Psychosocial stressors at work and musculoskeletal problems. *Scandivian Journal of Work Environmental Health,* 20, pp. 139-145.

Josephson, M. (1998). Workplace factors and care seeking for low back pain among female nursing personnel. *Scandivian Journal of Work Environmental Health,* 24, 6, pp. 465-472.

Korovessis, P., Repantis, T. & Baikousis, A. (2010). Factors affecting low back pain in adolescents. *Journal of Spinal Disorders and Techniques, 23*(8), 513-20.

Larsen, E. C., C. Wilken-Jensen, A. Hansen, D. V. Jensen, S. Johansen, H. Minck, M. Wormslev, M. Davidsen, & T. M. Hansen. (1999). Symptom-giving pelvic girdle relaxation in pregnancy. I: Prevalence and risk factors. *Acta Obstetricia et Gynecologica Scandinavica,* 78, 2, pp. 105-110.

Lundberg, U. (1999). Stress responses in low-status jobs and their relationship to health risks: Musculoskeletal disorders. *Annals New York Academy of Sciences,* 896, 1, pp. 162-172.

Luo, X., R. Pietrobon, S. X. Sun, G. G. Liu, & L. Hey. (2004). Estimates and patterns of direct health care expenditures among individuals with back pain in the United States. *Spine,* 29, 1, pp. 79-86.

MacGregor, A. J., T. Andrew, P. N. Sambrook, & T. D. Spector. (2004). Structural, psychological, and genetic influences on low back and neck pain: A study of adult female twins. *Arthritis and Rheumatism,* 51, 2, pp. 160-167.

Marras, W. S., K. G. Davis, C. A. Heaney, A. B. Maronitis, & W. G. Allread. (2000). The influence of psychsocial stress, gender, and personality on mechanical loading of the lumbar spine. *Spine,* 23, 3045-3054.

Mehling, W. E. & N. Krause. (2005). Are difficulties perceiving and expressing emotions associated with low-back pain? The relationship between lack of emotional awareness (alexithymia) and 12-month prevalence of low-back pain in 1180 urban public transit operators. *Journal of Psychosomatic Research,* 58, 1, pp. 73-81.

Meisler, j. g. (2003). Toward optimal health: The experts discuss management of low back pain in women. *Journal of Women's Health,* 12, 10, pp. 953-959.

Melzack, R. & P. Wall. (1965). Pain mechanisms: a new theory. *Science,* 150, 699, pp. 971-979.

Mens, J. M., A. Vleeming, R. Stoeckart, H. J. Stam, & C. J. Snijders. (1996). Understanding peripartum pelvic pain. Implications of a patient survey. *Spine,* 21, 11, pp. 1363-1369.

National Institute of Occupational Safety and Health. (1997). Workplace use of back belts: Review and recommendations. U.S.Department of Health and Human Services, Public Health Service, Centers for Disease Control and Prevention.

Olsson, C. & L. Nilsson-Wikmar. (2004). Health-related quality of life and physical ability among pregnant women with and without back pain in late pregnancy. *Acta Obstetricia et Gynecologica Scandinavica,* 83, 4, pp. 351-357.

Ostgaard, H. C. & G. B. Andersson. (1991). Previous back pain and risk of developing back pain in a future pregnancy. *Spine,* 16, 4, pp. 432-436.

Ostgaard, H. C., G. Zetherstrom, & E. Roos-Hansson. (1997). Back pain in relation to pregnancy: a 6-year follow-up. *Spine,* 22, 24, pp. 2945-2950.

Papageorgiou, A. C., G. J. Macfarlane, E. Thomas, P. R. Croft, M. I. Jayson, and A. J. Silman. (1997). Psychosocial factors in the workplace--do they predict new episodes of low back pain? Evidence from the South Manchester Back Pain Study. *Spine,* 22, 10, pp. 1137-1142.

Pincus, T., A. K. Burton, S. Vogel, & A. P. Field. (2002a). A systematic review of psychosocial factors as predictors of chronicity/ disability in prospective cohorts of low back pain. *Spine,* 27, pp. E209-E120.

Pincus, T., J. W. Vlaeyen, N. A. Kendall, M. R. Von Korff, D. A. Kalauokalani, & S. Reis. (2002b). Cognitive-behavioral therapy and psychosocial factors in low back pain: directions for the future. *Spine,* 27, 5, pp. E133-E138.

Punnett, L., A. Pruss-Utun, D. I. Nelson, M. A. Fingerhut, J. Leigh, S. Tak, & S. Phillips. (2005). Estimating the global burden of low back pain attributable to combined occupational exposures. *American Journal of Industrial Medicine,* 48, 6, pp. 459-469.

Quint, U., H. Hasenburg, T. Patsalis, & G. H. Franke. (1998). Psychological stress of inpatients with acute and chronic lumbar syndrome. *Zeitschrift fur Orthopadie und Ihre Grenzgebiete,* 136, 5, pp. 444-450.

Raak, R., K. Wikblas, A. Raak, M. Carlsson, & L. Wahren. (2002). Catastrophizing and health related quality of life - a 6- year follow-up of subjects suffering from chronic low back pain. *Rehabilitation Nursing Journal,* 27, 3, pp. 110-117.

Ramond, A., Bouton, C., Richard, I., Roquelaure, Y., Baufreton, C., Legrand, E. & Huez, J.-F. (2011). Psychosocial risk factors for chronic low back pain in primary care- a systematic review. *Family Practice, 28,* 12-21.

Robinson, M. E., E. A. Dannecker, S. Z. George, J. Otis, J. W. Atchison, & R. B. Fillingim. (2005). Sex differences in the associations among psychological factors and pain report: a novel psychophysical study of patients with chronic low back pain. *Journal of Pain and Symptom Management,* 6, 7, pp. 463-470.

Skillgate, E., E. Vingard, M. Josephson, T. Theorell, & L. Alfredsson. (2007). Life events and the risk of low back and neck/shoulder pain of the kind people are seeking care for: results from the MUSIC-Norrtalje case-control study. *Journal of Epidemiology and Community Health,* 61, 4, pp. 356-361.

Skovron, M. L., M. Szpalski, M. Nordin, C. Melot, & D. Cukier. (1994). Sociocultural factors and back pain: a population-based study in Belgian adults. *Spine,* 19, pp. 129-137.

Svensson, H. O., G. B. Andersson, A. Hagstad, & P. O. Jansson. (1990). The relationship of low-back pain to pregnancy and gynecologic factors. *Spine,* 15, pp. 371-375.

Thorbjornsson, C. O., L. Alfredsson, K. Fredriksson, M. Koster, H. Michelsen, E. Vingard, M. Torgen, & A. Kilbom. (1998). Psychosocial and physical risk factors associated with low back pain: a 24 year follow up among women and men in a broad range of occupations. *Occupational and Environmental Medicine,* 55, 2, pp. 84-90.

Truchon, M. (2001). Determinants of chronic disability related to low back pain: Towards an integrative biopsychosoical model. *Disability and Rehabilitation,* 23, 17, pp. 758-767.

Truchon, M. & L. Fillion. (2000b). Biopsychosocial determinants of chronic disability and low-back pain: a review. *Journal of Occupational Rehabilitation,* 10, pp. 117-142.

Vandergrift, J. L., Gold, J. E., Hanlon, A. & Punnett, L. (2012). Physical and psychological ergonomic risk factors for low back pain in automobile manufacturing workers. *Occupational Environmental Medicine, 69,* 29-34.

Vingard, E., L. Alfredsson, M. Hagberg, A. Kilbom, T. Theorell, & M. Waldenstrom. (2000). To what extent do current and past physical and psychosocial occupational factors explain care-seeking for low back pain in a working population? *Spine,* 25, 4, pp. 493-500.

Wu, W. H., O. G. Meijer, K. Uegaki, J. M. Mens, J. H. van Dieen, P. I. Wuisman, & H. C. Ostgaard. (2004). Pregnancy-related pelvic girdle pain (PPP), I: Terminology, clinical presentation, and prevalence. *European Spine Journal,* 13, pp. 575-589.

Yip, Y. B., S. C. Ho, & S. G. Chan. (2004). Identifying risk factors for low back pain (LBP) in Chinese middle-aged women: a case-control study. *Health Care for Women International,* 25, 4, pp. 358-369.

6

Evaluation and Management of Lower Back Pain in Oncological Patients

Joshua E. Schroeder[1], Yair Barzilay[1], Amir Hasharoni[1], Leon Kaplan[1], José E. Cohen[2] and Eyal Itshayek[2]

[1]*Departments of Orthopedic Surgery, Hadassah – Hebrew University Medical Center Jerusalem,*
[2]*Departments of Neurosurgery, Hadassah – Hebrew University Medical Center Jerusalem, Israel*

1. Introduction

Back pain is one of the most common complaints that brings patients to be examined by a physician (Moore, 2010). Pain may originate from a variety of tissues, including intervertebral disks, vertebrae, ligaments, neural structures, muscles, and fascia, or present as referred pain from adjacent pathology, such as peptic ulcers, pancreatitis, pyelonephritis, aortic aneurysm, and more (Henschke, et al., 2009). Nonspecific low back pain is typically managed with symptomatic care and physical therapy, with up to 90% of patients improving substantially over 3 months. It is such a common condition that the American College of Physicians has issued guidelines with a mandate against imaging patients for the first month after pain onset (Chou, 2010).

Serious, life-threatening diseases are uncommon causes of back pain; malignancy, ankylosing spondylitis, and infection together account for less than 5% of back pain cases in a typical primary care practice (Dagenais, et al., 2010). However, missing such a critical diagnosis represents a serious concern for every practitioner; thus complaints of back pain often lead to multiple imaging studies and consultations (Venkitaraman, et al., 2010).

The spine is one of the most common sites of metastasis with close to 20,000 cases of spine metastases arising each year in the United States (Sciubba, et al., 2009). The most common primary tumors in patients with metastases are breast, lung, prostate, and kidney cancer (Guillevin, et al., 2007). In close to 15% of oncology patients, the primary presenting symptoms of malignancy are related to spinal metastases. In these patients, the most common underlying pathology is lung cancer, followed by breast cancer in females and prostate cancer in male patients (Chamberlain & Kormanik, 1999).

When the patient's history is taken properly, a thorough physical examination is conducted, and appropriate diagnostic tests are performed, the physician can determine with a high level of accuracy whether an individual patient is suffering from nonspecific ("simple") back pain, or whether an underlying, potentially catastrophic disease is triggering the pain (Bach, et al., 1990).

In taking the patient's history, one should try to define specific characteristics of the pain. Is this radicular or axial back pain? Is the pain worse at night or in the morning? Is the pain mechanical in nature or constant? Is it progressing? The examining physician should also look for signs and symptoms of systemic disease, such as fatigue, night sweats, weight loss, and changes in bowel habits. Personal habits such as smoking, alcohol consumption, or drug use should be identified as a potential clue to the underlying pathology.

In a patient with a known history of cancer, the situation is quite different. Each new ache and bump might lead to the fear of a metastasis, and patients will thus be highly sensitive to changes. On the other hand, ignoring symptoms or assuming that they are normal side effects of medical treatment may lead a missed diagnosis and delayed treatment, with the potential of significantly shortening a patient's life expectancy or greatly reducing the quality of remaining life (Verbeeck, 2004). Sadly, although the awareness to the risk of spinal metastases is high, even in patients at high risk, progression, with catastrophic consequences can occur during a drawn out diagnostic process (Cole & Patchell, 2008; Hagelberg & Allan, 1990; Loblaw, et al., 2005). Although a high level of suspicion may lead to higher rates of imaging in these patients, any back pain in a patient with a history of malignancy should be considered as suspicious for spinal metastases and should be fully investigated.

We aimed to review the essential skills required for diagnosis of the etiology of back pain, and to outline basic elements of treatment in patients presenting with metastatic disease to the spine.

2. Making the diagnosis

2.1 Characteristics of benign back pain

Defining a patient's pain is a hard task. Pain is subjective. The way in which pain is experienced, tolerance for pain, the language used to describe it, and its impact on quality of life differ from patient to patient and from culture to culture. Pain can be affected by medications, comorbidities, prior treatments, and by the patient's life situation and mental state (Florence, 1981). With these points in mind, specific information can enable a skilled diagnostician to differentiate metastatic pain from benign back pain.

Benign back pain often arises from a specific event, is relieved by rest and lying down, and increases with activity such as lifting, sneezing, laughing, and the Valsalva maneuver (Lishchyna & Henderson, 2004). It is most commonly focal, with adjacent spasm of the lumbar muscles and buttocks. Benign back pain generally subsides several weeks after injury, and can be managed effectively with non-narcotic analgesics and physical therapy (Ladeira, 2011). Patients with benign pain may experience several relapses, but generally pain remits between attacks. In a portion of patients (up to 7%) benign pain becomes chronic, however a discussion of chronic back pain is beyond the scope of this chapter.

2.2 Characteristics of metastatic back pain

In contrast, pain caused by spinal metastases is typically persistent and progressive, and is not alleviated by rest. Often pain is worse at night, awakening the patient from sleep. This pain is typically focal at the level of the lesion, progresses over several days or weeks, and may be associated with neurological signs indicating pressure on the neural spinal elements (Bach, et al., 1990).

A spinal mass can cause one of several forms of back pain. Localized pain is confined to the region of the spine affected by the tumor. This type of pain presents when a metastasis that originally developed in the bone marrow extends to stretch the periosteum or invades soft tissues, triggering pain from the nerve roots or signaling instability in a specific spine segment (Cole & Patchell, 2008).

A second form of pain is radicular pain due to nerve root compression or invasion. This pain is also typically worse at night and when the patient is recumbent, due to lengthening of the spine and distension of the spinal epidural venous plexus. This pain is often made worse by a Valsalva maneuver or other stretching movement of the spine or lower limb. The pain is usually dermatome-linked and may be associated with weakness of the muscles innervated by this nerve root (Cole & Patchell, 2008). If more than one nerve root is involved, the pain might extend to more than one dermatome.

A third type of pain appears when a pathologic fracture is present. This pain is generally focal, associated with instability, and progressive. It will worsen with movement. In the case of a fracture, the patient may remember a specific event or time when the pain began. It can be debilitating, necessitating the use of large doses of narcotics or preventing the patient from sitting or walking (Smith, 2011). These fractures may also lead to neurological changes due to neural element compression (Shaw, et al., 1989).

Sadly, not all patients with spinal metastasis present with early back pain. Many times a metastasis, like other slowly evolving conditions, becomes symptomatic only when there is neural compression. In this case, the patient will come to medical care only when there is cord compression, with imminent risk of losing mobility and control of the bowels. This late presentation is associated with a lower probability of neurological recovery, and a high rate of morbidity and mortality (Sundaresan, et al., 1995).

2.3 Physical examination

As is the case with any diagnostic process, the physical examination begins when the patient walks through the door. The patient's general appearance, nutritional state, walking pattern, and general habitus can be assessed during the walk to the examination bed. It is thus important for the physician to be positioned to watch the patient's entry, and to be alert to these details.

A general examination should be performed, including clinical examination of the breasts, lungs, abdomen, thyroid, and prostate, in the appropriate setting. After the general examination, a thorough orthopedic examination should be performed to evaluate any limitation in movement or impairment due to metastatic disease in other locations. The spine should be examined to identify sites of focal tenderness and assess range of motion. Signs of spinal instability or neural compression should be evaluated, and a full neurological examination should be performed, including assessment of strength, deep and superficial sensation, and proprioception, as well as deep tendon and pathological reflexes (Winters, et al., 2006). In order to try to locate the specific location of the lesion, sensory and motor levels should be assessed and documented.

Physical examination should be repeated periodically to allow early detection of any pathological motor or sensory findings or abnormal reflexes or any signs suggesting spinal instability or pathological fracture (Bates & Reuler, 1988).

With findings from the physical examination in hand, the physician should have an understanding of the pathology, and whether it is pointing towards metastatic spread of the primary disease to the spine. The next steps in diagnosis or treatment are determined by these findings.

2.4 Blood work

If the history and physical examination leads to the suspicion that the patient's pain could originate from something more severe than common backache, blood work is mandated. If cancer is suspected, initial tests should include a complete blood count, a full chemistry panel including calcium and phosphate levels (de Mello, et al., 1983), evaluation of the erythrocyte sedimentation rate, and C-reactive protein levels (Elsberger, et al., 2011). If laboratory studies reveal anemia, thrombocytopenia hypercalcaemia, or elevated levels of alkaline phosphatase, concern should increase (Nieder, et al., 2010). Specific markers for prostate and breast cancer should be tested, as well as urine and blood protein electrophoresis if gammopathy (multiple myeloma or plasmacytoma) is suspected (Scharschmidt, et al., 2011).

A blood smear or a bone marrow biopsy may be indicated if hematological disease is suspected (Raje & Roodman, 2011).

2.5 Imaging

Metastases generally appear in more than one anatomic location. They can be in the brain, soft liver, lungs, or lymph nodes. In the skeleton, the third most common location for tumor spread, lesions may be found in the vertebrae, pelvis, proximal parts of the femur, ribs, proximal part of the humerus, and skull (Ratanatharathorn, et al., 1999). If a spinal metastasis is suspected, it is important to perform a full workup, both to identify the primary lesion and to detect other metastases.

2.5.1 X-ray

Simple X-rays of the spine are considered to be the first and most attainable imaging study. If obtained, they should be complimented with a chest X-ray for a preliminary search for lung involvement, however neither spinal nor lung tumors are well visualized on radiographs until the malignancy has advanced significantly (Nielsen, et al., 1991). Lateral X-ray may show vertebral body collapse. AP views may demonstrate pedicle erosion (the "winking owl" sign) or evidence of a paraspinal mass (Fehlings & Rao, 2000).

In cases where the physician has a high degree of suspicion towards metastatic spine disease, a more expeditious approach to diagnosis should be taken using advance imaging techniques such as CT or MRI (Black, et al., 1996). In these patients X-ray may be used as a complimentary study, since images obtained standing and dynamic X-rays can provide a better understanding of sagittal balance and stability of the diseased spine.

2.5.2 Computed Tomography

Computerized tomography (CT) has higher sensitivity and specificity than X-ray. Multislice CT (MDCT) systems facilitate a single rapid study of the chest-abdomen-pelvis. Osteolytic, sclerotic, and mixed lesions are depicted well on CT scans, as are lesions involving the viscera and vascular anatomy. However, while 16/64 row MDCT provides excellent image quality and clear assessment of bony structures, metastatic lesions without significant bone destruction may be missed (Buhmann-Kirchhoff, et al., 2009). CT scans are also associated with relatively high quantities of radiation, limiting the number of screening studies that should be performed, especially in a younger population (Huda & He, 2011). In addition, the differential diagnosis between a malignant process versus osteoporotic or degenerative disease can be challenging in the spinal column (Chassang, et al., 2007). CT findings that suggest metastatic disease include destruction of the anterolateral or posterior vertebral cortex, destruction of one or both pedicles, an epidural mass, and presence of a focal paraspinal soft tissue mass (Fehlings & Rao, 2000; Laredo, et al., 1995).

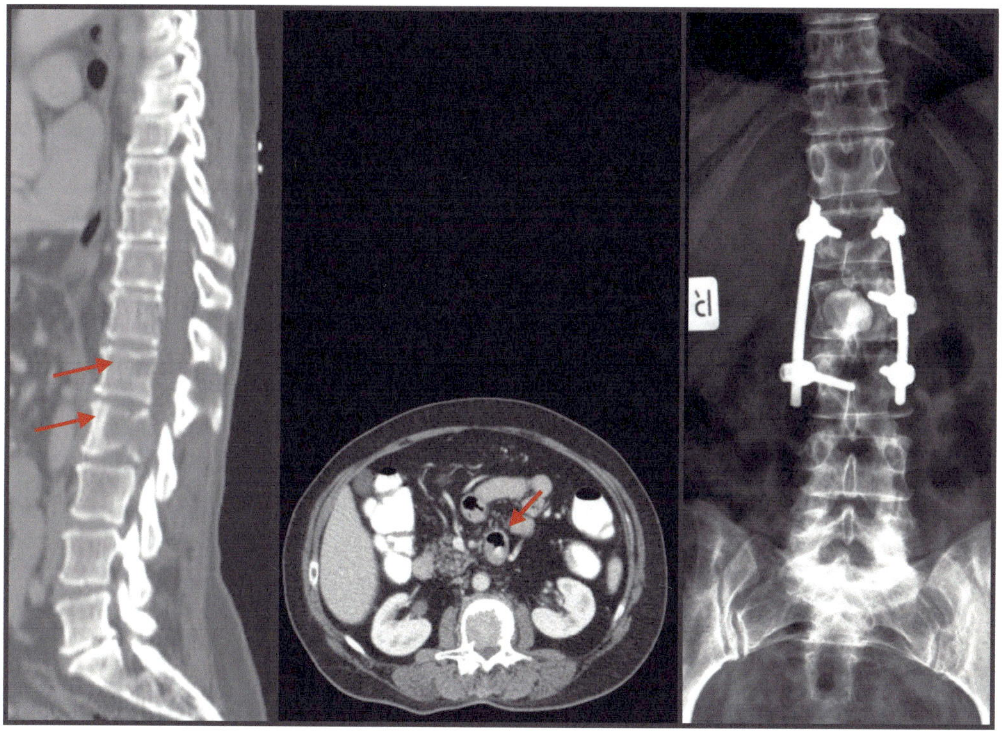

A B C

Fig. 1. **Spinal metastasis to the thoracic and lumbar spine.** A 74-year-old female with a history of gastrointestinal stromal tumor (GIST) presented with right leg sciatic pain and weakness. **(A)** Sagittal CT revealed an osteolytic metastasis in the vertebral body of T7, L1, and L2 (arrows). The L2 metastasis was most prominent, producing compression of the cauda equina. **(B)** Axial CT through the body of L2 demonstrating the soft tissue mass, which has created a cavity in the vertebral body, narrowed the spinal canal and created pressure on the cauda equina. **(C)** X-ray after transpedicular excision of the lesion shows reconstruction of the vertebral body with polymethylmethacrylate (PMMA) and L1-L3 posterolateral fixation with transpedicular screws and rods. The patient experienced immediate and sustained pain relief, with recovery of her previous strength. Two weeks after surgery she was treated with adjuvant EBRT.

2.5.3 Magnetic Resonance Imaging

Magnetic resonance imaging (MRI) is gold standard for evaluating spinal tumors. It depicts vertebral bone marrow infiltration by tumor cells as well as soft tissue masses in and around the spinal column. Bone marrow invaded by a neoplasm is characterized by increased cellularity, resulting in a decreased signal on T1-weighted images and a high signal on T2-weighted images, thus differentiating it from normal marrow tissue (Loblaw, et al., 2005). Intravenous gadolinium further increases the contrast between tumor and normal tissues (Loughrey, et al., 2000). MRI has been shown to detect up to 98.5% of vertebrae with metastatic disease, including both osteolyitic and osteoblastic lesions. This high level of detection is not compromised by osteoporosity (Buhmann-Kirchhoff, et al., 2009).

However, MRI is a costly, time consuming exam, limiting its efficacy in patients who have difficulty lying down without moving for long periods of time and those who are claustrophobic or morbid (Eshed, et al., 2007). In addition, it may be difficult to differentiate between osteoporotic compression fracture and metastatic disease, especially if a fracture co-exists. Signs that characterize malignant vertebral collapse include ill-defined vertebral margins, abnormal signal involvement of the pedicle, a marked and heterogeneous MR enhancement pattern, and the presence of an irregular nodular-type paraspinal vertebral lesion (Shih, et al., 1999). Using of different diffusion coefficients may assist correctly identifying this deferential diagnosis (Chan, et al., 2002).

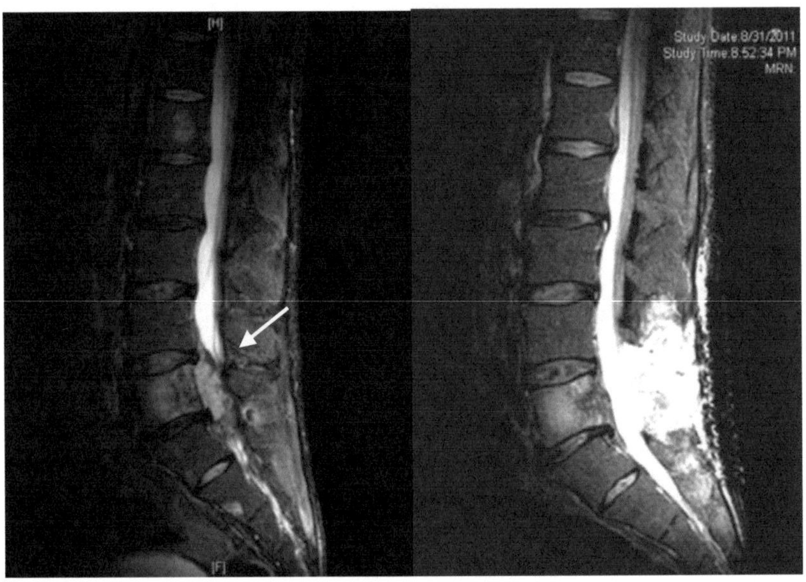

A B

Fig. 2. **MRI of the lumbar spine.** A 42-year-old male presented with cauda equina syndrome due to an epidural metastasis to the lumbar spine from a synovial sarcoma originating in the left lower limb. **(A)** Preoperative T2-weighted sagittal MRI demonstrating a metastasis to the L4 vertebral body invading the epidural space and compressing the thecal sac (arrow). The patient underwent laminectomy and resection of the epidural mass. (B) Postoperative T2-weighted sagittal MRI showing post-laminectomy decompression of the thecal sac.

2.5.4 ^{99}Tc bone scan

With the exception of purely lytic tumors such as myeloma, eosinophilic granuloma, and renal cell carcinoma, 99mTc-methylene diphosphonate (MDP) bone scan has good sensitivity to tumors, is widely available, and has a relatively low cost, with the ability to scan the entire skeleton in a single study. It can be performed as a flat two dimensional exam or, incorporating more advanced single-photon emission computed tomography (SPECT) technology, with true 3D information. Bone scans have moderate sensitivity in the spine and pelvis (Steinborn, et al., 1999). Sensitivity to detection of spine metastasis can be significantly improved by the combination of bone scan and SPECT imaging capabilities (Schirrmeister, et al., 2001). Bone scans combined with SPECT may depict spinal lesions as well as metastases in other bones or organs, and provide some indication of the site of the primary tumor in cases where this is not known. The main drawback to these studies is exposure to high levels of ionizing radiation and a lower rate of detection when compared with MRI (Sedonja & Budihna, 1999).

2.5.5 PET/CT

FDG-PET/CT studies in oncology utilize labeled glucose with the tracer fluorine-18 (F-18) fluorodeoxyglucose (FDG), in combination with MDCT, to provide the benefits of metabolic and anatomic imaging in a single study. FDG is a glucose analog that is taken up by glucose-using cells and phosphorylated by hexokinase in the mitochondria. Increased tumor [18F]-FDG uptake is depicted by positron emission tomography (PET), although proliferative activity is also broadly related to the number of viable tumor cells. When glucose is metabolized, the tissue incorporates the radioactive isomer, increasing local particle emission and producing a high intensity signal on PET/CT images (Young, et al., 1999).

PET-CT is a sensitive method for assessment of bone and bone marrow metastases, as well as vertebral and extravertebral skeletal masses. It is sensitive for osteolytic and osteoblastic metastasis and depicts early malignant bone-marrow infiltration (Kruger, et al., 2009). It provides precise localization and is sensitive to accompanying soft tissue metastases in the lung, liver, lymphatic system, and elsewhere (Metser, et al., 2004; Nguyen, et al., 2007). Low resolution PET scans are combined with the high resolution of thin slice CT, for exquisite depiction of even small metastases in early stages. PET/CT has higher sensitivity and specificity compared to bone scintigraphy or CT in the detection of skeletal metastases (Kruger, et al., 2009), however the rate of false-positive findings that require follow-up imaging with other modalities is also higher (Kuo & Cheng, 2005).

3. Cord compression

Discussions of metastatic epidural cord compression usually focus on the thoracic spine. Although the spinal cord usually ends in the lumbar spine between L1 and L2 in adults, up to 20% of metastatic cord compression occurs at these levels; thus lumbar cord compression must also be discussed.

Metastatic spinal cord compression (SCC) occurs in 5–10% of patients with cancer (Bach, et al., 1990). It is a true emergency, because delay in diagnosis and treatment may result in permanent neurological impairment. SCC is caused by direct compression of metastasis or primary tumors invading the vertebral bodies, breaking through the cortex, and compressing the vertebral canal and nerve roots (Bilsky, et al., 2000). The most commonly affected site is the thoracic spine, which is also the area in which the canal is the tightest, leaving little space for movement of the cord. Breast cancer is the most frequent primary malignancy associated with SCC, followed by lung, prostate, and renal cancers (Byrne, 1992). However, SCC is also common in patients with hematological tumors such as lymphoma and myeloma. SCC is the initial manifestation of a metastatic spread of a tumor in approximately 20% of cancer patients; in patients with lung cancer the rate of SCC at first presentation climbs to 30% (Chahal, et al., 2003). In most of the cases of SCC, back pain has been present for several months before detection, but was ignored by the patient and caretakers (Byrne, 1992).

The most important factor in determining the post-treatment outcome is pretreatment ambulatory function of the patient. Late presentation with neurological deficits, including bowel or bladder dysfunction, is often associated with irreversible paraplegia. In recent series, 74-100% of patients who were ambulatory before surgery retained the ability to walk after decompression, and 57-82% of nonambulatory patients regained ambulation (Bilsky, et al., 2009). However ambulation at surgery remains a key determinant of outcome, therefore, urgent investigations must be performed to facilitate treatment before function is lost.

4. Treatment

4.1 Medical management of pain

Basic pain control is achieved by the use of narcotics and non-narcotic analgesics. In patients with back pain due to metastases, pain control is one of the most important goals in management, as the life expectancy of the patients is limited (Padalkar & Tow, 2011). Sadly, despite this concept, most patients are undertreated for pain, which causes a significant reduction in quality of life (Cleeland, 2006). Two types of pain needed to be addressed when dealing with medical management of the metastatic patients, the constant pain of the metastasis and an acute episode of sharp pain, known as "breakthrough pain" (Lipton, 2011).

The first line of treatment in patients with mild low back pain is the use of nonsteroidal anti-inflammatory drugs (NSAIDs) and acetaminophen. These agents may provide adequate pain relief. There have been nine trials indicating that the use of NSAIDS provides pain relief in bone metastasis, including one showing that NSAIDS (specifically COX-2 inhibitors) assist in tumor control as well (Smith, 2011). Acetaminophen is preferred in patients with thrombocytopenia, renal dysfunction, those receiving nephrotoxic agents, and those at risk for gastrointestinal bleeding. In patients with liver dysfunction, NSAIDs are preferred for mild pain (Hitron & Adams, 2009)

In patients who need the next level of treatment for their low back pain, opioid therapy should be added. Common protocols begin with low doses of immediate-release short-acting agents (i.e. morphine), with reassessment for the level of effect every 1 to 2 hours. After 24 hours of pain control on a short-acting regimen, patients should be converted to a long-acting agent such as sustained-release morphine, oxycodone, fentanyl, or methadone for basal control. Patients who are opioid tolerant should begin with higher doses of short-acting agents, with the higher dose compensating for shorter duration of effect. If a patient is already using a long-acting product, this should be continued. A bowel regimen with a stimulant plus stool softener should be initiated to prevent constipation from opioid use. Care must be taken to avoid misuse and abuse of opioids, even in cancer patients (Manchikanti, et al., 2010).

Adjunct agents such as anticonvulsants (e.g., gabapentin, lamotrigine, topiramate), tricyclic antidepressants (e.g., amitriptyline, imipramine, desipramine, nortriptyline), venlafaxine, duloxetine, or topical analgesics (e.g., lidocaine or capsaicin) may also help to reduce neuropathic pain caused by nerve compression (Hitron & Adams, 2009).

In cases of breakthough pain, oral transmucosal fentanyl and transdermal patches have been shown to be an effective treatment for pain including sharp pain episodes, yielding a more rapid onset of relief. In cases where patients developed tolerance towards opioid narcotics, the use of anesthetics, such as ketamine can provide adequate pain relief (Chazan, et al., 2008).

The use of cannabinoids as an alternative or to augment narcotics has become popular, primarily for light-to-intermediate pain. Three cannabinoids, all a combination of the delta-9-tetra-hydrocannabinol and cannabidiol cannabinoids from the *Cannabis sativa* plant, are available on the market. Sativex, a cannabinoid medication, has recently received approval as an adjuvant medication for the treatment of cancer pain in North America. It is a sublingual (mouth) spray that can be titrated up to the most effective dose (Bonneau, 2008). Other forms of cannabinoids treatment include smoking marijuana, or ingestion of oils or even cookies with insertion of active cannabinoid components. In our experience, "cannabis cookies" have provided relief or reduction in chronic pain levels for some cancer patients. The use of cannabinoids is indicated in cancer pain with a neuropathic component that is not adequately controlled with opioids. Cannabinoids are not suitable as a single medication in spine metastasis.

4.2 Steroids

Steroids are commonly prescribed in patients with metastatic bone disease. Steroids reduce edema and have been shown to reduce the size of metastases from tumors of hematological origin, and occasionally breast cancer (Cole & Patchell, 2008). They have also been shown to have a rapid analgesic effect (Bonneau, 2008). However, they are associated with side effects, such as wound dehiscence, gastric ulcers, rectal bleeding, psychosis, and diabetes mellitus, and may increase susceptibility to infections (Shih & Jackson, 2007).

An early study (Greenberg, et al., 1980) demonstrated quick and significant pain reduction in patients treated with steroids. A subsequent randomized trial (Vecht, et al., 1989) compared the effect of 10 mg IV dexamethasone versus 100 mg IV followed by 16 mg daily orally and found no differences between the conventional and high-dose group on pain, ambulation, or bladder function. Both the conventional and high-dose regimens provided significant pain relief.

A randomized, controlled trial (Sorensen, et al., 1994) studied the administration of 96 mg of IV dexamethasone followed by 96 mg orally for 3 days and then tapered in 10 days, with subsequent radiotherapy, versus radiotherapy as a single modality in the treatment of SCC. Steroid treatment provided a statistically significant improvement in ambulation at 3 to 6 months, albeit with increased side effects, including psychoses and gastric ulcers requiring surgery. The effect of dexamethasone on pain reduction was not addressed.

Currently there are no absolute guidelines for steroid treatment in patients with back pain due to spinal metastases. The decision of whether to treat, as well as steroid dose are determined by the treating physician. In cases of SCC, steroid treatment is given as an adjuvant to surgery or to chemo- or radiotherapy, and not as sole treatment modality.

4.3 Bisphosphonates

Bisphosphonates are pyrophosphate analogues. The bisphosphonates strongly bind hydroxyapatite crystals in bone, preventing the creation of the ruffle border of the osteoclasts that prevent bone resorption. In addition, bisphosphonates induce osteoclast apoptosis (Li, et al., 2011).

The rationale for using drugs such as bisphosphonates in patients with metastatic cancer is that osteoclasts, as the mediators of bone absorption, are often activated by the tumors and thus allow metastatic invasion into the medulla of the vertebrae. Preventing such activity may reduce metastatic invasion and spread (Orita, et al., 2011). In addition, by inhibiting bone resorption, a secondary correction of hypercalcaemia and hypercalciuria will be achieved (Woodward & Coleman, 2010).

Most importantly, the use of bisphosphonates reduces pain and the occurrence of fracture, as well as the development of new osteolytic lesions and, as a consequence, improves patients' quality of life (Fleisch, 1991). Compounds that are commercially available for use in tumor-induced bone disease are, in order of increasing potency, etidronate, clodronate, pamidronate, and alendronate.

Van Holten-Verzantvoort et al found a significant reduction in morbidity from bone metastases in pamidronate-treated breast cancer patients, with a 30–50% reduction of pain and lower rate of new pathological fractures (van Holten-Verzantvoort, et al., 1991). Thurlimann et al showed that pain relief was achieved in about 30% of patients who received pamidronate every 4 weeks (Thurlimann, et al., 1994).

Side effects from bisphosphonates, including transient low grade fever, nausea, myalgia, gastrointestinal side-effects, bone pain, and mild infusion-site reactions, are usually minimal. More rarely, osteonecrosis of the jaw and long bone "frozen bone" fractures due to long term use have been reported (Kim, et al., 2011; Mercadante, 1997).

A meta-analysis (Fulfaro, et al., 1998) concluded that bisphosphonates, and in particular IV pamidronate, are an important therapeutic tool in association with other therapeutic modalities for the treatment of metastatic bone disease with marked osteolysis, such as multiple myeloma and breast cancer. The authors did not evaluate the impact on back pain, but looked at the general effect of bisphosphonates on bone metastases.

4.4 Chemotherapy

As is the case with metastatic disease in other locations, the long-term control of spine metastases entails systemic chemotherapy. The type of treatment and its duration largely depend on the tumor histology and specific tumor receptors. The full treatment plan should be determined by a team of oncologists, spine surgeons, and radiotherapists, balancing the different modalities of treatment with the side effects and complications.

Hoy et al showed that many patients experience pain relief with chemotherapy, even in cases where there is no objective tumor response (Hoy, 1989). Although palliative, the exact benefit of chemotherapy is difficult to measure. Potential indications for palliative chemotherapy include symptoms due to metastatic disease, hypercalcemia, bone marrow infiltration, and clinical conditions known to be associated with bone or spine metastases. As patients are mainly treated with a combination of therapies, the exact effect of the response in the bone is more difficult to assess (Lote, et al., 1986).

In tumors that react quickly to chemotherapy, as in testicular cancers, lymphoma, or myeloma, administration of chemotherapy can have a pronounced effect in reducing back pain (Samoladas, et al., 2008). In tumors where the response to chemotherapy is slower, as in pancreatic cancer, pain reduction parallels the decrease in pressure exerted on spinal structures as the tumor slowly diminishes in size (Takuma, et al., 2006). However, in very advanced cancer patients there is little justification for using chemotherapy because of its toxicity (Bruera, 1993).

Hormonal therapy, based on the principle of depriving tumor cells of the growth stimulus induced by hormones that change the hormonal environment, may be beneficial in breast, prostate, and endometrial cancers (Wood, 1993). This treatment has a lower toxicity profile compared to chemotherapy, and should be the first therapeutic modality if possible. Dearnaley showed that hormonal therapy provides sustained symptom relief in patients with widespread painful bone metastases from prostatic cancer (Dearnaley, 1994). Cresenda et al showed that the combination of hormonal therapy with radiation for spinal metastasis reduces pain in 77% of patients (Cereceda, et al., 2003). There are no data regarding the impact on pain from the isolated use of hormonal therapy in patients with spinal metastasis.

4.5 Radiation-based treatment

Radiotherapy has served as a cornerstone for the treatment of bone metastases since it was introduced in the 1950s. Its benefits include reduction of pain and neurological complications arising from spinal cord compression. Radiation-based treatment is undergoing a revolution, with new stereotactic techniques for single-shot or fractionated treatment showing good results. However the mainstay of radiation-based treatment is external beam radiation therapy (EBRT) (Cole & Patchell, 2008; Harel & Angelov, 2010).

4.5.1 External beam radiation therapy

The exact pathophysiology of tumor response to radiotherapy is not fully understood, however EBRT can lead to rapid reduction in the tumor size, sometimes within 24 hours, decreasing local periosteal nerve stimulation. EBRT is typically given from a posterior field to the affected vertebra(e), with radiation delivered to a treatment area including one vertebral body above and one below to compensate for daily variations in patient setup.

A meta-analysis (Agarawal, et al., 2006), reported that over 40% of patients with bone metastases treated with radiation achieved at least a 50% reduction in pain, however fewer than 30% experienced complete pain relief at 1 month.

In cases of SCC, when surgery is not performed, the use of radiation combined with steroids has been shown to provide improved pain management and improved outcome when compared to radiation alone (Loblaw, et al., 2005). In cases were surgery is planned, the use of preoperative EBRT is associated with an increase in local wound-related complications (Bilsky, et al., 2000; Itshayek, et al., 2010; Itshayek, et al., In press).

Numerous EBRT protocols may be employed in the management of painful bone metastasis, including fractionated and single-fraction regimens. The dose ranges from 8 Gy delivered in a single fraction, up to a cumulative dose of 30–40 Gy administered over 3–4 weeks (Gerrard & Franks, 2004; Harel & Angelov, 2010). In a retrospective evaluation of treatment, comparing a single 8 Gy dose with several hypofractionated regimens in 1304 patients with spine metastases, there was no statistical difference for the various protocols in terms of patient survival, pain control, or side effects, however there were higher recurrence rates in patients treated with a lower cumulative dose (Rades, et al., 2005).

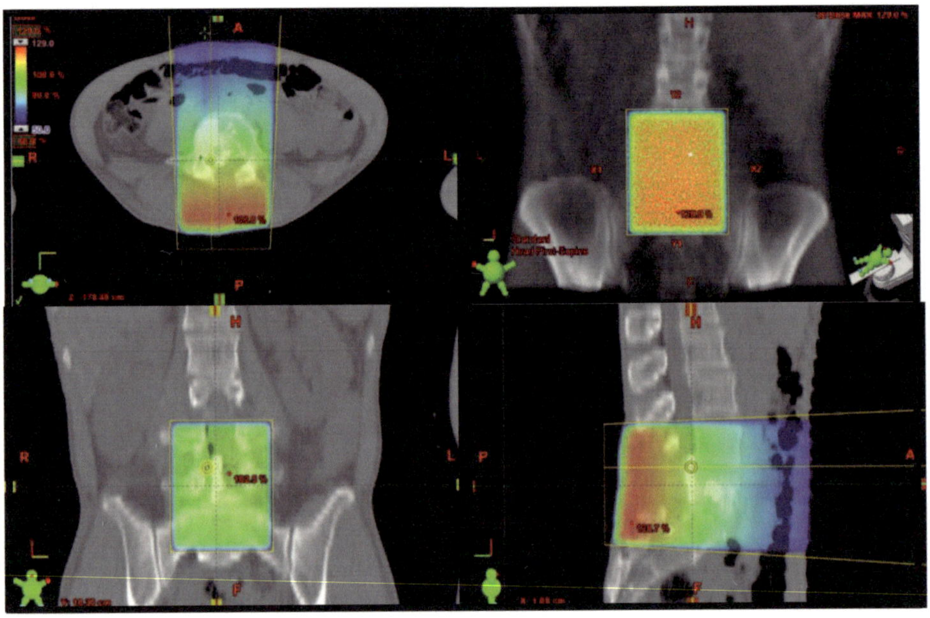

Fig. 3. **Treatment plan showing a map of the radiation fields for external beam radiation therapy.** A 42-year-old male presented with low back pain and cauda equina syndrome due to metastatic synovial sarcoma to L4 (Fig. 2). Following laminectomy and resection, he suffered persistent axial back pain and proceeded to vertebral augmentation (Fig. 5). After augmentation ERBT was administered for tumor control.

4.5.2 Stereotactic radiosurgery

Stereotactic radiosurgery (SRS) is a new modality that is used in patients with spinal metastases in some centers. The use of stereotaxy enables deployment of high doses of radiation to inactivate or eradicate defined targets in the head or spine, even tumor found in close proximity to the spinal cord or other critical structures. SRS uses sophisticated computer algorithms to divide the lethal doses of radiation into multiple projections, focusing a very high dose in the target zone with rapid dose fall off, and thus minimal collateral damage to adjacent tissues (Moulding, et al., 2010). These systems are extremely accurate, with less than 1 mm deviation from the target.

The optimal SRS protocol is highly individual, and should take into account the tumor's size, its pathological composure, and distance from the cord, as well as other radiosensitive organs. Data accumulating from an increasing number of studies suggests that SRS is a relatively safe, quick method to control spine metastases (Bilsky, et al., 2009). It provides excellent pain relief in most patients. In one study, significant pain relief was reported 6 months after treatment in 91% of patients treated with the Cyberknife (Accuray, Sunnyvale CA, USA) (Chang, et al., 2009). Another study reported improved pain control and improved quality of life 1 year after treatment (Degen, et al., 2005).

A possible side affect of spinal SRS is compression fracture, which may be attributed to rapid tumor lysis after radiation. These fractures can be managed conservatively or with pre-radiation vertebral augmentation (Gerszten, et al., 2005).

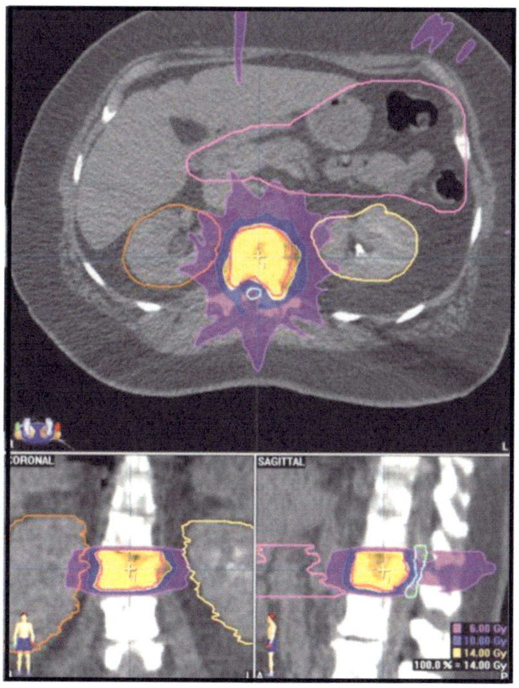

Fig. 4. **Treatment plan showing radiation fields for a stereotactic radiosurgery procedure.** A 65-year-old male with renal cell carcinoma that was metastatic to L2 that was treated with single-shot SRS.

4.6 Surgery

4.6.1 Tumor decompression and instrumental stabilization

The surgical treatment of spinal metastases has evolved significantly over the last 20 years, with mounting evidence of the value of surgery in the treatment of metastasis (Bilsky, et al., 2009; Cole & Patchell, 2008; Harel & Angelov, 2010; Tomita, et al., 1997). Surgery was once limited to posterior laminectomy for the management of neurological decompression in SCC {Roy-Camille, 1990 #219}. With this philosophy, outcomes were mixed since most metastases are anterior to the cord and thus not directly accessible with posterior surgical approaches. With advances in technique, new surgical approaches, and spinal instrumentation, patient outcomes have significantly improved, due to a combination of anterior tumor decompression and instrumental stabilization. The surgical approach is tailored, depending on the anatomic locations of metastases, aiming if possible to reach en block total resection of the tumor, the best possible outcome for the patient {Tomita, 2006 #220}. Tumors anterior to the spinal cord are usually approached with a purely anterior approach, especially in the cervical spine, or through a posterolateral or lateral approach in the thoracic and lumbar spine. These approaches allow for anterior decompression with anterior column support, as well as posterior instrumented fusion for improved stability. Although laminectomy can help to increase the diameter of the spinal canal at the affected levels, it does not achieve immediate cord decompression and can lead to instability {Black, 1996 #213; Camins, 2004 #221}.

As in most procedures for the management of metastatic disease, the aim is not to provide a cure but rather to relieve pain, stabilize the spine, prevent neurological deterioration, and at times provide a pathological diagnosis (Choi, et al., 2010). In select cases, complete resection of a spinal metastasis can be performed, which can provide the best opportunity for long-term local control and palliation in patients with isolated metastases or radioresistant tumors (Gallo & Donington, 2007). In cases where total resection cannot be achieved, stabilization and decompression provide a window of opportunity for other treatment modalities to take their course without jeopardizing the patients' health. In two large cohorts of patients, surgery reduced pain, and improved physical function and quality of life in patients with symptomatic vertebral metastases at 12 month follow up and beyond, with an acceptable complication rate (Falicov, et al., 2006; Quan, et al., 2011).

In cases of SCC, treatment should include corticosteroids, surgery, and radiotherapy along with aggressive systemic chemotherapy. Surgical decompression followed by radiotherapy was shown to be superior to radiotherapy alone in a study that was halted early after interim analysis demonstrated superior results in the surgery and radiotherapy group compared to the radiotherapy alone group (Patchell, et al., 2005). Patients treated with surgery plus radiotherapy had a median survival of 126 days with ambulation for 122 days, compared to a median 100 days survival and only 13 days ambulation in patients treated with EBRT alone. These findings were confirmed in a retrospective study conducted at our center (unpublished results). Patients who were operated had longer periods of ambulation relative to patients treated solely by radiotherapy. They also required lower doses and lighter pain medications.

4.6.2 Vertebral augmentation using vertbroplasty and kyphoplasty

Over the past two decades, percutaneous cement augmentation techniques have been developed for the treatment of spinal metastases and fractures. The two most common techniques are vertebroplasty and kyphoplasty. Vertebroplasty is a technique in which cement is injected into the vertebral body; kyphoplasty uses a balloon to create a void in the vertebrae into which the cement is injected. In a recent review comparing these techniques, pain scores for both the vertebroplasty and the kyphoplasty groups decreased significantly from preoperative values at 6-month follow-up (Liu, et al., 2010). The National Institute for Health and Clinical Excellence (NICE) guidelines (N.H.S., 2003; 2006) state that vertebroplasty or kyphoplasty should be considered for patients with vertebral metastases when their pain is refractory to analgesia, or when there is evidence of vertebral body collapse, in cases where is no evidence of cord compression or spinal instability.

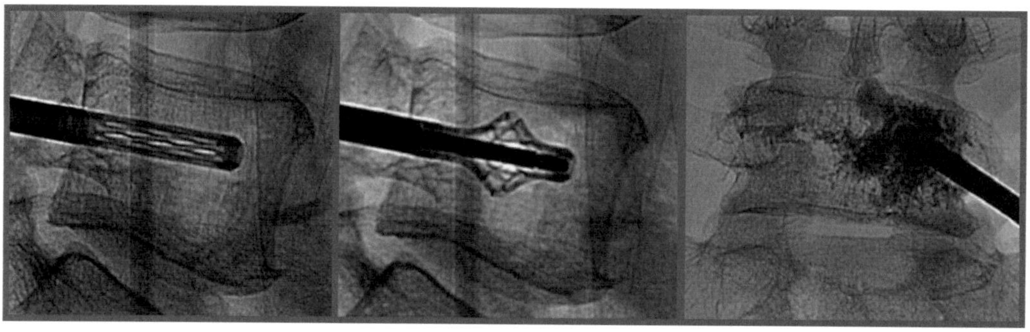

A B C

Fig. 5. **Vertebral augmentation.** A 47-year-old male presented with persistent axial low back pain that was refractory to medical management following laminectomy and resection of an epidural mass from an L4 metastasis of a synovial sarcoma (see Figs. 2, 3). The patient underwent vertebral augmentation. **(A)** An OsseoFix implant (Alphatech Spine, Carsbad CA, USA) is inserted in an unexpanded configuration into the L4 vertebral body. **(B)** Once proper placement is confirmed on fluoroscopy, the implant is expanded. **(C)** The implant is filled with PMMA bone cement. Following augmentation the patient is pain free.

5. Conclusions

Back pain is one of the most prevalent medical problems, and is usually benign. However, in patients with a history of cancer, one must take extreme precautions to make sure that a metastatic spread of the tumor is not missed. Metastases may present as mechanical pain, radicular pain or weakness, or with myelopathy. Pain is usually unremitting and progressive.

MRI is the gold standard for image-based diagnosis, with excellent depiction of soft tissue, including the spinal cord and nerve roots; CT provides optimal visualization of bony structures.

Pain is the most common and earliest complaint, and sadly, it is undertreated in most patients. Most studies are not focused on pain control, but rather on the neurological status of the patient, while pain control is a secondary objective.

Management of back pain in metastatic patients may involve a combination of pain medications, chemotherapy, radiation-based treatment, and surgery. Patient management thus requires the combined efforts of a multidisciplinary team, including oncologists, radiation oncologists, and spine surgeons. Treatment should be tailored to the needs of specific patients, based on their prognosis, neurological status, age, and primary pathology.

6. Acknowledgements

The authors wish to thank Shifra Fraifeld, a research associate in the Department of Neurosurgery, for her editorial assistance in the preparation of this material.

7. References

Agarawal, J.P., Swangsilpa, T., van der Linden, Y., Rades, D., Jeremic, B. & Hoskin, P.J. (2006). The role of external beam radiotherapy in the management of bone metastases. *Clin Oncol (R Coll Radiol),* Vol. 18, pp.747-60.

Bach, F., Larsen, B.H., Rohde, K., Borgesen, S.E., Gjerris, F., Boge-Rasmussen, T., Agerlin, N., Rasmusson, B., Stjernholm, P. & Sorensen, P.S. (1990). Metastatic spinal cord compression. Occurrence, symptoms, clinical presentations and prognosis in 398 patients with spinal cord compression. *Acta Neurochir (Wien),* Vol. 107, pp.37-43.

Bates, D.W. & Reuler, J.B. (1988). Back pain and epidural spinal cord compression. *J Gen Intern Med,* Vol. 3, pp.191-7.

Bilsky, M.H., Boland, P., Lis, E., Raizer, J.J. & Healey, J.H. (2000). Single-stage posterolateral transpedicle approach for spondylectomy, epidural decompression, and circumferential fusion of spinal metastases. *Spine (Phila Pa 1976),* Vol. 25, pp.2240-9,discussion 250.

Bilsky, M.H., Laufer, I. & Burch, S. (2009). Shifting paradigms in the treatment of metastatic spine disease. *Spine (Phila Pa 1976),* Vol. 34, pp.S101-7.

Black, P., SNair, S. & Giannakopoulos, G. (1996). Spinal epidural tumors, In *Neurosurgery, 2nd ed., Vol. II* (pp. 1791-1804) R. H. Wilkins and S. S. Rengachary (Ed.),^(Eds.), McGraw-Hill, New York.

Bonneau, A. (2008). Management of bone metastases. *Can Fam Physician,* Vol. 54, pp.524-7.

Buhmann-Kirchhoff, S., Becker, C., Duerr, H.R., Reiser, M. & Baur-Melnyk, A. (2009). Detection of osseous metastases of the spine: comparison of high resolution multi-detector-CT with MRI. *Eur J Radiol,* Vol. 69, pp.567-73.

Byrne, T.N. (1992). Spinal cord compression from epidural metastases. *N Engl J Med,* Vol. 327, pp.614-9.

Cereceda, L.E., Flechon, A. & Droz, J.P. (2003). Management of vertebral metastases in prostate cancer: a retrospective analysis in 119 patients. *Clin Prostate Cancer,* Vol. 2, pp.34-40.

Chahal, S., Lagera, J.E., Ryder, J. & Kleinschmidt-DeMasters, B.K. (2003). Hematological neoplasms with first presentation as spinal cord compression syndromes: a 10-year retrospective series and review of the literature. *Clin Neuropathol,* Vol. 22, pp.282-90.

Chamberlain, M.C. & Kormanik, P.A. (1999). Epidural spinal cord compression: a single institution's retrospective experience. *Neuro Oncol,* Vol. 1, pp.120-3.

Chan, J.H., Peh, W.C., Tsui, E.Y., Chau, L.F., Cheung, K.K., Chan, K.B., Yuen, M.K., Wong, E.T. & Wong, K.P. (2002). Acute vertebral body compression fractures: discrimination between benign and malignant causes using apparent diffusion coefficients. *Br J Radiol,* Vol. 75, pp.207-14.

Chang, U.K., Youn, S.M., Park, S.Q. & Rhee, C.H. (2009). Clinical results of cyberknife(r) radiosurgery for spinal metastases. *J Korean Neurosurg Soc,* Vol. 46, pp.538-44.

Chassang, M., Grimaud, A., Cucchi, J.M., Novellas, S., Amoretti, N., Chevallier, P. & Bruneton, J.N. (2007). Can low-dose computed tomographic scan of the spine replace conventional radiography? An evaluation based on imaging myelomas, bone metastases, and fractures from osteoporosis. *Clin Imaging,* Vol. 31, pp.225-7.

Chazan, S., Ekstein, M.P., Marouani, N. & Weinbroum, A.A. (2008). Ketamine for acute and subacute pain in opioid-tolerant patients. *J Opioid Manag,* Vol. 4, pp.173-80.

Choi, D., Crockard, A., Bunger, C., Harms, J., Kawahara, N., Mazel, C., Melcher, R. & Tomita, K. (2010). Review of metastatic spine tumour classification and indications for surgery: the consensus statement of the Global Spine Tumour Study Group. *Eur Spine J,* Vol. 19, pp.215-22.

Chou, R. (2010). Low back pain (chronic). *Clin Evidence,* BMJ Brit Med J. Retrieved from http://clinicalevidence.bmj.com/ceweb/conditions/msd/1116/1116-get.pdf

Cleeland, C.S. (2006). The measurement of pain from metastatic bone disease: capturing the patient's experience. *Clin Cancer Res,* Vol. 12, pp.6236s-6242s.

Cole, J.S. & Patchell, R.A. (2008). Metastatic epidural spinal cord compression. *Lancet Neurol,* Vol. 7, pp.459-66.

Dagenais, S., Tricco, A.C. & Haldeman, S. (2010). Synthesis of recommendations for the assessment and management of low back pain from recent clinical practice guidelines. *Spine J,* Vol. 10, pp.514-29.

de Mello, J., Struthers, L., Turner, R., Cooper, E.H. & Giles, G.R. (1983). Multivariate analyses as aids to diagnosis and assessment of prognosis in gastrointestinal cancer. *Br J Cancer,* Vol. 48, pp.341-8.

Dearnaley, D.P. (1994). Cancer of the prostate. *BMJ,* Vol. 308, pp.780-4.

Degen, J.W., Gagnon, G.J., Voyadzis, J.M., McRae, D.A., Lunsden, M., Dieterich, S., Molzahn, I. & Henderson, F.C. (2005). CyberKnife stereotactic radiosurgical treatment of spinal tumors for pain control and quality of life. *J Neurosurg Spine,* Vol. 2, pp.540-9.

Elsberger, B., Lankston, L., McMillan, D.C., Underwood, M.A. & Edwards, J. (2011). Presence of tumoural C-reactive protein correlates with progressive prostate cancer. *Prostate Cancer Prostatic Dis,* Vol. 14, pp.122-8.

Eshed, I., Althoff, C.E., Hamm, B. & Hermann, K.G. (2007). Claustrophobia and premature termination of magnetic resonance imaging examinations. *J Magn Reson Imaging,* Vol. 26, pp.401-4.

Falicov, A., Fisher, C.G., Sparkes, J., Boyd, M.C., Wing, P.C. & Dvorak, M.F. (2006). Impact of surgical intervention on quality of life in patients with spinal metastases. *Spine (Phila Pa 1976),* Vol. 31, pp.2849-56.

Fehlings, M.G. & Rao, S.C. (2000). Spinal cord and spinal column tumors, In *Neuro-oncology: the essentials* (pp. 445-464) M. Bernstein and M. S. Berger (Ed.),^(Eds.), Thieme Medical Publishers, Inc., New York.

Fleisch, H. (1991). Bisphosphonates. Pharmacology and use in the treatment of tumour-induced hypercalcaemic and metastatic bone disease. *Drugs,* Vol. 42, pp.919-44.

Florence, D.W. (1981). The chronic pain syndrome: a physical and psychologic challenge. *Postgrad Med,* Vol. 70, pp.217-9, 222-3, 226-8.

Fulfaro, F., Casuccio, A., Ticozzi, C. & Ripamonti, C. (1998). The role of bisphosphonates in the treatment of painful metastatic bone disease: a review of phase III trials. *Pain,* Vol. 78, pp.157-69.

Gallo, A.E. & Donington, J.S. (2007). The role of surgery in the treatment of stage III non-small-cell lung cancer. *Curr Oncol Rep,* Vol. 9, pp.247-54.

Gerrard, G.E. & Franks, K.N. (2004). Overview of the diagnosis and management of brain, spine, and meningeal metastases. *J Neurol Neurosurg Psychiatry,* Vol. 75 Suppl 2, pp.ii37-42.

Gerszten, P.C., Germanwala, A., Burton, S.A., Welch, W.C., Ozhasoglu, C. & Vogel, W.J. (2005). Combination kyphoplasty and spinal radiosurgery: a new treatment paradigm for pathological fractures. *J Neurosurg Spine,* Vol. 3, pp.296-301.

Greenberg, H.S., Kim, J.H. & Posner, J.B. (1980). Epidural spinal cord compression from metastatic tumor: results with a new treatment protocol. *Ann Neurol,* Vol. 8, pp.361-6.

Guillevin, R., Vallee, J.N., Lafitte, F., Menuel, C., Duverneuil, N.M. & Chiras, J. (2007). Spine metastasis imaging: review of the literature. *J Neuroradiol,* Vol. 34, pp.311-21.

Hagelberg, C. & Allan, D. (1990). Restricted diffusion of integral membrane proteins and polyphosphoinositides leads to their depletion in microvesicles released from human erythrocytes. *Biochem J,* Vol. 271, pp.831-4.

Harel, R. & Angelov, L. (2010). Spine metastases: current treatments and future directions. *Eur J Cancer,* Vol. 46, pp.2696-707.

Henschke, N., Maher, C.G., Refshauge, K.M., Herbert, R.D., Cumming, R.G., Bleasel, J., York, J., Das, A. & McAuley, J.H. (2009). Prevalence of and screening for serious spinal pathology in patients presenting to primary care settings with acute low back pain. *Arthritis Rheum,* Vol. 60, pp.3072-80.

Hitron, A. & Adams, V. (2009). The pharmacological management of skeletal-related events from metastatic tumors. *Orthopedics,* Vol. 32, pp.188.

Hoy, A.M. (1989). Radiotherapy, chemotherapy and hormone therapy: treatment for pain, In *Textbook of Pain* (pp. 966-978) P. D. Wall and R. Melzack (Ed.),^(Eds.), Churchill Livinstone, Edinburgh.

Huda, W. & He, W. (2011). Estimating cancer risks to adults undergoing body CT examinations. *Radiat Prot Dosimetry,* Vol.

Itshayek, E., Yamada, J., Bilsky, M., Schmidt, M., Shaffrey, C., Gerszten, P., Polly, D., Gokaslan, Z., Varga, P.P. & Fisher, C.G. (2010). Timing of surgery and radiotherapy in the management of metastatic spine disease: a systematic review. *Int J Oncol,* Vol. 36, pp.533-44.

Itshayek, E., Yamada, J., Mahgerefteh, S., Cohen, J.E. & Fisher, C.G. (In press). Sequence of surgery, radiotherapy, and seterotactic radiosurgery in the treatment of metastatic spine disease: effects on wound healing, In *Spinal tumors Part 1, Vol. 6* M. A. Hayat (Ed.),^(Eds.), Springer-Verlag, London.

Kim, S.Y., Schneeweiss, S., Katz, J.N., Levin, R. & Solomon, D.H. (2011). Oral bisphosphonates and risk of subtrochanteric or diaphyseal femur fractures in a population-based cohort. *J Bone Miner Res,* Vol. 26, pp.993-1001.

Kruger, S., Buck, A.K., Mottaghy, F.M., Hasenkamp, E., Pauls, S., Schumann, C., Wibmer, T., Merk, T., Hombach, V. & Reske, S.N. (2009). Detection of bone metastases in patients with lung cancer: 99mTc-MDP planar bone scintigraphy, 18F-fluoride PET or 18F-FDG PET/CT. *Eur J Nucl Med Mol Imaging,* Vol. 36, pp.1807-12.

Kuo, P.H. & Cheng, D.W. (2005). Artifactual spinal metastases imaged by PET/CT: a case report. *J Nucl Med Technol,* Vol. 33, pp.230-1.

Ladeira, C.E. (2011). Evidence based practice guidelines for management of low back pain: physical therapy implications. *Rev Bras Fisioter,* Vol. 15, pp.190-9.

Laredo, J.D., Lakhdari, K., Bellaiche, L., Hamze, B., Janklewicz, P. & Tubiana, J.M. (1995). Acute vertebral collapse: CT findings in benign and malignant nontraumatic cases. *Radiology,* Vol. 194, pp.41-8.

Li, B., Ling Chau, J.F., Wang, X. & Leong, W.F. (2011). Bisphosphonates, specific inhibitors of osteoclast function and a class of drugs for osteoporosis therapy. *J Cell Biochem,* Vol. 112, pp.1229-42.

Lipton, A. (2011). New strategies to prevent and manage bone complications in cancer. *Clin Adv Hematol Oncol,* Vol. 9, pp.42-4.

Lishchyna, N. & Henderson, S. (2004). Acute onset-low back pain and hip pain secondary to metastatic prostate cancer: a case report. *J Can Chiropr Assoc,* Vol. 48, pp.5-12.

Liu, J.T., Liao, W.J., Tan, W.C., Lee, J.K., Liu, C.H., Chen, Y.H. & Lin, T.B. (2010). Balloon kyphoplasty versus vertebroplasty for treatment of osteoporotic vertebral compression fracture: a prospective, comparative, and randomized clinical study. *Osteoporos Int,* Vol. 21, pp.359-64.

Loblaw, D.A., Perry, J., Chambers, A. & Laperriere, N.J. (2005). Systematic review of the diagnosis and management of malignant extradural spinal cord compression: the Cancer Care Ontario Practice Guidelines Initiative's Neuro-Oncology Disease Site Group. *J Clin Oncol,* Vol. 23, pp.2028-37.

Lote, K., Walloe, A. & Bjersand, A. (1986). Bone metastasis. Prognosis, diagnosis and treatment. *Acta Radiol Oncol,* Vol. 25, pp.227-32.

Loughrey, G.J., Collins, C.D., Todd, S.M., Brown, N.M. & Johnson, R.J. (2000). Magnetic resonance imaging in the management of suspected spinal canal disease in patients with known malignancy. *Clin Radiol,* Vol. 55, pp.849-55.

Manchikanti, L., Fellows, B., Ailinani, H. & Pampati, V. (2010). Therapeutic use, abuse, and nonmedical use of opioids: a ten-year perspective. *Pain Physician,* Vol. 13, pp.401-35.

Mercadante, S. (1997). Malignant bone pain: pathophysiology and treatment. *Pain,* Vol. 69, pp.1-18.

Metser, U., Lerman, H., Blank, A., Lievshitz, G., Bokstein, F. & Even-Sapir, E. (2004). Malignant involvement of the spine: assessment by 18F-FDG PET/CT. *J Nucl Med,* Vol. 45, pp.279-84.

Moore, J.E. (2010). Chronic low back pain and psychosocial issues. *Phys Med Rehabil Clin N Am,* Vol. 21, pp.801-15.

Moulding, H.D., Elder, J.B., Lis, E., Lovelock, D.M., Zhang, Z., Yamada, Y. & Bilsky, M.H. (2010). Local disease control after decompressive surgery and adjuvant high-dose single-fraction radiosurgery for spine metastases. *J Neurosurg Spine,* Vol. 13, pp.87-93.

N.H.S. (2003). Percutaneous vertebroplasty. National Institute for Clinical Excellence (NICE), National Health Service (NHS). Retrieved from http://www.nice.org.uk/nicemedia/live/11058/30792/30792.pdf (Accessed October 3, 2011)

N.H.S. (2006). Balloon kyphoplasty for vertebral compression fractures. National Institute for Health and Clinical Excellence (NICE), National Health Service (NHS). Retrieved from http://www.nice.org.uk/guidance/IPG166. (Accessed September 26, 2011)

Nguyen, N.C., Chaar, B.T. & Osman, M.M. (2007). Prevalence and patterns of soft tissue metastasis: detection with true whole-body F-18 FDG PET/CT. *BMC Med Imaging,* Vol. 7, pp.8.

Nieder, C., Haukland, E., Pawinski, A. & Dalhaug, A. (2010). Anaemia and thrombocytopenia in patients with prostate cancer and bone metastases. *BMC Cancer,* Vol. 10, pp.284.

Nielsen, O.S., Munro, A.J. & Tannock, I.F. (1991). Bone metastases: pathophysiology and management policy. *J Clin Oncol,* Vol. 9, pp.509-24.

Orita, Y., Sugitani, I., Toda, K., Manabe, J. & Fujimoto, Y. (2011). Zoledronic acid in the treatment of bone metastases from differentiated thyroid carcinoma. *Thyroid,* Vol. 21, pp.31-5.

Padalkar, P. & Tow, B. (2011). Predictors of survival in surgically treated patients of spinal metastasis. *Indian J Orthop,* Vol. 45, pp.307-13.

Patchell, R.A., Tibbs, P.A., Regine, W.F., Payne, R., Saris, S., Kryscio, R.J., Mohiuddin, M. & Young, B. (2005). Direct decompressive surgical resection in the treatment of spinal cord compression caused by metastatic cancer: a randomised trial. *Lancet,* Vol. 366, pp.643-8.

Quan, G.M., Vital, J.M., Aurouer, N., Obeid, I., Palussiere, J., Diallo, A. & Pointillart, V. (2011). Surgery improves pain, function and quality of life in patients with spinal metastases: a prospective study on 118 patients. *Eur Spine J,* Vol.

Rades, D., Stalpers, L.J., Veninga, T., Schulte, R., Hoskin, P.J., Obralic, N., Bajrovic, A., Rudat, V., Schwarz, R., Hulshof, M.C., Poortmans, P. & Schild, S.E. (2005). Evaluation of five radiation schedules and prognostic factors for metastatic spinal cord compression. *J Clin Oncol,* Vol. 23, pp.3366-75.

Raje, N. & Roodman, G.D. (2011). Advances in the biology and treatment of bone disease in multiple myeloma. *Clin Cancer Res,* Vol. 17, pp.1278-86.

Ratanatharathorn, V., Powers, W.E., Moss, W.T. & Perez, C.A. (1999). Bone metastasis: review and critical analysis of random allocation trials of local field treatment. *Int J Radiat Oncol Biol Phys,* Vol. 44, pp.1-18.

Samoladas, E.P., Anbar, A.S., Lucas, J.D., Fotiadis, H. & Chalidis, B.E. (2008). Spinal cord compression by a solitary metastasis from a low grade leydig cell tumour: a case report and review of the literature. *World J Surg Oncol,* Vol. 6, pp.75.

Scharschmidt, T.J., Lindsey, J.D., Becker, P.S. & Conrad, E.U. (2011). Multiple myeloma: diagnosis and orthopaedic implications. *J Am Acad Orthop Surg,* Vol. 19, pp.410-9.

Schirrmeister, H., Glatting, G., Hetzel, J., Nussle, K., Arslandemir, C., Buck, A.K., Dziuk, K., Gabelmann, A., Reske, S.N. & Hetzel, M. (2001). Prospective evaluation of the clinical value of planar bone scans, SPECT, and (18)F-labeled NaF PET in newly diagnosed lung cancer. *J Nucl Med,* Vol. 42, pp.1800-4.

Sciubba, D.M., Nguyen, T. & Gokaslan, Z.L. (2009). Solitary vertebral metastasis. *Orthop Clin North Am,* Vol. 40, pp.145-54, viii.

Sedonja, I. & Budihna, N.V. (1999). The benefit of SPECT when added to planar scintigraphy in patients with bone metastases in the spine. *Clin Nucl Med,* Vol. 24, pp.407-13.

Shaw, B., Mansfield, F.L. & Borges, L. (1989). One-stage posterolateral decompression and stabilization for primary and metastatic vertebral tumors in the thoracic and lumbar spine. *J Neurosurg,* Vol. 70, pp.405-10.

Shih, A. & Jackson, K.C., 2nd. (2007). Role of corticosteroids in palliative care. *J Pain Palliat Care Pharmacother,* Vol. 21, pp.69-76.

Shih, T.T., Huang, K.M. & Li, Y.W. (1999). Solitary vertebral collapse: distinction between benign and malignant causes using MR patterns. *J Magn Reson Imaging,* Vol. 9, pp.635-42.

Smith, H.S. (2011). Painful osseous metastases. *Pain Physician,* Vol. 14, pp.E373-403.

Sorensen, S., Helweg-Larsen, S., Mouridsen, H. & Hansen, H.H. (1994). Effect of high-dose dexamethasone in carcinomatous metastatic spinal cord compression treated with radiotherapy: a randomised trial. *Eur J Cancer,* Vol. 30A, pp.22-7.

Steinborn, M.M., Heuck, A.F., Tiling, R., Bruegel, M., Gauger, L. & Reiser, M.F. (1999). Whole-body bone marrow MRI in patients with metastatic disease to the skeletal system. *J Comput Assist Tomogr,* Vol. 23, pp.123-9.

Sundaresan, N., Sachdev, V.P., Holland, J.F., Moore, F., Sung, M., Paciucci, P.A., Wu, L.T., Kelligher, K. & Hough, L. (1995). Surgical treatment of spinal cord compression from epidural metastasis. *J Clin Oncol,* Vol. 13, pp.2330-5.

Takuma, Y., Kawai, D., Makino, Y., Saito, S., Tanaka, S., Ogata, M., Ohta, T., Murakami, I. & Nishiura, T. (2006). Vertebral metastasis of intraductal papillary mucinous tumor of the pancreas. *Pancreas,* Vol. 33, pp.206-8.

Thurlimann, B., Morant, R., Jungi, W.F. & Radziwill, A. (1994). Pamidronate for pain control in patients with malignant osteolytic bone disease: a prospective dose-effect study. *Support Care Cancer,* Vol. 2, pp.61-5.

Tomita, K., Kawahara, N., Baba, H., Tsuchiya, H., Fujita, T. & Toribatake, Y. (1997). Total en bloc spondylectomy. A new surgical technique for primary malignant vertebral tumors. *Spine (Phila Pa 1976),* Vol. 22, pp.324-33.

van Holten-Verzantvoort, A.T., Zwinderman, A.H., Aaronson, N.K., Hermans, J., van Emmerik, B., van Dam, F.S., van den Bos, B., Bijvoet, O.L. & Cleton, F.J. (1991). The effect of supportive pamidronate treatment on aspects of quality of life of patients with advanced breast cancer. *Eur J Cancer,* Vol. 27, pp.544-9.

Vecht, C.J., Haaxma-Reiche, H., van Putten, W.L., de Visser, M., Vries, E.P. & Twijnstra, A. (1989). Initial bolus of conventional versus high-dose dexamethasone in metastatic spinal cord compression. *Neurology,* Vol. 39, pp.1255-7.

Venkitaraman, R., Sohaib, S.A., Barbachano, Y., Parker, C.C., Huddart, R.A., Horwich, A. & Dearnaley, D. (2010). Frequency of screening magnetic resonance imaging to detect occult spinal cord compromise and to prevent neurological deficit in metastatic castration-resistant prostate cancer. *Clin Oncol (R Coll Radiol),* Vol. 22, pp.147-52.

Verbeeck, A. (2004). Bone metastases from breast cancer: guidelines for diagnosis. *J Manipulative Physiol Ther,* Vol. 27, pp.211-5.

Winters, M.E., Kluetz, P. & Zilberstein, J. (2006). Back pain emergencies. *Med Clin North Am,* Vol. 90, pp.505-23.

Wood, B.C. (1993). Hormone treatments in the common 'hormone-dependent' carcinomas. *Palliat Med,* Vol. 7, pp.257-72.

Woodward, E.J. & Coleman, R.E. (2010). Prevention and treatment of bone metastases. *Curr Pharm Des,* Vol. 16, pp.2998-3006.

Young, H., Baum, R., Cremerius, U., Herholz, K., Hoekstra, O., Lammertsma, A.A., Pruim, J. & Price, P. (1999). Measurement of clinical and subclinical tumour response using [18F]-fluorodeoxyglucose and positron emission tomography: review and 1999 EORTC recommendations. European Organization for Research and Treatment of Cancer (EORTC) PET Study Group. *Eur J Cancer,* Vol. 35, pp.1773-82.

Part 2

Conservative Treatment

7

Pharmacotherapy for Chronic Low Back Pain

John H. Peniston
Feasterville Family Health Care Center, USA

1. Introduction

Chronic low back pain is a common, debilitating, and costly health problem (Centers for Disease Control and Prevention, 2001; Collins et al., 2005; Mapel et al., 2004; Vogt et al., 2005). The prevalence of chronic low back pain is higher in women and in whites versus blacks (Andersson, 1999; Mapel et al., 2004). Although the risk of chronic low back pain increases with age (Mapel et al., 2004), back pain remains the most common cause of disability in adults aged <45 years (Andersson, 1999).

Treatment guidelines issued in the United States (Chou et al., 2007) and Europe (Airaksinen et al, 2006) both state that back pain becomes chronic if it persists for ≥12 weeks. United States guidelines also distinguish between acute (<4 weeks) and subacute (4–<12 weeks) back pain (Chou et al., 2007). The 12-week threshold for classifying back pain as chronic makes sense given data suggesting that 80% to 95% of patients with disabling back pain can return to normal activities within 12 weeks, with less certain outcomes thereafter (Andersson, 1999).

Treatments for chronic low back pain include nonpharmacologic therapies (eg, exercise, lifestyle modification), which are discussed elsewhere in this book. First-line pharmacotherapies include acetaminophen, nonsteroidal anti-inflammatory drugs (NSAIDs), weak opioids, and strong opioids. Each of these therapies, including opioids, may be initiated during the acute phase of back pain, depending on the severity of pain (Chou et al., 2007). However, the course of chronic low back pain may be protracted (Andersson, 1999), meaning patients may require treatment over years or decades. It is therefore important that pharmacological treatments be effective and as safe as possible both in the short term and in patients requiring long-term treatment. Acetaminophen is a well-tolerated first-line pharmacotherapy but has limited efficacy (Zhang et al., 2010) and high-dose, long-term use is associated with hepatic toxicity (Watkins et al., 2006). NSAIDs have better efficacy than acetaminophen but have well-known risks of gastrointestinal (Boers et al., 2007; Gabriel et al., 1991; Hippisley-Cox et al., 2005), cardiovascular (Antman et al., 2007; Caldwell et al., 2006; Hippisley-Cox & Coupland, 2005; Kearney et al., 2006; Motsko et al., 2006), and other systemic adverse effects that increase with age, dose, and duration of use.

The serotonin norepinephrine reuptake inhibitor (SNRI) duloxetine has received US Food and Drug Administration (FDA) approval for the treatment of musculoskeletal pain such as chronic low back pain (Cymbalta® Delayed-Release Capsules, 2010). Duloxetine has only modest efficacy (Skljarevski et al., 2010a; Skljarevski et al., 2009), is associated with systematic adverse events (Skljarevski et al., 2010a; Skljarevski et al., 2009), and may also interact with other analgesics, most notably certain opioids (Smith, 2009).

Opioids have the greatest efficacy of any oral therapy for relieving pain but also have significant risk of adverse events and potential for pharmacokinetic drug interactions (Malhotra et al., 2001; Tulner et al., 2008) and carry a substantial risk of addiction and abuse. Chronic opioid therapy requires long-term monitoring for treatment compliance.

This chapter will review guideline recommendations and clinical evidence for pharmacotherapies available for the management of chronic low back pain.

2. Review of guidelines for chronic low back pain and clinical evidence for pharmacotherapies

2.1 Multimodal approach to therapy

United States guidelines for the management of chronic low back pain recommend that effective pharmacotherapy be administered in conjunction with self-care options (eg, application of heat, continued activity, adoption of physical fitness regimens, use of a medium-firm mattress, lifestyle modifications) (Airaksinen et al., 2006; Chou et al., 2007). Nonpharmacologic therapies are discussed in detail elsewhere in this book. For patients not responding to medication and self-care, clinicians should employ other nonpharmacologic options, including intensive interdisciplinary rehabilitation, exercise therapy, acupuncture, massage therapy, spinal manipulation, yoga, cognitive behavioral therapy, and progressive relaxation (Airaksinen et al., 2006; Chou et al., 2007).

2.2 Acetaminophen

United States Guidelines for the management of chronic low back pain recommend acetaminophen as a first-line therapy for mild to moderate pain (Chou et al., 2007); the recommendations are based on modest efficacy and overall favorable tolerability in patients with osteoarthritis. European guidelines do not recommend acetaminophen (Airaksinen et al., 2006).

Randomized controlled trials of acetaminophen for chronic low back pain are lacking. Both US and European guidelines acknowledge that acetaminophen has been less effective than oral NSAIDs for osteoarthritis pain (Airaksinen et al., 2006; Chou et al., 2007), and effect sizes for acetaminophen calculated by the Osteoarthritis Research Society International are below the threshold for a clinically meaningful analgesia (Zhang et al., 2010). Acetaminophen is not as safe as it was once believed to be, particularly for a chronic condition such as chronic low back pain that may require treatment for many years and may also require dose escalation as the disease progresses over time. At doses >4 g/d, acetaminophen is known to cause liver enzyme increases >3-fold greater than the upper limit of normal in healthy volunteers (Watkins et al., 2006). The FDA is currently considering lowering the maximum recommended dose to 3.25 g/day (Kuehn, 2009). In the United States, acetaminophen is the leading cause of acute liver failure (Chun et al., 2009).

Overdose of acetaminophen may be accidental (Camidge et al., 2003; Larson et al., 2005), resulting from patients taking too much without realizing there is a dosage ceiling, or from patients taking a product that combines a narcotic with acetaminophen and then augmenting it with additional acetaminophen (Larson et al., 2005). Overdose may also be deliberate (Camidge et al., 2003; Larson et al., 2005). Despite its reputation as a safe drug, acetaminophen is among the drugs implicated most frequently in suicide attempts (Camidge et al., 2003; Larson et al., 2005). Also, thought not typically believed to carry substantial potential for pharmacokinetic interactions, acetaminophen-associated liver failure is significantly associated with alcohol abuse and antidepressant use (Larson et al., 2005).

2.3 Nonsteroidal anti-inflammatory drugs

A Cochrane review of 65 randomized controlled trials found evidence that NSAIDs are effective for the management of chronic low back pain pain, but effect sizes were indicative of only a small treatment effect (Roelofs et al., 2008). The evidence presented in this Cochrane review suggested that different NSAIDs, including cyclooxygenase-2 inhibitors, show similar efficacy in patients with chronic low back pain.

Evidence from meta-analyses provide evidence that NSAIDs are moderately more effective than acetaminophen (Lee et al., 2004; Roelofs et al., 2008), and chronic low back pain guidelines also state that NSAIDs are more effective (Airaksinen et al., 2006; Chou et al., 2007). In patients with osteoarthritis, effect sizes for NSAIDs are substantially higher than with acetaminophen but nonetheless correspond to a small treatment effect (Zhang et al., 2010).

Nonsteroidal anti-inflammatory drugs are associated with well-known risks of gastrointestinal (Boers et al., 2007; Gabriel et al., 1991; Hippisley-Cox et al., 2005), cardiovascular (Antman et al., 2007; Caldwell et al., 2006; Hippisley-Cox & Coupland, 2005; Juni et al., 2004; Kearney et al., 2006; Motsko et al., 2006), and renal (Barkin & Buvanendran, 2004; Evans et al., 1995) adverse events. Hepatic adverse events have been reported infrequently (Rostom et al., 2005; Rubenstein & Laine, 2004), and are certainly less common than with acetaminophen. NSAIDs have been associated with potentially clinically meaningful increases in blood pressure (Forman et al., 2005; Johnson et al., 1994; Pavlicevic et al., 2008), probably both through direct effects (Forman et al., 2005; Johnson et al., 1994) and adverse pharmacokinetic interactions with diuretics and angiotensin-converting enzyme inhibitors (Pavlicevic et al., 2008). Drug interactions of NSAIDs with antihypertensives (angiotensin-converting enzyme inhibitors, diuretics), selective serotonin reuptake inhibitors (SSRIs), and corticosteroids (American Geriatrics Society, 2009; Malhotra et al., 2001; Tulner et al., 2008) are common. Ibuprofen can nullify the cardioprotective effects of aspirin (Gengo et al., 2008). Gastrointestinal bleeding events may occur in patients receiving an NSAID with warfarin (Cheetham et al., 2009), an SSRI, or a corticosteroid (American Geriatrics Society, 2009). Proton pump inhibitors administered as gastroprotection in NSAID-treated patients can block the cardioprotective effects of clopidogrel (Ho et al., 2009; Juurlink et al., 2009).

Cyclooxygenase-2 inhibitors have been associated with fewer adverse events than nonselective agents (Roelofs et al., 2008), and in particular have shown improved gastrointestinal tolerability (Singh et al., 2006). In a 12-week, 13,000 patient trial, the occurrence of abdominal pain, dyspepsia, diarrhea, headache, and nausea was reported in

2% to 5% of patients treated with celecoxib and 3% to 6% of those treated with naproxen or diclofenac (Singh et al., 2006). Confirmed gastrointestinal bleeding events were also less frequent with celecoxib compared with nonselective agents, but occurred in 1% of patients compared with 2.1% treated with the nonselective agents.

Given their risk/benefit ratio, guidelines for the treatment of chronic low back pain (Airaksinen et al., 2006; Chou et al., 2007) and osteoarthritis (Zhang et al., 2007; Zhang et al., 2008) recommend that oral NSAIDs be administered for the shortest period possible at the lowest effective dose. Guidelines for the management of chronic pain in the elderly suggest that regular oral NSAID use should be avoided if possible (American Geriatrics Society, 2009).

2.4 Topical therapies

United States guidelines for the management of chronic low back pain do not recommend topical therapies. European guidelines recommend only capsaicin plaster for short-term therapy (<3 wk) in patients with chronic low back pain (Airaksinen et al., 2006).

2.4.1 The lidocaine 5% patch

The lidocaine 5% patch has US approval only for the management of postherpetic neuralgia (LIDODERM®, 2010), particularly with allodynia (Attal et al., 2010). However, the lidocaine 5% patch has also shown efficacy in patients with chronic lower back pain (Galer et al., 2004; Gimbel et al., 2005). In a 6-week open-label trial, 71 patients with acute/subacute (n=11), short-term chronic (n=17), or long-term chronic (n=43) nonradicular back pain applied up to 4 patches to the maximal area of pain (Galer et al., 2004). All patient subgroups reported significant improvements in scores on the Neuropathic Pain Scale. In a second 6-week open-label study, 131 patients with subacute (n=21), short-term chronic (n=33), and long-term chronic (n=77) back pain reported significant mean improvements in scores on the Brief Pain Inventory (Gimbel et al., 2005) and significant improvements in the Brief Pain Inventory item for interference with quality of life. Beck Depression Inventory scores also improved significantly. Approximately two thirds of patients and investigators reported being Satisfied or Very Satisfied with treatment.

Application of up to 4 lidocaine 5% patches daily results in peak systemic lidocaine concentrations <200 ng/mL, or >7-fold less than the concentration of >1500 ng/mL required to produce cardiac effects and 20-fold less than the threshold concentration for toxic effects (Gammaitoni et al., 2003). Consistent with low systemic exposure, the most common adverse events reported in a trial of the lidocaine 5% patch in patients with chronic low back pain were mild to moderate skin and subcutaneous tissue reactions (eg, rash, pruritus, dermatitis, erythema, edema). Treatment-related systemic adverse events were relatively uncommon and included dizziness (4%), headache (2%), and nausea (2%-3%) (Gammaitoni et al., 2003).

2.4.2 The capsaicin 8% patch

In a double-blind, randomized trial, 320 patients with chronic low back pain were assigned to apply the capsaicin 8% patch or placebo patch once daily for 3 weeks (Frerick et al., 2003).

The capsaicin 8% patch was associated with a 42% reduction in pain assessed using the Arhus low back rating scale versus a 31% reduction with placebo. Sixty-seven percent of the capsaicin 8% patch group and 49% of the placebo group met response criteria, defined as a ≥30% reduction in pain. Though statistically significant, these differences compared with placebo are very modest. These results were similar to those obtained with a similar capsaicin 8% patch in a similarly designed trial (Keitel et al., 2001). The most common adverse events were application site warmth or itching, although inflammatory contact eczema, urticaria, small hemorrhagic spots, vesiculation, and dermatitis have been reported in a small number of patients (Frerick et al., 2003; Keitel et al., 2001). Given modest efficacy and generally good tolerability, the capsaicin 8% patch may only be useful as adjunctive therapy or as monotherapy in a small subgroup of patients.

2.4.3 Topical nonsteroidal anti-inflammatory drugs

Topical NSAIDs produce dramatically lower systemic NSAID concentrations compared with oral NSAIDs (Kienzler et al., 2010) and are recommended as a first-line therapy in patients with osteoarthritis of the knees or hands (Zhang et al., 2007; Zhang et al., 2008), particularly the elderly (American Geriatrics Society, 2009). However, no chronic low back pain guidelines recommend topical NSAIDs and chronic low back pain is not an approved indication for either of the two topical NSAID formulations used in the United States: diclofenac sodium 1% gel (Voltaren® Gel 1%, 2007) and diclofenac sodium 1.5% in 45.5% dimethyl sulfoxide solution (Pennsaid®, 2009). Neither formulation has been evaluated in patients with chronic low back pain.

Topical diclofenac diethylamine 1.16% gel and an indomethacin plaster were evaluated in 64 patients with mild to moderate, nonsurgical chronic low back pain (Waikakul et al., 1996). Both formulations provided statistically significant improvements in pain and function; however, topical NSAIDs have not been evaluated in clinical trials since this 1996 study.

2.5 Serotonin-norepinephrine reuptake inhibitors

Duloxetine has shown modest efficacy in two 12-week trials (Skljarevski et al., 2010a; Skljarevski et al., 2009) and in a 41-week open-label study (Skljarevski et al., 2010b). Active treatment was superior to placebo in both 12-week trials when a 30% improvement in pain was defined as a positive response (Skljarevski et al., 2010a; Skljarevski et al., 2009) but only in 1 trial when a 50% improvement was defined as a positive response (Skljarevski et al., 2010a). Given modest efficacy, duloxetine may be a helpful adjunctive therapy (discussed below).

2.6 Weak opioids

The weak opioids codeine, hydrocodone, and propoxyphene have not been studied extensively in patients with chronic low back pain. A single 1976 study of propoxyphene showed efficacy in patients with chronic low back pain (Baratta, 1976) and a 1998 study of an extended release formulation of codeine combined with acetaminophen demonstrated efficacy similar to tramadol (Muller et al., 1998). Hydrocodone has not been studied in patient with chronic low back pain. Nonetheless, a meta-analysis showed that weak opioids do not have a clear efficacy advantage compared with oral NSAIDs or tricyclic antidepressants (Furlan et al., 2006). Several safety issues limit use of weak opioids. These agents are commonly combined with acetaminophen, imposing a dosage ceiling on their use (Victor et al., 2009). Codeine has been associated with a high rate of constipation. Moreover, though a weak opioid, propoxyphene is associated with more potent respiratory and cardiovascular depression than are strong opioids (Barkin et al., 2006; Ulens et al., 1999).

2.7 Combination weak opioid-norepinephrine reuptake inhibitors

A Cochrane review of 3 trials found that tramadol is more effective than placebo for treating pain and functional impairment (Deshpande et al., 2007). It should be observed that 2 of the 3 trials analyzed in the Cochrane review evaluated products combining tramadol with acetaminophen (Peloso et al., 2004; Ruoff et al., 2003). Although pure opioids theoretically have no analgesic dosage ceiling, the presence of acetaminophen imposes a dosage ceiling on the product because acetaminophen is hepatotoxic above a daily dose of 4 g (Watkins et al., 2006). Tramadol is not a pure opioid agonist but rather also has serotonergic effects that limit the recommended daily dose to 300 mg (Ultram ER®, 2009). Serotonergic effects of tramadol increase the risk of serotonin syndrome in patients receiving other serotonergic drugs, and must not be combined with SNRIs, SSRIs, and tricyclic antidepressants (Ultram ER®, 2009).

Two 6-week trials found that tramadol monotherapy was less effective than the selective cyclooxygenase-2 inhibitor celecoxib (O'Donnell et al., 2009). Likewise, tramadol was found to be no more effective than extended-release codeine, even though tramadol has the additional noradrenergic mechanism of action. The most common adverse events reported in the 3 trials of tramadol with or without acetaminophen were typical opioid effects, including nausea (8.7%–13%), headache (4.7%–6.6%), somnolence (9%–12.4%), dizziness (7.5%–10.8%), dry mouth (6%–8.1%), vomiting (6%), and constipation (10.2%–11.2%) (Peloso et al., 2004; Ruoff et al., 2003; Schnitzer et al., 2000).

Tramadol relies on the cytochrome P450 enzyme 2D6 for metabolism to its active metabolite, *O*-desmethyltramadol, giving it potential for pharmacokinetic interaction with drugs metabolized via this pathway. Five to 15% of the white population is classified as having either rapid or slow cytochrome P450 enzyme 2D6 metabolizers, meaning that some patients will have insufficient or increased sensitivity to both analgesic and adverse effects of the drug (Smith, 2009). Collectively, tramadol appears to offer no efficacy advantage compared with weak opioids despite its secondary norepinephrine reuptake inhibitor activity.

A second weak opioid-SNRI combination, tapentadol, differs from tramadol in several important respects. Tapentadol is metabolized via glucuronidation and therefore lacks the potential for pharmacokinetic interaction and variability response observed with tramadol (Kneip et al., 2008). Its noradrenergic effects are more potent than its serotonergic effects (Tzschentke et al., 2007), theoretically reducing the risk of serotonergic adverse events, although in the United States tapentadol has an FDA "black box" warning against its combined use with SSRIs or SNRIs. Analgesic effects of SNRIs are attributed primarily to the noradrenergic effects (Max et al., 1992); hence, reduced activity with respect to serotonergic function decreases adverse events without compromising analgesic efficacy.

In a prospective, randomized, double-blind, trial, 981 patients with chronic low back pain received tapentadol 100 to 250 mg/d, oxycodone controlled release, or placebo for a 3-week titration period followed by a 12-week maintenance period. Tapentadol and oxycodone produced similar, statistically significant improvements in pain compared with placebo. Tapentadol was associated with a lower occurrence compared with oxycodone of total adverse events, constipation, nausea, and vomiting. A pooled analysis of this study and two 12-week studies of patients with osteoarthritis (n=2010) demonstrated that tapentadol was as effective as oxycodone controlled release in improving mean pain intensity, but more people treated with tapentadol experienced 30% and 50% reductions in pain (Lange et al., 2010). Only tapentadol was associated with significant improvements in quality of life, which may reflect superior tolerability with tapentadol. Tapentadol exhibited a superior gastrointestinal tolerability compared with oxycodone (Lange et al., 2010). A subsequent subanalyses of these trials found that the lower incidence of constipation with tapentadol (13%–18%) compared with oxycodone (27%–36%) should be associated with a lower rate of work absenteeism and productivity loss in tapentadol-treated patients, based on previous research correlating constipation with productivity (Cepeda et al., 2011).

A limitation of studies comparing tapentadol with oxycodone controlled release was the dosage ranges used. Patients assigned to oxycodone received a dose of 20 to 50 mg twice daily (Lange et al., 2010), which is well below the effective dose for many patients. All patients, including opioid-experienced patients, were started on oxycodone 10 mg. The maximum dose of 50 mg is much less than the maximum doses reported in oxycodone trials (Rauck et al., 2006a). Nonetheless, if patients were underdosed with respect to analgesia, they also would be expected to have more adverse events at higher, more effective doses. Thus, at optimal oxycodone doses with respect to analgesia, differences in tolerability favoring tapentadol might be even more pronounced.

Given statistically significant efficacy compared with placebo, a trial of a weak opioid-SNRI combination drug before prescribing a strong opioid may be a valuable option in many patients.

2.8 Strong opioids: Efficacy

Despite increasing use in patients with chronic noncancer pain in recent years, opioids have not been studied extensively in patients with chronic low back pain. A 2007 Cochrane review of opioids for chronic low back pain (Deshpande et al., 2007) included only 1 trial of a strong opioid (Jamison et al., 1998). This 1998 trial found that oxycodone immediate release and oxycodone combined with morphine extended release were both significantly superior to naproxen in patients with moderate to severe chronic low back pain (Jamison et al., 1998). Similarly a 2006 meta-analysis of opioids in chronic noncancer pain included the same single study of oxycodone in chronic low back pain and reached the same conclusion (Furlan et al., 2006). A 2011 meta-analysis included 4 additional studies, one of tramadol, 2 of oxymorphone extended release, and 1 of oxycodone with ultra-low-dose naltrexone (Kuijpers et al., 2011).

Oxymorphone extended release was evaluated in two 12-week randomized controlled trials in opioid-naive and opioid-experienced patients with moderate to severe chronic low back pain (Hale et al., 2007; Katz et al., 2007). Results from these studies demonstrated that use of a flexible, individualized titration schedule will allow the majority (60% in the 2 studies combined) of patients to be titrated to an effective, generally well-tolerated opioid dose

(Peniston & Gould, 2009). The trial in opioid-naïve patients demonstrated that the lowest available doses of oxymorphone extended release can be administered safety in patients who have never received an opioid. In contrast, some long-acting opioids are not appropriate as a starting medication for opioid-naïve patients; for example, the fentanyl transdermal patch and hydromorphone OROS (discussed following) (Duragesic®, 2009; EXALGO complete prescribing information, 2010). The trial of oxymorphone extended release in opioid-experienced patients demonstrated that patients who grow tolerant to one opioid formulation can be successfully transitioned to another opioid with ostensibly the same mechanism of action and experience adequate pain relief with acceptable tolerability.

This finding is important given the clinical observation that patients requiring long-term opioid therapy typically require successive trials of different opioids to find one that works, and that even after finding the right opioid, will frequently grow tolerant to the initial opioid and need to be rotated to another opioid to maintain analgesia or avoid adverse events that emerge with dose escalation (Grilo et al., 2002; Mercadante & Bruera, 2006; Quang-Cantagrel et al., 2000). It is important to have multiple effective opioids from which to choose an initial agent and have options for switching as tolerance develops (Slatkin, 2009). It remains to be determined whether after switching from an initial opioid to another, responsiveness to the original opioid will return over time to allow a switch back if tolerance to the second opioid emerges. In both opioid-naïve and –experienced patients, oxymorphone extended release analgesia and functional improvement remained significantly superior to placebo for the entire 12-week treatment period without dose escalation (Hale et al., 2007; Katz et al., 2007). Of patients successfully titrated and randomized to oxymorphone extended release, 69% completed 12 weeks of treatment (Peniston & Gould, 2009).

Oxycodone has been formulated with ultra-low-dose naltrexone in order to minimize adverse events (eg, respiratory depression, constipation, physical dependence) (Amass et al., 2000; Chindalore et al., 2005; Webster et al., 2006). In a double-blind trial, 719 patients with moderate to severe chronic low back pain received placebo, oxycodone immediate release 4 times daily, oxycodone with naltrexone 4 times daily, or oxycodone plus naltrexone twice daily for 12 weeks (Webster et al., 2006). All patients initiated treatment with a total daily oxycodone dose of 10 mg and were titrated over a period of 1 to 6 weeks to a dose that provided effective analgesia with acceptable tolerability or to a maximum daily dose of 80 mg. All oxycodone regimens provided significantly superior analgesia compared with placebo. The mean daily oxycodone dose was 12% lower with the oxycodone plus naltrexone regimens than with oxycodone. Compared with oxycodone administered 4 times daily, oxycodone plus naltrexone twice daily was associated with 44% less constipation, 33% less somnolence, and 51% less pruritus. Upon abrupt opioid discontinuation, the severity of withdrawal symptoms assessed using the Short Opiate Withdrawal Scale was 55.8% less with oxycodone plus naltrexone twice daily compared with oxycodone 4 times daily (Webster et al., 2006).

In a randomized, open-label trial, 392 patients with chronic low back pain were randomized to an effective, generally well-tolerated dose of once-daily morphine extended release or twice-daily oxycodone controlled release (Rauck et al., 2006a). Patients then entered an 8-week evaluation phase comparing the 2 treatments. As in the previous trials, use of a flexible titration schedule allowed 67.9% of patients (morphine extended release, 65.0%;

oxycodone controlled release, 70.9%) to be titrated successful. Once-daily morphine extended-release capsules provided significantly better analgesia with a lower morphine-equivalent daily dose compared with oxycodone twice daily (Rauck et al., 2006a). Sleep quality was significantly better with once-daily morphine, and rescue ibuprofen use was significantly less compared with twice-daily oxycodone. During a 4-month extension phase, patients receiving once-daily morphine continued to experience lower pain intensity and greater improvements in sleep quality (Rauck et al., 2006b).

A second once-daily opioid formulation, hydromorphone OROS, was also associated with statistically significant improvements in pain, quality of life measures, and sleep quality during a 6-week open-label trial (Wallace et al., 2007) and 6-month open-label follow-up (Wallace & Thipphawong, 2010). A clinical trial of hydromorphone OROS demonstrated that administration with ethanol does not result in premature release of hydromorphone from the capsule, a phenomenon referred to as "dose-dumping" (Sathyan et al., 2008). This is important because a previous formulation of hydromorphone extended release exhibited dose-dumping, resulting in its withdrawal from the US market (FDA asks Purdue Pharma to withdraw Palladone, 2005).

The efficacy and tolerability of fentanyl transdermal patch were demonstrated in open-label trials lasting 1 month (Simpson et al., 1997), 3 months (Lee et al., 2011), and 13 months (Allan et al., 2005), respectively. In the 13-month study (Allan et al., 2005), 673 patients with moderate to severe chronic low back pain ingested morphine sustained release once every 12 hours or applied a single fentanyl patch every 72 hours. Both treatments provided similar, significant analgesia throughout the study. Each opioid formulation was associated with typical opioid-associated adverse events such as constipation, nausea, vomiting, dizziness, diarrhea, and somnolence. Constipation was significantly less frequent with the fentanyl patch (52%) than with morphine sustained release (65%). The 3-month trial enrolled 1576 patients with severe, treatment-refractory chronic low back pain. Fentanyl provided significant improvements in measures of pain, function, and sleep quality. It should be noted, however, that <1% of patients enrolled in this trial had received a strong opioid as their prior medication; hence, the results support the use of transdermal fentanyl in patients with severe refractory pain but do not distinguish fentanyl from other opioids. Of note, the use of fentanyl transdermal patch for opioid-naïve patients in this study is contrary to the products' approved use.(Duragesic®, 2009) In contrast, studies of oxymorphone (Hale et al., 2007), oxycodone controlled release, and morphine extended release (Rauck et al., 2006a), and hydromorphone OROS (Wallace et al., 2007) have demonstrated efficacy in patients with chronic low back pain that was moderate to severe despite previous opioid therapy. Fentanyl may offer advantages over oral opioids for patients with compliance issues, including those with problems chewing or swallowing. Fentanyl clearance appears to be affected very little by renal failure (Dean, 2004) or hepatic cirrhosis (Haberer et al., 1982) impairment, making it an option that may be used with caution in these populations. However, fentanyl patch should not be administered to opioid-naïve patients because there is a significant risk of respiratory depression even with the lowest available fentanyl patch dose, 25 µg/hour (Duragesic®, 2009).

In a randomized, double-blind crossover study, 79 patients with chronic low back pain applied a 7-day buprenorphine transdermal patch or placebo patch for 4 weeks of treatment before crossing over to the alternate therapy for an additional 4 weeks (Gordon et al., 2010).

Patients began treatment with the 5 μg/h patch and were titrated with each successive application to the maximum tolerated dose (5, 10, or 20 μg/h). Fifty-four of 73 (74%) patients completed 4 weeks of treatment with buprenorphine patch. Mean reduction in pain (100-mm Visual Analog Scale) was 39.5% in the buprenorphine group and 29.8% in the placebo group, which was statistically significant. Improvements in Pain Disability Index scores were significant compared with baseline but not compared with placebo.

2.8.1 Summary of efficacy data

Flexible titration allows for successful titration in the majority of opioid-treated patients. The availability of formulations administered twice daily, daily, every 3 days, or once weekly allows for individualized treatment of patients to optimize treatment compliance. Studies of long-acting oxymorphone, oxycodone, morphine, hydromorphone, fentanyl, and buprenorphine confirm the analgesic efficacy in patients with chronic low back pain. Patients who do not respond or who grow tolerant to an opioid therapy can be transitioned to a second opioid with the same mechanism of action. The impact of opioids on functional status is variable.

2.8.2 Summary of safety data

The most common adverse events during opioid therapy include constipation (41%), nausea (32%), somnolence/sedation (29%), vomiting (15%), dizziness (20%), itching (15%), dry mouth (13%), and headache (8%) (Kalso et al., 2004). Approximately one quarter of patients discontinue clinical trials owing to adverse events (Kalso et al., 2004). Flexible titration reduces adverse events and allows the majority of patients to be titrated to an effective dose (Peniston & Gould, 2009; Rauck et al., 2006a). In 2 trials of oxymorphone extended release, nearly two thirds (348 of 575; 60.5%) of patients were successfully titrated to a well-tolerated oxymorphone extended release dose. Opioid-naïve patients were started at a low dose (10 mg/d) and opioid-experienced patients were transitioned from their previous opioid using published conversion ratios. Titration was gradual (5–10 mg every 3–7 days) and at investigators discretion. Discontinuations due to adverse events during titration (18.4%) were lower than typically observed in opioid trials (Peniston & Gould, 2009).

Similarly, in a trial comparing morphine extended release with oxycodone controlled release, use of a flexible titration schedule allowed 67.9% of patients (morphine extended release, 65.0%; oxycodone controlled release, 70.9%) to be titrated successfully (Rauck et al., 2006a). Discontinuations due to adverse events during the titration period were reported in 18.7% of patients treated with morphine extended release and 14.3% treated with oxycodone controlled release.

All opioids have a potential for clinically significant pharmacodynamic drug interactions, particularly when combined with drugs that cause respiratory depression (eg, barbiturates) or central nervous system depression (eg, alcohol, benzodiazepines), but opioids differ in their potential for pharmacokinetic drug interactions. The majority of opioids are metabolized by the cytochrome P450 system and have potential for pharmacokinetic interaction with many drugs (Smith, 2009). Exceptions to this are morphine (Coffman et al., 1997), oxymorphone (Adams et al., 2005; OPANA® ER, 2010), and hydromorphone (Hydromorphone-HP Injection 10 mg 2008), each of which primarily undergo phase 2 metabolism by glucuronidation. Selection of an opioid that is not metabolized via cytochrome P450 enzymes is important when there is a known cytochrome P450 pharmacokinetic interaction with a concurrent medication but may also be desirable in patients likely to be prescribed multiple medications, such as the elderly, patients with psychiatric illness, and patients with multiple medical problems (Smith & Bruckenthal, 2010).

The labels for all strong opioids recommend against concurrent use with alcohol because of the potential for clinically meaningful additive pharmacodynamic effects (Dilaudid® Oral Liquid and Dilaudid® Tablets, 2008; Duragesic®, 2009; Kadian®, 2007; OPANA® ER, 2010; OxyContin®, 2010). Additive effects when opioids are administered with alcohol are to be distinguished from the phenomenon of dose-dumping described previously with reference to hydromorphone extended release. Clinical trials have demonstrated the integrity of hydromorphone OROS (Sathyan et al., 2008) and morphine extended release (Barkin et al., 2009; Johnson et al., 2008) when administered with 240 mL of 4% to 40% ethanol. Coadministration of oxymorphone extended release with 240 mL of 4% and 20% ethanol, corresponding to modest and moderate alcohol consumption, did not cause premature release of oxymorphone from the extended release tablet (Fiske et al., 2011). Oxymorphone extended release administered with a 240 mL solution of 40% ethanol resulted in potentially clinically meaningful increases in oxymorphone maximum concentration (Fiske et al., 2011). It should be noted, however, that the 40% ethanol solution corresponds to heavy alcohol intake, which is to be avoided in any opioid-treated patient.

2.8.3 Minimize abuse potential

Before initiating opioid therapy, physicians should take a thorough patient history that includes a history of prior or current substance abuse or psychiatric disease, as well as accurate documentation of current prescribed medications. Patients should undergo a physical examination sufficient to confirm the diagnosis of chronic low back pain and determine pain intensity and the extent of disability. This will help establish that a request for an analgesic is legitimate rather than drug-seeking behavior (Cone & Caplan, 2009). Patients should also be screened for abuse risk using validated screening instruments such as the Opioid Risk Tool (Webster & Webster, 2005), Current Opioid Misuse Measure (Butler et al., 2007), and the revised Screener and Opioid Assessment for Patients with Pain (Butler et al., 2008).

Opioid therapy should be conducted using a plan that includes controlled substance agreements, frequent random urine drug testing, and confirmatory urine drug testing (Cone & Caplan, 2009; Manchikanti et al., 2008). Screening should be conducted in all patients, regardless of perceived risk of abuse (Gourlay et al., 2005). In patients believed to be at increased risk, selection of an opioid that does not produce metabolites that are identical to other opioids may simplify urine drug screening (Cone & Caplan, 2009). Positive test results for patients prescribed buprenorphine, fentanyl, oxymorphone, tapentadol, and tramadol will not be confused by the presence of multiple opioids.

Drug companies are currently attempting to develop formulations that incorporate obstacles to abuse. Most of these technologies are designed to resist common methods by which abusers tamper with an oral opioid formulation to facilitate intranasal or intravenous abuse

(Katz, 2008). Tamper-resistant opioids include formulations with physical barriers to crushing, chewing, and dissolution; formulations with sequestered antagonists that neutralize opioid effects if the formulation is chewed or crushed; and formulations with sequestered noxious components that are released if the opioid is chewed or crushed.

At present, 2 tamper-resistant formulations have received FDA approval. OXECTA® is a short-acting formulation of oxycodone with sequestered ingredients that are released if the opioid is chewed or crushed (OXECTA®, 2011). In addition to oxycodone, the formulation contains several ingredients that may cause irritation if swallowed or inhaled, including colloidal silicon dioxide (Amorim et al., 2010), crospovidone (Lowe et al., 2006), microcrystalline cellulose (Teshima et al., 2002), and sodium lauryl sulfate (Engel et al., 2008).

Oxycodone controlled release has also been reformulated in a polymer matrix that resists crushing or chewing (US Food and Drug Administration, 2010) and becomes a viscous gel if the product is immersed in fluid (Schneider et al., 2010). The value of tamper-resistant formulations of oxycodone and other opioids is likely to be limited unless all opioids are reformulated with this purpose in mind. Otherwise, the availability of one opioid in a tamper-resistant formulation may simply divert abusers to another opioid that has not been reformulated. Since the approval of reformulated oxycodone, oxycodone has seen a decline in abuse, with a corresponding increase in oxymorphone abuse (Goodnough, 2011; Nassau County executive: abuse of painkiller Opana is growing, 2011). Oxymorphone extended release is not currently available in a tamper resistant formulation, although such a formulation is in development.

2.9 Adjunctive therapies

Tricyclic antidepressants are recommended in both US and European guidelines for the management of chronic low back pain. Two systematic reviews concluded that tricyclic antidepressants have modest beneficial effects on pain but limited effects on functional status (Salerno et al., 2002; Staiger et al., 2003).

Duloxetine can be administered safely with NSAIDs, acetaminophen, and opioids that do not undergo cytochrome P450 metabolism, such as oxymorphone, hydromorphone, fentanyl, or morphine (Smith, 2009). Because it is a substrate and moderate inhibitor of the cytochrome P450 enzyme 2D6, pharmacokinetic interactions with oxycodone, tramadol, and codeine are possible (Cymbalta® Delayed-Release Capsules, 2010). Given its modest efficacy (discussed previously), addition of duloxetine to ongoing opioid or NSAID therapy may be a useful alternative to switching to another medication or increasing the dose of an opioid or NSAID. Duloxetine may be of particular value in chronic pain patients with comorbid depression. In an open-label pilot study (Karp et al., 2010), duloxetine administered to patients with comorbid chronic low back pain and major depression improved both pain and depression.

Guidelines in the United States state that gabapentin has modest efficacy in patients with chronic low back pain (Chou et al., 2007), whereas the European guidelines state that there is no convincing evidence (Airaksinen et al., 2006). Other anticonvulsants are not recommended. Benzodiazepines have been recommended for use as muscle relaxants for short-term therapy in both US and European guidelines (Airaksinen et al., 2006; Chou et al.,

2007). These anxiolytics have a high risk of addiction and abuse, and their combined use with opioids is of particular concern (Rich & Webster, 2011).

2.10 Invasive therapies

In US guidelines, epidural steroid injection is recommended for patients with persistent chronic low back pain and signs of radiculopathy or spinal stenosis (Chou et al., 2007). European guidelines state that epidural injections have insufficient evidence to support a recommendation (Airaksinen et al., 2006). Intrathecal opioids (morphine or methadone) have been administered for severe chronic low back pain, almost always following failed back surgery (Noble et al., 2010). Though effective, intrathecal delivery of opioids has been associated with granuloma formation (Allen et al., 2006). Neither intrathecal nor epidural opioid infusion is recommended in US or European treatment guidelines.

United States guidelines state that transcutaneous nerve stimulation has not been proven effective for chronic low back pain (Chou et al., 2007). European guidelines state that clinicians may consider percutaneous electrical nerve stimulation or neuroflexotherapy (Airaksinen et al., 2006). European guidelines also recommend against local facet nerve blocks, trigger point injections, and spinal cord stimulation (Airaksinen et al., 2006). Surgery is not recommended unless other treatment options have been unsuccessful for 1 year (US) (Chou et al., 2007) or 2 years (Europe) (Airaksinen et al., 2006).

3. Treatment algorithm

Treatment recommendations based on the clinical data provided previously are summarized in **Figure 1**. Patients with mild to moderate chronic low back pain should begin pharmacotherapy with acetaminophen or an NSAID. Acetaminophen has limited efficacy, poses risks of liver failure with cumulative/escalating use, is not recommended in patients with hepatic impairment, and should be given cautiously to patients with depression or at risk of substance abuse because of its documented use by patients wishing to inflict self-harm. Patients with dementia may have difficulties reporting pain, so clinicians should be sure that patients are not in need of more potent analgesia.

Oral NSAIDs are more effective than acetaminophen but still have only modest efficacy, making their use best reserved for mild to moderate pain that does not respond to acetaminophen. NSAIDs should be administered for the shortest period of time at the lowest effective dose, and are not generally recommended in older patients and those with current or a history of gastrointestinal bleeding or significant cardiovascular or renal comorbidities. Caution is recommended when administering NSAIDs in patients taking SSRIs, corticosteroids, warfarin, cardioprotective aspirin (ibuprofen only), diuretics and angiotensin-converting enzyme inhibitors. An oral NSAID with a gastroprotective proton pump inhibitor should not be administered to patients receiving clopidogrel.

Topical lidocaine patch is recommended for the treatment of chronic low back pain in Europe but not the United States. Lidocaine patch has shown efficacy in patients with chronic low back pain. Capsaicin plasters have shown only very modest efficacy and cause localized burning that limits tolerability. Capsaicin patches are not recommended, and their localized numbing effect is better achieved using a lidocaine patch. Topical NSAIDs have few data supporting their use for chronic low back pain and require further investigation of their potential value for this indication.

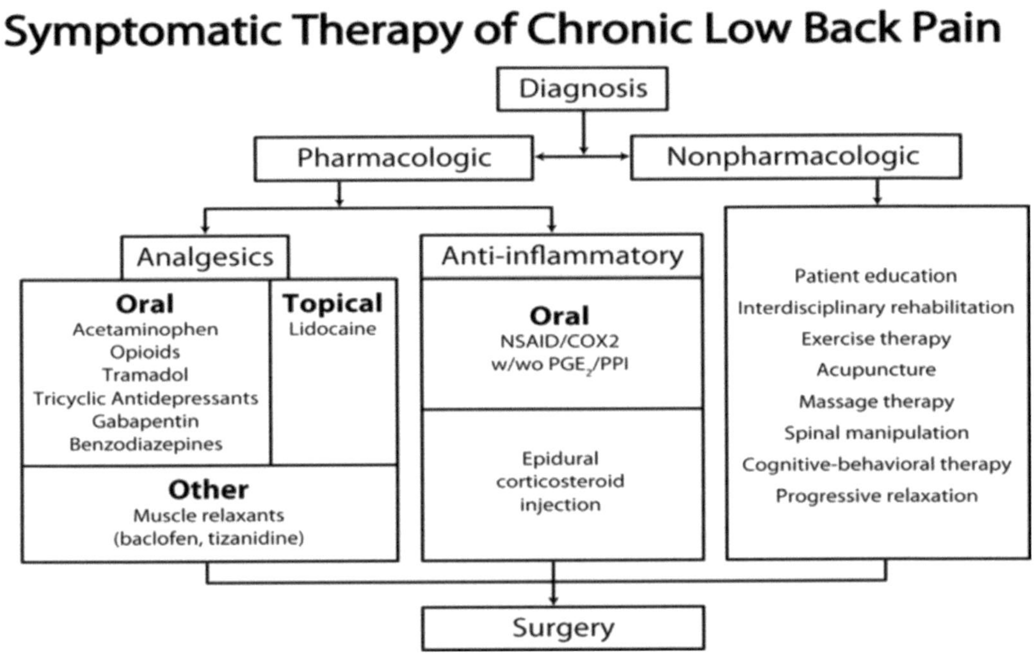

Reprinted from PostGrad Med, 122, Altman RD, Smith HS. Opioid Therapy for Osteoarthritis and Chronic Low Back Pain, 87-97, 2010, with permission from JTE Multimedia.

Fig. 1. Treatment algorithm for patients with chronic low back pain. COX2= cyclooxygenase -2, NSAID=nonsteroidal anti-inflammatory drug, PGE_2= prostaglandin E_2, PPI=proton pump inhibitor.

There are limited data supporting the use of tramadol in patients with chronic low back pain. Tramadol may be administered in combination products that include acetaminophen. This imposes a dosage ceiling on the product that may limit use of tramadol. There are more robust data for tapentadol; however, it is unclear whether this is because of a difference in efficacy or to it having been studied more rigorously. Both tramadol and tapentadol are associated with typical opioid adverse events, but gastrointestinal adverse events are less frequent than with strong opioids. Though less potent than strong opioids, tapentadol may be a safer option for patients who may benefit from an opioid but want to reduce risks. As noradrenergic agents may be administered as adjunctive therapy for chronic low back pain, tapentadol affords an opportunity to utilize 2 analgesic mechanisms in a single pill.

Opioids have a growing body of evidence supporting their use in patients with chronic low back pain. Much of these data have been published after the most recent US and European guidelines were developed. Trials of long-acting formulations of oxycodone, morphine, oxymorphone, hydromorphone, and buprenorphine have shown good efficacy with acceptable tolerability in patients with chronic low back pain. Starting with a low dose in opioid-naïve patients and titrating slowly based on individual response increases the likelihood that treatment will be tolerated. Opioid-experienced patients can be transitioned to a new opioid at a dose slightly lower than that considered equipotent based on established conversion ratios. Titration should proceed slowly based on individual response. Opportunities for twice-daily, once-daily, or once-weekly dosing allow clinicians to tailor treatment in order to maximum compliance. In patients receiving multiple medications, it may be advisable to prescribe an opioid that is not metabolized by the cytochrome P450 enzyme system. Opioid therapy needs to be predicated on a full history, screening for abuse potential, and agreement to regular monitoring for treatment compliance and signs of abuse. Drugs that do not produce other opioids as metabolites will simplify urine drug screening.

Epidural corticosteroid injections should be reserved for patients with persistent pain that does not respond adequately to rehabilitation and pharmacotherapy. Surgery should be reserved for patients with prolonged, disabling, nonresponsive pain. Discussions of these more invasive therapies are reserved for elsewhere in this book.

4. Conclusion

Though ideally chronic low back pain would respond to self-management techniques, pharmacotherapy remains an essential component of treatment without which patients cannot maintain their levels of activity. There are clinical data that suggest uncontrolled pain can interfere with physical therapy, whereas improvements in pain are accompanied by improvements in function and response to therapy (Gross et al., 2008; Schein et al., 2008; Soin et al., 2008; Teske et al., 2008). Sufficient analgesia may be necessary to allow patients to effectively participate in a physical therapy program. The perceived choice facing clinicians is between medications with limited efficacy, but relatively benign adverse events profiles, and more potent analgesics with more substantial risks of adverse events or drug abuse.

Acetaminophen and NSAIDs have modest efficacy, but acetaminophen has significant risks of hepatic adverse events and NSAIDs with gastrointestinal, cardiovascular, renal and other systemic adverse events. Duloxetine has moderate efficacy comparable to acetaminophen or NSAIDs and is useful as an adjunctive medication. Tramadol and tapentadol offer advantages of opioid agonism paired with norepinephrine reuptake inhibition, but should not be combined with SNRIs or SSRIs. Opioids offer the greatest efficacy of available pharmacologic therapies, and, although opioids have significant risks of abuse, overdose, and distressing gastrointestinal adverse effects (nausea, constipation), they have a safety advantage over NSAIDs with respect to cardiovascular adverse events and gastrointestinal bleeding. The American Heart Association has stated that opioids may be preferred over NSAIDs in patients with significant cardiovascular risk (Antman et al., 2007).

5. Acknowledgments

The authors would like to acknowledge Jeff Coleman, MA, and Robert Gatley, MD, of Complete Healthcare Communications, Inc. (Chadds Ford, PA, USA), who provided medical writing and editorial support with funding by Endo Pharmaceuticals Inc., Chadds Ford, PA, USA.

6. References

Adams, M., Pieniaszek, H. J., Jr, Gammaitoni, A. R. & Ahdieh, H. (2005). Oxymorphone extended release does not affect CYP2C9 or CYP3A4 metabolic pathways. *Journal of Clinical Pharmacology*. Vol.45, No.3, pp. 337-345

Airaksinen, O., Brox, J. I., Cedraschi, C., Hildebrandt, J., Klaber-Moffett, J., Kovacs, F., Mannion, A. F., Reis, S., Staal, J. B., Ursin, H. & Zanoli, G. (2006). Chapter 4. European guidelines for the management of chronic nonspecific low back pain. *European Spine Journal*. Vol.15 No.(suppl)2, pp. S192-300

Allan, L., Richarz, U., Simpson, K. & Slappendel, R. (2005). Transdermal fentanyl versus sustained release oral morphine in strong-opioid naive patients with chronic low back pain. *Spine*. Vol.30, No.22, pp. 2484-2490

Allen, J. W., Horais, K. A., Tozier, N. A. & Yaksh, T. L. (2006). Opiate pharmacology of intrathecal granulomas. *Anesthesiology*. Vol.105, No.3, pp. 590-598

Amass, L., Kamien, J. B. & Mikulich, S. K. (2000). Efficacy of daily and alternate-day dosing regimens with the combination buprenorphine-naloxone tablet. *Drug and Alcohol Dependence*. Vol.58, No.1-2, pp. 143-152

American Geriatrics Society. (2009). Pharmacological management of persistent pain in older persons. *Journal of the American Geriatrics Society*. Vol.57, No.8, pp. 1331-1346

Amorim, C., Couto, A. G., Netz, D. J., de Freitas, R. A. & Bresolin, T. M. (2010). Antioxidant idebenone-loaded nanoparticles based on chitosan and N-carboxymethylchitosan. *Nanomedicine*. Vol.6, No.6, pp. 745-752

Andersson, G. B. (1999). Epidemiological features of chronic low-back pain. *Lancet*. Vol.354, No.9178, pp. 581-585

Antman, E. M., Bennett, J. S., Daugherty, A., Furberg, C., Roberts, H. & Taubert, K. A. (2007). Use of nonsteroidal antiinflammatory drugs: an update for clinicians: a scientific statement from the American Heart Association. *Circulation*. Vol.115, No.12, pp. 1634-1642

Attal, N., Cruccu, G., Baron, R., Haanpaa, M., Hansson, P., Jensen, T. S. & Nurmikko, T. (2010). EFNS guidelines on the pharmacological treatment of neuropathic pain: 2010 revision. *European Journal of Neurology*. Vol.17, No.9, pp. 1113-e1188

Baratta, R. R. (1976). A double-blind comparative study of carisoprodol, propoxyphene, and placebo in the management of low back syndrome. *Curr Ther Res Clin Exp*. Vol.20, No.3, pp. 233-240

Barkin, R. L., Barkin, S. J. & Barkin, D. S. (2006). Propoxyphene (dextropropoxyphene): a critical review of a weak opioid analgesic that should remain in antiquity. *American Journal of Therapeutics*. Vol.13, No.6, pp. 534-542

Barkin, R. L. & Buvanendran, A. (2004). Focus on the COX-1 and COX-2 agents: renal events of nonsteroidal and anti-inflammatory drugs-NSAIDs. *American Journal of Therapeutics*. Vol.11, No.2, pp. 124-129

Barkin, R. L., Shirazi, D. & Kinzler, E. (2009). Effect of ethanol on the release of morphine sulfate from Oramorph SR tablets. *American Journal of Therapeutics*. Vol.16, No.6, pp. 482-486

Boers, M., Tangelder, M. J., van Ingen, H., Fort, J. G. & Goldstein, J. L. (2007). The rate of NSAID-induced endoscopic ulcers increases linearly but not exponentially with age: a pooled analysis of 12 randomised trials. *Annals of the Rheumatic Diseases*. Vol.66, No.3, pp. 417-418

Butler, S. F., Budman, S. H., Fernandez, K. C., Houle, B., Benoit, C., Katz, N. & Jamison, R. N. (2007). Development and validation of the Current Opioid Misuse Measure. *Pain.* Vol.130, No.1-2, pp. 144-156

Butler, S. F., Fernandez, K., Benoit, C., Budman, S. H. & Jamison, R. N. (2008). Validation of the revised Screener and Opioid Assessment for Patients with Pain (SOAPP-R). *Journal of Pain.* Vol.9, No.4, pp. 360-372

Caldwell, B., Aldington, S., Weatherall, M., Shirtcliffe, P. & Beasley, R. (2006). Risk of cardiovascular events and celecoxib: a systematic review and meta-analysis. *Journal of the Royal Society of Medicine.* Vol.99, No.3, pp. 132-140

Camidge, D. R., Wood, R. J. & Bateman, D. N. (2003). The epidemiology of self-poisoning in the UK. *British Journal of Clinical Pharmacology.* Vol.56, No.6, pp. 613-619

Centers for Disease Control and Prevention. Prevalence of disabilities and associated health conditions - United States, 1999. (2001). *MMWR; Morbidity and Mortality Weekly Report.* Vol.50, No.7, pp. 120-125

Cepeda, M. S., Sutton, A., Weinstein, R. & Kim, M. (2012). Effect of tapentadol extended release on productivity: results from an analysis combining evidence from multiple sources. *Clinical Journal of Pain.* Vol.28,No.1, pp.8-13.

Cheetham, T. C., Levy, G., Niu, F. & Bixler, F. (2009). Gastrointestinal safety of nonsteroidal antiinflammatory drugs and selective cyclooxygenase-2 inhibitors in patients on warfarin. *Annals of Pharmacotherapy.* Vol.43, No.11, pp. 1765-1773

Chindalore, V. L., Craven, R. A., Yu, K. P., Butera, P. G., Burns, L. H. & Friedmann, N. (2005). Adding ultralow-dose naltrexone to oxycodone enhances and prolongs analgesia: a randomized, controlled trial of Oxytrex. *Journal of Pain.* Vol.6, No.6, pp. 392-399

Chou, R., Qaseem, A., Snow, V., Casey, D., Cross, J. T., Jr., Shekelle, P. & Owens, D. K. (2007). Diagnosis and treatment of low back pain: a joint clinical practice guideline from the American College of Physicians and the American Pain Society. *Annals of Internal Medicine.* Vol.147, No.7, pp. 478-491

Chun, L. J., Tong, M. J., Busuttil, R. W. & Hiatt, J. R. (2009). Acetaminophen hepatotoxicity and acute liver failure. *Journal of Clinical Gastroenterology.* Vol.43, No.4, pp. 342-349

Coffman, B. L., Rios, G. R., King, C. D. & Tephly, T. R. (1997). Human UGT2B7 catalyzes morphine glucuronidation. *Drug Metabolism and Disposition.* Vol.25, No.1, pp. 1-4

Collins, J. J., Baase, C. M., Sharda, C. E., Ozminkowski, R. J., Nicholson, S., Billotti, G. M., Turpin, R. S., Olson, M. & Berger, M. L. (2005). The assessment of chronic health conditions on work performance, absence, and total economic impact for employers. *Journal of Occupational and Environmental Medicine.* Vol.47, No.6, pp. 547-557

Cone, E. J. & Caplan, Y. H. (2009). Urine toxicology testing in chronic pain management. *Postgraduate Medicine.* Vol.121, No.4, pp. 91-102

Cymbalta® Delayed-Release Capsules. Duloxetine hydrochloride. Indianapolis, IN: Eli Lilly and Company; 2010

Dean, M. (2004). Opioids in renal failure and dialysis patients. *Journal of Pain and Symptom Management.* Vol.28, No.5, pp. 497-504

Deshpande, A., Furlan, A., Mailis-Gagnon, A., Atlas, S. & Turk, D. (2007). Opioids for chronic low-back pain. *Cochrane Database of Systematic Reviews.* No.3, pp. CD004959

Dilaudid® Oral Liquid and Dilaudid® Tablets (hydromorphone hydrochloride). Full Prescribing Information, Purdue Pharma, LP, Stamford, CT, 2008

Duragesic® (fentanyl transdermal system). Full Prescribing Information, Ortho-McNeil-Janssen Pharmaceuticals, Inc., Raritan, NJ, 2009

Engel, K., Reuter, J., Seiler, C., Schulte Monting, J., Jakob, T. & Schempp, C. M. (2008). Anti-inflammatory effect of pimecrolimus in the sodium lauryl sulphate test. *Journal of the European Academy of Dermatology and Venereology*. Vol.22, No.4, pp. 447-450

Evans, J. M., McGregor, E., McMahon, A. D., McGilchrist, M. M., Jones, M. C., White, G., McDevitt, D. G. & MacDonald, T. M. (1995). Non-steroidal anti-inflammatory drugs and hospitalization for acute renal failure. *QJM*. Vol.88, No.8, pp. 551-557

EXALGO (hydromorphone ER). complete prescribing information. Hazelwood, MO: Mallinckrodt; 2010

FDA asks Purdue Pharma to withdraw Palladone. (2005). *FDA Consumer*. Vol.39, No.(5), pp. 7

Fiske, W. D., Jobes, J., Xiang, Q., Chang, S. & Benedek, I. H. (2011). The effects of ethanol on the bioavailability of oxymorphone extended-release tablets. *Annals of Pharmacotherapy*. In press

Forman, J. P., Stampfer, M. J. & Curhan, G. C. (2005). Non-narcotic analgesic dose and risk of incident hypertension in US women. *Hypertension*. Vol.46, No.3, pp. 500-507

Frerick, H., Keitel, W., Kuhn, U., Schmidt, S., Bredehorst, A. & Kuhlmann, M. (2003). Topical treatment of chronic low back pain with a capsicum plaster. *Pain*. Vol.106, No.1-2, pp. 59-64

Furlan, A. D., Sandoval, J. A., Mailis-Gagnon, A. & Tunks, E. (2006). Opioids for chronic noncancer pain: a meta-analysis of effectiveness and side effects. *Canadian Medical Association Journal*. Vol.174, No.11, pp. 1589-1594

Gabriel, S. E., Jaakkimainen, L. & Bombardier, C. (1991). Risk for serious gastrointestinal complications related to use of nonsteroidal anti-inflammatory drugs. A meta-analysis. *Annals of Internal Medicine*. Vol.115, No.10, pp. 787-796

Galer, B. S., Gammaitoni, A. R., Oleka, N., Jensen, M. P. & Argoff, C. E. (2004). Use of the lidocaine patch 5% in reducing intensity of various pain qualities reported by patients with low-back pain. *Current Medical Research and Opinion*. Vol.20, No.(suppl)2, pp. S5-12

Gammaitoni, A. R., Alvarez, N. A. & Galer, B. S. (2003). Safety and tolerability of the lidocaine patch 5%, a targeted peripheral analgesic: a review of the literature. *Journal of Clinical Pharmacology*. Vol.43, No.2, pp. 111-117

Gengo, F. M., Rubin, L., Robson, M., Rainka, M., Gengo, M. F., Mager, D. E. & Bates, V. (2008). Effects of ibuprofen on the magnitude and duration of aspirin's inhibition of platelet aggregation: clinical consequences in stroke prophylaxis. *Journal of Clinical Pharmacology*. Vol.48, No.1, pp. 117-122

Gimbel, J., Linn, R., Hale, M. & Nicholson, B. (2005). Lidocaine patch treatment in patients with low back pain: results of an open-label, nonrandomized pilot study. *American Journal of Therapeutics*. Vol.12, No.4, pp. 311-319

Goodnough, A. Drug is harder to abuse, but users persevere. *New York Times*. June 15, 2011.

Gordon, A., Rashiq, S., Moulin, D. E., Clark, A. J., Beaulieu, A. D., Eisenhoffer, J., Piraino, P. S., Quigley, P., Harsanyi, Z. & Darke, A. C. (2010). Buprenorphine transdermal system for opioid therapy in patients with chronic low back pain. *Pain Research and Management*. Vol.15, No.3, pp. 169-178

Gourlay, D. L., Heit, H. A. & Almahrezi, A. (2005). Universal precautions in pain medicine: a rational approach to the treatment of chronic pain. *Pain Medicine*. Vol.6, No.2, pp. 107-112

Grilo, R. M., Bertin, P., Scotto di Fazano, C., Coyral, D., Bonnet, C., Vergne, P. & Treves, R. (2002). Opioid rotation in the treatment of joint pain. A review of 67 cases. *Joint, Bone, Spine*. Vol.69, No.5, pp. 491-494

Gross, D. P., Bhambhani, Y., Haykowsky, M. J. & Rashiq, S. (2008). Acute opioid administration improves work-related exercise performance in patients with chronic back pain. *Journal of Pain.* Vol.9, No.9, pp. 856-862

Haberer, J. P., Schoeffler, P., Couderc, E. & Duvaldestin, P. (1982). Fentanyl pharmacokinetics in anaesthetized patients with cirrhosis. *British Journal of Anaesthesia.* Vol.54, No.12, pp. 1267-1270

Hale, M. E., Ahdieh, H., Ma, T. & Rauck, R. (2007). Efficacy and safety of OPANA ER (oxymorphone extended release) for relief of moderate to severe chronic low back pain in opioid-experienced patients: a 12-week, randomized, double-blind, placebo-controlled study. *Journal of Pain.* Vol.8, No.2, pp. 175-184

Hippisley-Cox, J. & Coupland, C. (2005). Risk of myocardial infarction in patients taking cyclo-oxygenase-2 inhibitors or conventional non-steroidal anti-inflammatory drugs: population based nested case-control analysis. *BMJ: British Medical Journal.* Vol.330, No.7504, pp. 1366

Hippisley-Cox, J., Coupland, C. & Logan, R. (2005). Risk of adverse gastrointestinal outcomes in patients taking cyclo-oxygenase-2 inhibitors or conventional non-steroidal anti-inflammatory drugs: population based nested case-control analysis. *BMJ: British Medical Journal.* Vol.331, No.7528, pp. 1310-1316

Ho, P. M., Maddox, T. M., Wang, L., Fihn, S. D., Jesse, R. L., Peterson, E. D. & Rumsfeld, J. S. (2009). Risk of adverse outcomes associated with concomitant use of clopidogrel and proton pump inhibitors following acute coronary syndrome. *Journal of the American Medical Association.* Vol.301, No.9, pp. 937-944

Hydromorphone-HP Injection 10 mg (hydromorphone hydrochloride). Full Prescribing Information, Abbott Laboratories, North Chicago, IL, 2008

Jamison, R. N., Raymond, S. A., Slawsby, E. A., Nedeljkovic, S. S. & Katz, N. P. (1998). Opioid therapy for chronic noncancer back pain. A randomized prospective study. *Spine.* Vol.23, No.23, pp. 2591-2600

Johnson, A. G., Nguyen, T. V. & Day, R. O. (1994). Do nonsteroidal anti-inflammatory drugs affect blood pressure? A meta-analysis. *Annals of Internal Medicine.* Vol.121, No.4, pp. 289-300

Johnson, F., Wagner, G., Sun, S. & Stauffer, J. (2008). Effect of concomitant ingestion of alcohol on the in vivo pharmacokinetics of KADIAN (morphine sulfate extended-release) capsules. *Journal of Pain.* Vol.9, No.4, pp. 330-336

Juni, P., Nartey, L., Reichenbach, S., Sterchi, R., Dieppe, P. A. & Egger, M. (2004). Risk of cardiovascular events and rofecoxib: cumulative meta-analysis. *Lancet.* Vol.364, No.9450, pp. 2021-2029

Juurlink, D. N., Gomes, T., Ko, D. T., Szmitko, P. E., Austin, P. C., Tu, J. V., Henry, D. A., Kopp, A. & Mamdani, M. M. (2009). A population-based study of the drug interaction between proton pump inhibitors and clopidogrel. *Canadian Medical Association Journal.* Vol.180, No.7, pp. 713-718

Kadian® (morphine sulfate sustained release capsules). Full Prescribing Information, Alpharma, Piscataway, NJ, 2007

Kalso, E., Edwards, J. E., Moore, R. A. & McQuay, H. J. (2004). Opioids in chronic non-cancer pain: systematic review of efficacy and safety. *Pain.* Vol.112, No.3, pp. 372-380

Karp, J. F., Weiner, D. K., Dew, M. A., Begley, A., Miller, M. D. & Reynolds, C. F., 3rd. (2010). Duloxetine and care management treatment of older adults with comorbid major depressive disorder and chronic low back pain: results of an open-label pilot study. *International Journal of Geriatric Psychiatry.* Vol.25, No.6, pp. 633-642

Katz, J., Rauck, R., Ahdieh, H., Ma, T., Gerritsen van der Hoop, R., Kerwin, R. & Podolsky, G. (2007). A 12-week randomized placebo-controlled trial assessing the safety and efficacy of oxymorphone extended release for opioid-naive patients with chronic low back pain. *Current Medical Research and Opinion*. Vol.23, No.1, pp. 117-128

Katz, N. (2008). Abuse-deterrent opioid formulations: are they a pipe dream? *Current Rheumatology Reports*. Vol.10, No.1, pp. 11-18

Kearney, P. M., Baigent, C., Godwin, J., Halls, H., Emberson, J. R. & Patrono, C. (2006). Do selective cyclo-oxygenase-2 inhibitors and traditional non-steroidal anti-inflammatory drugs increase the risk of atherothrombosis? Meta-analysis of randomised trials. *British Medical Journal. (Clinical Research Ed.)*. Vol.332, No.7553, pp. 1302-1308

Keitel, W., Frerick, H., Kuhn, U., Schmidt, U., Kuhlmann, M. & Bredehorst, A. (2001). Capsicum pain plaster in chronic non-specific low back pain. *Arzneimittel-Forschung*. Vol.51, No.11, pp. 896-903

Kienzler, J., Gold, M. & Nollevaux, F. (2010). Systemic bioavailability of topical diclofenac sodium gel 1% versus oral diclofenac sodium in healthy volunteers. *Journal of Clinical Pharmacology and New Drugs*. Vol.50, pp. 50-61

Kneip, C., Terlinden, R., Beier, H. & Chen, G. (2008). Investigations into the drug-drug interaction potential of tapentadol in human liver microsomes and fresh human hepatocytes. *Drug Metab Lett*. Vol.2, No.1, pp. 67-75

Kuehn, B. M. (2009). FDA focuses on drugs and liver damage: labeling and other changes for acetaminophen. *Journal of the American Medical Association*. Vol.302, No.4, pp. 369-371

Kuijpers, T., van Middelkoop, M., Rubinstein, S. M., Ostelo, R., Verhagen, A., Koes, B. W. & van Tulder, M. W. (2011). A systematic review on the effectiveness of pharmacological interventions for chronic non-specific low-back pain. *European Spine Journal*. Vol.20, No.1, pp. 40-50

Lange, B., Kuperwasser, B., Okamoto, A., Steup, A., Haufel, T., Ashworth, J. & Etropolski, M. (2010). Efficacy and safety of tapentadol prolonged release for chronic osteoarthritis pain and low back pain. *Advances in Therapy*. Vol.27, No.6, pp. 381-399

Larson, A. M., Polson, J., Fontana, R. J., Davern, T. J., Lalani, E., Hynan, L. S., Reisch, J. S., Schiodt, F. V., Ostapowicz, G., Shakil, A. O. & Lee, W. M. (2005). Acetaminophen-induced acute liver failure: results of a United States multicenter, prospective study. *Hepatology*. Vol.42, No.6, pp. 1364-1372

Lee, C., Straus, W. L., Balshaw, R., Barlas, S., Vogel, S. & Schnitzer, T. J. (2004). A comparison of the efficacy and safety of nonsteroidal antiinflammatory agents versus acetaminophen in the treatment of osteoarthritis: a meta-analysis. *Arthritis and Rheumatism*. Vol.51, No.5, pp. 746-754

LIDODERM® (lidocaine patch 5%). Full Prescribing Information, ENDO Pharmaceuticals, Chadds Ford, PA, 2010

Lowe, D. O., Knowles, S. R., Weber, E. A., Railton, C. J. & Shear, N. H. (2006). Povidone-iodine-induced burn: case report and review of the literature. *Pharmacotherapy*. Vol.26, No.11, pp. 1641-1645

Malhotra, S., Karan, R. S., Pandhi, P. & Jain, S. (2001). Drug related medical emergencies in the elderly: role of adverse drug reactions and non-compliance. *Postgraduate Medical Journal*. Vol.77, No.913, pp. 703-707

Manchikanti, L., Atluri, S., Trescot, A. M. & Giordano, J. (2008). Monitoring opioid adherence in chronic pain patients: tools, techniques, and utility. *Pain Physician.* Vol.11, No.2(suppl), pp. S155-180

Mapel, D. W., Shainline, M., Paez, K. & Gunter, M. (2004). Hospital, pharmacy, and outpatient costs for osteoarthritis and chronic back pain. *Journal of Rheumatology.* Vol.31, No.3, pp. 573-583

Max, M. B., Lynch, S. A., Muir, J., Shoaf, S. E., Smoller, B. & Dubner, R. (1992). Effects of desipramine, amitriptyline, and fluoxetine on pain in diabetic neuropathy. *New England Journal of Medicine.* Vol.326, No.19, pp. 1250-1256

Mercadante, S. & Bruera, E. (2006). Opioid switching: a systematic and critical review. *Cancer Treatment Reviews.* Vol.32, No.4, pp. 304-315

Motsko, S. P., Rascati, K. L., Busti, A. J., Wilson, J. P., Barner, J. C., Lawson, K. A. & Worchel, J. (2006). Temporal relationship between use of NSAIDs, including selective COX-2 inhibitors, and cardiovascular risk. *Drug Safety.* Vol.29, No.7, pp. 621-632

Muller, F. O., Odendaal, C. L., Muller, F. R., Raubenheimer, J., Middle, M. V. & Kummer, M. (1998). Comparison of the efficacy and tolerability of a paracetamol/codeine fixed-dose combination with tramadol in patients with refractory chronic back pain. *Arzneimittel-Forschung.* Vol.48, No.6, pp. 675-679

Nassau County executive: abuse of painkiller Opana is growing. *CBS News New York.* May 9, 2011.

Noble, M., Treadwell, J. R., Tregear, S. J., Coates, V. H., Wiffen, P. J., Akafomo, C. & Schoelles, K. M. (2010). Long-term opioid management for chronic noncancer pain. *Cochrane Database of Systematic Reviews.* No.1, CD006605

O'Donnell, J. B., Ekman, E. F., Spalding, W. M., Bhadra, P., McCabe, D. & Berger, M. F. (2009). The effectiveness of a weak opioid medication versus a cyclo-oxygenase-2 (COX-2) selective non-steroidal anti-inflammatory drug in treating flare-up of chronic low-back pain: results from two randomized, double-blind, 6-week studies. *Journal of International Medical Research.* Vol.37, No.6, pp. 1789-1802

OPANA® ER (oxymorphone hydrochloride extended-release tablets). Full Prescribing Information, Endo Pharmaceuticals Inc., Chadds Ford, PA, 2010

OXECTA® (oxycodone hydrochloride). Full Prescribing Information, King Pharmaceuticals Inc., Bristol, TN, 2011

OxyContin® (oxycodone HCl controlled-release tablets). Full Prescribing Information, Purdue Pharma L.P., Stamford, CT, 2010

Pavlicevic, I., Kuzmanic, M., Rumboldt, M. & Rumboldt, Z. (2008). Interaction between antihypertensives and NSAIDs in primary care: a controlled trial. *Canadian Journal of Clinical Pharmacology.* Vol.15, No.3, pp. e372-382

Peloso, P. M., Fortin, L., Beaulieu, A., Kamin, M. & Rosenthal, N. (2004). Analgesic efficacy and safety of tramadol/ acetaminophen combination tablets (Ultracet) in treatment of chronic low back pain: a multicenter, outpatient, randomized, double blind, placebo controlled trial. *Journal of Rheumatology.* Vol.31, No.12, pp. 2454-2463

Peniston, J. H. & Gould, E. (2009). Oxymorphone extended release for the treatment of chronic low back pain: a retrospective pooled analysis of enriched-enrollment clinical trial data stratified according to age, sex, and prior opioid use. *Clinical Therapeutics.* Vol.31, No.2, pp. 347-359

Quang-Cantagrel, N. D., Wallace, M. S. & Magnuson, S. K. (2000). Opioid substitution to improve the effectiveness of chronic noncancer pain control: a chart review. *Anesthesia and Analgesia.* Vol.90, No.4, pp. 933-937

Rauck, R. L., Bookbinder, S. A., Bunker, T. R., Alftine, C. D., Ghalie, R., Negro-Vilar, A., de Jong, E. & Gershon, S. (2006a). The ACTION study: a randomized, open-label, multicenter trial comparing once-a-day extended-release morphine sulfate capsules (AVINZA) to twice-a-day controlled-release oxycodone hydrochloride tablets (OxyContin) for the treatment of chronic, moderate to severe low back pain. *Journal of Opioid Management*. Vol.2, No.3, pp. 155-166

Rauck, R. L., Bookbinder, S. A., Bunker, T. R., Alftine, C. D., Ghalie, R., Negro-Vilar, A., de Jong, E. & Gershon, S. (2006b). A randomized, open-label study of once-a-day AVINZA (morphine sulfate extended-release capsules) versus twice-a-day OxyContin (oxycodone hydrochloride controlled-release tablets) for chronic low back pain: the extension phase of the ACTION trial. *Journal of Opioid Management*. Vol.2, No.6, pp. 325-328, 331-333

Rich, B. A. & Webster, L. R. (2011). A review of forensic implications of opioid prescribing with examples from malpractice cases involving opioid-related overdose. *Pain Medicine*. Vol.12, No.(suppl)2, pp. S59-65

Roelofs, P. D., Deyo, R. A., Koes, B. W., Scholten, R. J. & van Tulder, M. W. (2008). Nonsteroidal anti-inflammatory drugs for low back pain: an updated Cochrane review. *Spine (Phila Pa 1976)*. Vol.33, No.16, pp. 1766-1774

Rostom, A., Goldkind, L. & Laine, L. (2005). Nonsteroidal anti-inflammatory drugs and hepatic toxicity: a systematic review of randomized controlled trials in arthritis patients. *Clinical Gastroenterology and Hepatology*. Vol.3, No.5, pp. 489-498

Rubenstein, J. H. & Laine, L. (2004). Systematic review: the hepatotoxicity of non-steroidal anti-inflammatory drugs. *Alimentary Pharmacology and Therapeutics*. Vol.20, No.4, pp. 373-380

Ruoff, G. E., Rosenthal, N., Jordan, D., Karim, R. & Kamin, M. (2003). Tramadol/acetaminophen combination tablets for the treatment of chronic lower back pain: a multicenter, randomized, double-blind, placebo-controlled outpatient study. *Clinical Therapeutics*. Vol.25, No.4, pp. 1123-1141

Salerno, S. M., Browning, R. & Jackson, J. L. (2002). The effect of antidepressant treatment on chronic back pain: a meta-analysis. *Archives of Internal Medicine*. Vol.162, No.1, pp. 19-24

Sathyan, G., Sivakumar, K. & Thipphawong, J. (2008). Pharmacokinetic profile of a 24-hour controlled-release OROS formulation of hydromorphone in the presence of alcohol. *Current Medical Research and Opinion*. Vol.24, No.1, pp. 297-305

Schein, J. R., Kosinski, M. R., Janagap-Benson, C., Gajria, K., Lin, P. & Freedman, J. D. (2008). Functionality and health-status benefits associated with reduction of osteoarthritis pain. *Current Medical Research and Opinion*. Vol.24, No.5, pp. 1255-1265

Schneider, J. P., Matthews, M. & Jamison, R. N. (2010). Abuse-deterrent and tamper-resistant opioid formulations: what is their role in addressing prescription opioid abuse? *CNS Drugs*. Vol.24, No.10, pp. 805-810

Schnitzer, T. J., Gray, W. L., Paster, R. Z. & Kamin, M. (2000). Efficacy of tramadol in treatment of chronic low back pain. *Journal of Rheumatology*. Vol.27, No.3, pp. 772-778

Simpson, R. K., Jr., Edmondson, E. A., Constant, C. F. & Collier, C. (1997). Transdermal fentanyl as treatment for chronic low back pain. *Journal of Pain and Symptom Management*. Vol.14, No.4, pp. 218-224

Singh, G., Fort, J. G., Goldstein, J. L., Levy, R. A., Hanrahan, P. S., Bello, A. E., Andrade-Ortega, L., Wallemark, C., Agrawal, N. M., Eisen, G. M., Stenson, W. F. & Triadafilopoulos, G. (2006). Celecoxib versus naproxen and diclofenac in osteoarthritis patients: SUCCESS-I Study. *American Journal of Medicine*. Vol.119, No.3, pp. 255-266

Skljarevski, V., Desaiah, D., Liu-Seifert, H., Zhang, Q., Chappell, A. S., Detke, M. J., Iyengar, S., Atkinson, J. H. & Backonja, M. (2010a). Efficacy and safety of duloxetine in patients with chronic low back pain. *Spine (Phila Pa 1976)*. Vol.35, No.13, pp. E578-585

Skljarevski, V., Ossanna, M., Liu-Seifert, H., Zhang, Q., Chappell, A., Iyengar, S., Detke, M. & Backonja, M. (2009). A double-blind, randomized trial of duloxetine versus placebo in the management of chronic low back pain. *European Journal of Neurology*. Vol.16, No.9, pp. 1041-1048

Skljarevski, V., Zhang, S., Chappell, A. S., Walker, D. J., Murray, I. & Backonja, M. (2010b). Maintenance of effect of duloxetine in patients with chronic low back pain: a 41-week uncontrolled, dose-blinded study. *Pain Medicine*. Vol.11, No.5, pp. 648-657

Slatkin, N. E. (2009). Opioid switching and rotation in primary care: implementation and clinical utility. *Current Medical Research and Opinion*. Vol.25, No.9, pp. 2133-2150

Smith, H. & Bruckenthal, P. (2010). Implications of opioid analgesia for medically complicated patients. *Drugs and Aging*. Vol.27, No.5, pp. 417-433

Smith, H. S. (2009). Opioid metabolism. *Mayo Clinic Proceedings*. Vol.84, No.7, pp. 613-624

Soin, A., Cheng, J., Brown, L., Moufawad, S. & Mekhail, N. (2008). Functional outcomes in patients with chronic nonmalignant pain on long-term opioid therapy. *Pain Practice*. Vol.8, No.5, pp. 379-384

Staiger, T. O., Gaster, B., Sullivan, M. D. & Deyo, R. A. (2003). Systematic review of antidepressants in the treatment of chronic low back pain. *Spine (Phila Pa 1976)*. Vol.28, No.22, pp. 2540-2545

Teshima, D., Yamauchi, A., Makino, K., Kataoka, Y., Arita, Y., Nawata, H. & Oishi, R. (2002). Nasal glucagon delivery using microcrystalline cellulose in healthy volunteers. *International Journal of Pharmaceutics*. Vol.233, No.1-2, pp. 61-66

Teske, W., Anastasiadis, A., Kramer, J. & Theodoridis, T. (2008). Effective pain relief facilitates exercise therapy: results of a multicenter study with controlled-release oxycodone in patients with movement pain. [article in German] *Orthopade*. Vol.37, No.12, pp. 1210-1216

Tulner, L. R., Frankfort, S. V., Gijsen, G. J., van Campen, J. P., Koks, C. H. & Beijnen, J. H. (2008). Drug-drug interactions in a geriatric outpatient cohort: prevalence and relevance. *Drugs and Aging*. Vol.25, No.4, pp. 343-355

Tzschentke, T. M., Christoph, T., Kogel, B., Schiene, K., Hennies, H. H., Englberger, W., Haurand, M., Jahnel, U., Cremers, T. I., Friderichs, E. & De Vry, J. (2007). (-)-(1*R*,2*R*)-3- (3-dimethylamino-1-ethyl-2-methyl-propyl)- phenol hydrochloride (tapentadol HCl): a novel mu-opioid receptor agonist/norepinephrine reuptake inhibitor with broad-spectrum analgesic properties. *Journal of Pharmacology and Experimental Therapeutics*. Vol.323, No.1, pp. 265-276

Ulens, C., Daenens, P. & Tytgat, J. (1999). Norpropoxyphene-induced cardiotoxicity is associated with changes in ion-selectivity and gating of HERG currents. *Cardiovascular Research*. Vol.44, No.3, pp. 568-578

Ultram ER® (tramadol hydrochloride). Full Prescribing Information, Ortho-McNeil-Janssen Pharmaceuticals, Inc., Raritan, NJ, 2009

US Food and Drug Administration. FDA approves new formulation for OxyContin [press release]. Available at: http://www.fda.gov/NewsEvents/Newsroom/PressAnnouncements/ucm207480.htm. Accessed February 14, 2011

Victor, T. W., Alvarez, N. & Gould, E. (2009). Opioid prescribing practices in chronic pain management: guidelines do not sufficiently influence clinical practice. *Journal of Pain.* Vol.10, No.10, pp. 1051-1057

Vogt, M. T., Kwoh, C. K., Cope, D. K., Osial, T. A., Culyba, M. & Starz, T. W. (2005). Analgesic usage for low back pain: impact on health care costs and service use. *Spine.* Vol.30, No.9, pp. 1075-1081

Voltaren® Gel 1% (diclofenac sodium topical gel). Full Prescribing Information, Novartis Consumer Health, Parsippany, NJ, 2007

Waikakul, S., Danputipong, P. & Soparat, K. (1996). Topical analgesics, indomethacin plaster and diclofenac emulgel for low back pain: a parallel study.*Journal of The Medical Association of Thailand.* Vol.79, No.8, pp. 486-490

Wallace, M., Rauck, R. L., Moulin, D., Thipphawong, J., Khanna, S. & Tudor, I. C. (2007). Once-daily OROS hydromorphone for the management of chronic nonmalignant pain: a dose-conversion and titration study. *International Journal of Clinical Practice.* Vol.61, No.10, pp. 1671-1676

Wallace, M. & Thipphawong, J. (2010). Open-label study on the long-term efficacy, safety, and impact on quality of life of OROS hydromorphone ER in patients with chronic low back pain. *Pain Medicine.* Vol.11, No.10, pp. 1477-1488

Watkins, P. B., Kaplowitz, N., Slattery, J. T., Colonese, C. R., Colucci, S. V., Stewart, P. W. & Harris, S. C. (2006). Aminotransferase elevations in healthy adults receiving 4 grams of acetaminophen daily: a randomized controlled trial. *Journal of the American Medical Association.* Vol.296, No.1, pp. 87-93

Webster, L. R., Butera, P. G., Moran, L. V., Wu, N., Burns, L. H. & Friedmann, N. (2006). Oxytrex minimizes physical dependence while providing effective analgesia: a randomized controlled trial in low back pain. *Journal of Pain.* Vol.7, No.12, pp. 937-946

Webster, L. R. & Webster, R. M. (2005). Predicting aberrant behaviors in opioid-treated patients: preliminary validation of the Opioid Risk Tool. *Pain Medicine.* Vol.6, No.6, pp. 432-442

Zhang, W., Doherty, M., Leeb, B. F., Alekseeva, L., Arden, N. K., Bijlsma, J. W., Dincer, F., Dziedzic, K., Hauselmann, H. J., Herrero-Beaumont, G., Kaklamanis, P., Lohmander, S., Maheu, E., Martin-Mola, E., Pavelka, K., Punzi, L., Reiter, S., Sautner, J., Smolen, J., Verbruggen, G. & Zimmermann-Gorska, I. (2007). EULAR evidence based recommendations for the management of hand osteoarthritis: report of a Task Force of the EULAR Standing Committee for International Clinical Studies Including Therapeutics (ESCISIT). *Annals of the Rheumatic Diseases.* Vol.66, No.3, pp. 377-388

Zhang, W., Moskowitz, R. W., Nuki, G., Abramson, S., Altman, R. D., Arden, N., Bierma-Zeinstra, S., Brandt, K. D., Croft, P., Doherty, M., Dougados, M., Hochberg, M., Hunter, D. J., Kwoh, K., Lohmander, L. S. & Tugwell, P. (2008). OARSI recommendations for the management of hip and knee osteoarthritis, part II: OARSI evidence-based, expert consensus guidelines. *Osteoarthritis and Cartilage.* Vol.16, No.2, pp. 137-162

Zhang, W., Nuki, G., Moskowitz, R. W., Abramson, S., Altman, R. D., Arden, N. K., Bierma-Zeinstra, S., Brandt, K. D., Croft, P., Doherty, M., Dougados, M., Hochberg, M., Hunter, D. J., Kwoh, K., Lohmander, L. S. & Tugwell, P. (2010). OARSI recommendations for the management of hip and knee osteoarthritis: part III: changes in evidence following systematic cumulative update of research published through January 2009. *Osteoarthritis and Cartilage.* Vol.18, No.4, pp. 476-499

8

Application of Radiofrequency in Low Back Pain Treatment

Hsi-Kai Tsou and Ting-Hsien Kao
Department of Neurosurgery, Taichung Veterans General Hospital, Taichung, Taiwan, Republic of China

1. Introduction

Chronic low back pain (LBP) has become a main cause of absenteeism and disability in industrialized societies and is a major health problem with enormous economic and costs (Andersson, 1999). As many as 80% of adults experience at least one episode of LBP during their lifetime. Only 5% of patients suffering from chronic LBP can find a specific cause such as disc herniation, spondylolisthesis, discitis or spondylitis. No definite evident causes were found in 95% patients with low back pain (Schwarze et al., 1995a, 1995b).

At present, the treatment of low back pain consists of therapies, both conservative and invasive, that are aimed at symptomatic relief. As the evidence-based medicine developed over the years, there are now much more accumulated data that inform us how to treat patients with chronic LBP. Unfortunately, many of the treatments used today are not strongly effective (Carragee 2005).

1.1 Application of radiofrequency in medicine

Radiofrequency (RF) is a minimally invasive, target-selective technique that has been in clinical use for more than decades and has been demonstrated to be successful for treating cardiac arrhythmias (Baszko et al., 2002), dysplasia (Shahee et al., 2009) and reducing pain in several chronic pain conditions including trigeminal neuralgia, chronic LBP, postherpetic neuralgia, complex regional pain syndrome, ischemic pain, cervicobrachialgia, postthoracotomy pain, occipital neuralgia, and cervical or lumbar radicular pain (Chao et al., 2008; Navani et al., 2006; Racz & Ruiz-Lopez 2006; Zhang et al., 2011). Focusing on pain management, RF can not only reach directly to the source of pain but also modulate the pain signal transmission.

1.2 Principles and mechanisms of radiofrequency

There are two major mechanisms of radiofrequency (RF) treatment: thermal (continuous) RF and non-thermal (pulsed) RF (PRF). Thermal RF caused by continuous current within frequencies between 300 Hz and 300 GHz generates both current and heat on exposed tissues. Since temperatures above 45°C result in nonselective destruction of both myelinated and nonmyelinated nerve fibers (Smith et al., 1981), the thermal RF procedure has limited applications and caused some adverse effects. Unlike continuous RF, PRF generates intermittent pulsed current which lowers the target tissue temperature to below 45°C and causes different neurobiological effects (Cahana et al., 2003).

For example, a marker for neuronal activity in the dorsal horn, c-Fos, has been reported to be expressed immediately and up to 7 days after PRF treatment in rat models (Higuchi et al., 2002; Van Zundert et al., 2005). The long lasting effect of c-Fos expression was caused by non thermal RF but inhibition of excitatory C-fiber responses was seen in long-term depression (Richebé et al., 2005). Continuous RF creates a longer blockade of synaptic transmission than PRF in an in vitro model even under similar temperature. Both RF and PRF treatments induce distance dependent tissue destruction under the stimulating needle, but the effect was more pronounced in the RF group (Cahana et al., 2003). Moreover, morphological study showed no pathological findings in the control and sham-operated groups, minimal morphological changes in the PRF group, and neuro-destruction in the continuous RF group (Erdine et al., 2005). All these findings together indicate that the use of PRF promises to be a safer, reversible and nondestructive approach to various chronic pain conditions.

2. Application of radiofrequency treatment in low back pain from different origins and mechanisms

Low back pain can originate from several sources, such as discs, ligaments, muscles, and sacroiliac joints, and another cause can be lumbar facet joint degeneration (Deyo & Weinstein 2001). Since all the pain signals were transmitted by nerves, applying treatment targeting the neuronal transmission pathway can reasonably relieve the pain. Here we review the anatomy, possible biomechanical mechanisms, clinical presentation and physical examination findings of different sources of low back pain. The diagnostic tools and image findings are discussed as well.

2.1 Discogenic low back pain

Considering the diagnosis of LBP, pure discogenic pain is thought to be less than 10% (Deyo & Weinstein 2001). However, in chronic persistent LBP patients, intervertebral disc (IVD) degeneration seemed to be the initial step and played the most important role (Carragee 2005). After IVD degeneration, the biomechanical status of the vertebral column changes, and the possibility of facet joint degeneration, spondylosis, spondylolisthesis and spinal stenosis increases as well.

2.1.1 Anatomy and pathogenesis of disc and its degeneration

The IVD is composed of a tough outer ring, the annular fibrosus (AF), a gelatinous inner core, the nucleous pulposus (NP), and the adjacent vertebral endplate (VE). The axial loading force of IVD was support by posterior two facet joints. The healthy IVD is avascular and its nutritional supply depends on diffusion via the AF and VE. Symptomatic degeneration of the IVD is thought to be the leading causes of chronic back pain.

In a normal disc, the NP is devoid of nerve fibers, while the outer AF and VE contain nerve fibers. The nerve supply of the IVD is from branches of sympathetic trunk and sinuvertebral nerves (Fig. 1). The sinuvertebral nerves run ventral to the nerve root, back to the spinal canal and divide the posterior longitudinal ligament and ventral dural branches. The anterior part of IVD was supplied by branches through the anterior longitudinal ligament which is from the sympathetic trunk. The lateral and ventral aspects of IVD are supplied by branches of rami communicantes (Fig. 1).

Because the IVD was supply by the sympathetic trunk, somatosensory nerves and their communicating network through multiple segments, discogenic back pain is always hard to localize and seemed to be a visceral pain (Bogduk et al., 1981) and RF applying to the target nerves of IVD is much more complicated than other parts of the spinal column (Bogduk et al., 1981; Brown et al., 1997; Edgar 2007).

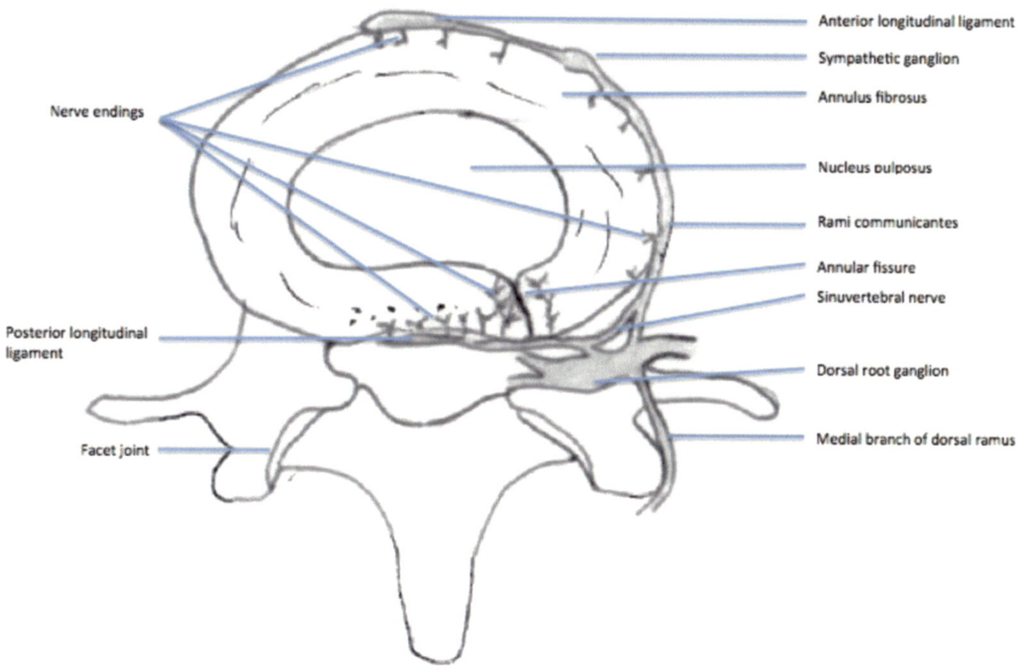

Fig. 1. Schematic illustration of the lumbosacral intervertebral disc innervation.

During IVD degeneration, increase neuronal activity is found in inner NP which is the possible mechanism of painful disc (Freemont et al., 1997; Coppes et al., 1997; Hurri & Karppinen 2004; Peng et al., 2006; Peng et al., 2009; Freemont et al., 2002; Freemont 2009). Nociceptive neuropeptides just like calcitonin gene-related peptide and substance P, which are present within the nerve fibers in the outer AF and dorsal root ganglion (DRG), may likely play a role in discogenic pain transmission (Brown et al., 1997; Ohtori et al., 2002). It is believed that most afferent fibers from the low lumbar discs travel in the sinuvertebral nerve, pass through the ramus communicantes and lumbar sympathetic chain, and finally enter the spinal cord through the L2 ramus communicantes and L2 spinal nerve roots (Nakamura et al., 1996a; 1996b).

Since most nerve fibers that innervate the disc emanate from the sympathetic nervous system (Bogduk et al., 1981; Nakamura et al., 1996a; 1996b), RF targeting discogenic low back pain can reasonably apply through two locations. One is thermal RF lesioning causing nerve fiber destruction within the inner disc and the other is sensory modulation targeting the DRG of L2, the level at which the sympathetic nerve fibers leave the spinal cord.

2.1.2 Clinical presentation, physical examination

There are no typical physical examination findings of painful IVD and most of the findings appear in other types of LBP. Diagnosis of discogenic LBP is based on these non-specific past histories and physical examinations as well as the image study. Generally speaking, most of the discogenic back pain is not localized. It seemed to be visceral pain because of its characteristics and nerve supply (Bogduk et al., 1981). Most patients experience typical features including persistent nociceptive low back (more than six months), groin with or without leg pain which worsens with axial loading or flexion of painful segment, and pain relief when lying down. Moreover, there are nerve roots lying just posteriorly to the disc margin. So some patients with discogenic LBP experience some referred pain. For example, painful IVDs for upper lumbar segment typically cause referred pain to the anterior aspect of thigh and lower lumbar segments and sometimes cause referred pain down to the posterior thigh and leg (Ohnmeiss et al, 1999a; 1999b)

2.1.3 Image diagnosis and discography

Before the development of computed tomography (CT) and magnetic resonance imaging (MRI), discography had been used to diagnose possible disc pathology (Lindblom, 1951). However, it was thought to be obsolete because of its complications and efficacy for diagnosis after the invention of CT and MRI (Walsh et al., 1990). There are many benefits of CT and MRI for disc pathology diagnosis. They provide clear three dimensional images of the spinal column and can be reconstructed to view different aspects. Moreover, some unusual pathology can be found during CT and MRI study combined with contrast enhancement (Maus, 2010).

Considering radiation exposure, MRI seems better than CT except for bony structure evaluation and determination of pre-existing metal material inside the body. T2-weighted sequence MRI can provide detailed information of disc pathology included disc height and morphology change, herniation of nucleous pulposus, spinal canal or neuroforamen stenosis, hydration of disc and with gadolinium enhancement, some inflammation pathology or neogrowth can be detected (Maus, 2010).

However, the clinical symptoms and outcome of disc degeneration cannot be predicted even if MRI can identify signal changes in the discs themselves and surrounding soft tissues (Keller et al., 2011). Moreover, MRI also provides adjacent vertebrae end plate signal change and annular tear which is thought to be strongly associated with disogenic LBP (Carragee & Hannibal 2004; Zhou & Abdi 2006).

Because painful IVD was found to increase neuronal activity in its inner layer (Freemont et al., 1997; Coppes et al., 1997; Freemont et al., 2002), a direct increase in intradiscal pressure may cause more pain. Besides, during degeneration or trauma, the tough annular ring AF becomes weaker and then tears as fissure formation. When the intradiscal pressure increases, the force is transferred to the outer area of AF through the fissure which is always located in the posterior part of the disc and causes pain; so, provocative discography is thought to be useful to find the "exact" pain source. In summary, discography provokes pain through the following mechanisms: 1) Increase in intradiscal pressure during the injection of contrast material. Mechanical stretching of the annular fibers of the painful disc may stimulate the overgrowth of nerve endings. 2) Chemical irritation of the surrounding nerve endings within the disc.

The interpretation of discogram findings includes the degree of pain generation during the procedure in each disc, and the appearance of contrast medium in the disc. Sachs et al. described the Dallas grading system using the CT discogram appearance (Sachs et al., 1987). The 'Modified Dallas Discogram Description' was finalized in the 1990's and is the 'Gold Standard' for the CT classification of anular tears (Fig. 2). Based on their article, they divided the CT discogram finding into six degrees: Grade 0: the contrast medium within the inner NP margin. Grade 1 to grade 3 indicates the contrast medium leaking to inner, middle, or outer layer of AF. Grade 4 indicates the circumferential spread greater than 30 degrees. Grade 5 tear describes either a grade 3 or grade 4 radial tear that has completely ruptured the outer layers of the disc and is 'leaking' contrast medium from the disc into the epidural space (Fig. 2).

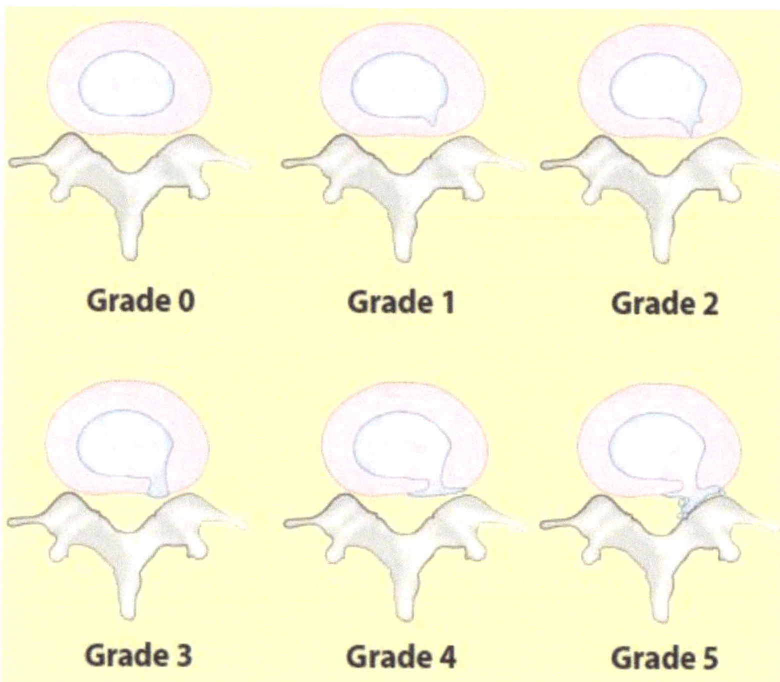

Fig. 2. The Modified Dallas Discogram

The degree of pain mainly depends on the Numeric Rating Scale (NRS), the reproduction of concordant pain and comparison of normal adjacent discs. According to the guidelines of the International Association for the Study of Pain (IASP) and the International Spine

Intervention Society (ISIS), they suggest at least two adjacent levels should be tested as controls during the provocative discography procedures. The criteria of discogenic pain is listed in table 1 (Kallewaard et al., 2010). However, because the discography is done without direct visualization of the disc structure and multiple combined pathologies of the lower back in most of the patients suffering from chronic LBP, the diagnosis made by provocative discography is controversial (Wichman, 2007). Wichman summarized three reasons causing controversy about discography for the diagnosis of discogenic LBP, namely, techniques, the disc pathology itself and symptom interpretation (Wichman, 2007).

Diagnosis	Diagnostic criteria
Absolute discogenic pain	Reproduce concordant pain in diseased level during procedures NRS at least 7 The pain develops less than 15 psi above the opening pressure Stimulation of the two adjacent discs is not painful
Highly probable discogenic pain	Reproduce concordant pain in diseased level during procedures NRS at least 7 The pain develops less than 15 psi above the opening pressure Stimulation of the one of the adjacent discs is not painful
Discogenic pain	Reproduce concordant pain in diseased level during procedures NRS at least 7 The pain develops less than 50 psi above the opening pressure Stimulation of the two adjacent discs is not painful
Possible discogenic pain	Reproduce concordant pain in diseased level during procedures NRS at least 7 The pain develops less than 50 psi above the opening pressure Stimulation of the one of the adjacent discs is not painful, and stimulation of another disc is painful at a pressure greater than 50 psi above the opening pressure, and the pain is discordant

(summary from the article by Kallewaard et al., 2010)

Table 1. Diagnostic criteria of discogenic pain by discography

2.1.4 Treatment choices of radiofrequency applying to discogenic low back pain

RF applying to the disc itself included transdiscal biacuplasty (Baylis Medical Inc., Montreal, Canada), intradiscal electrothermal therapy (IDET) with spinecath (OratecInterventions, Inc., Menlo Park, CA) and disctrode (Radionics RFG-3C,Valleylab, Tyco Healthcare Group LP 5920 Longbow Drive, Boulder, Colorado 80301-3299 USA) (Karasek & Bogduk 2000; Saal JA & Saal JS, 2000, 2002; Davis et al., 2004; Kapural & Mekhail 2006; Andersson et al., 2006; Kvarstein et al., 2009; Tsou et al., 2010). The spinecath and disctrode are flexible catheters with a distal thermocoil. When it is introduced into the annulus, the distal part should ideally be along the internal aspect of the posterior annulus. Local denervation effect is caused by heating of the distal portion of the catheter. The other two major mechanisms of IDET are intradiscal pressure reduction and enhancement of the annular healing process (Saal JA & Saal JS, 2000). However, two prospective studies have shown the opposite efficacy results using IDET for discogenic LBP patients (Pauza et al., 2004; Freeman et al., 2005). There are limited prospective studies mentioning transdiscal RF annuloplasty and their effects are uncertain (Helm et al., 2009).

The other choice of the RF target for discogenic LBP is the L2 DRG, which is based on the natural history of discogenic LBP and was not easily confirmed. With regard to discogenic pain, Nakamura et al. proposed that the main afferent pathway of pain from the lower intervertebral discs is through the L2 spinal nerve root, presumably via sympathetic afferents from the sinuvertebral nerve (Helm et al., 2009). Therefore, discogenic pain should be regarded as visceral pain due to its neural pathway. Nakamura et al. believe that the nerve fibers in the sinuvertebral nerve originate from the rami communicans of the sympathetic nerves (Nakamura et al., 1996b). RF lesioning to cervical DRG has been proved effective in a select group of patients with cervical brachialgia (van Kleef et al., 1996; Van Boxem et al., 2011). However, because of the side effects and possible complications of neuropathic pain, PRF was applied to cervical DRG treatment. In contrast, DRG lesioning for lumbosacral radicular pain was thought to be less effective (Geurts et al., 2003). Since the DRG of L2 was thought to be the main trunk of the lower back sensory afferent pathway, some clinical retrospective studies of PRF showed effectiveness for chronic LBP with highly suspect disc origins or unspecific LBP with or without radicular pain (Tsou et al., 2010; Nagda et al., 2011).

2.2 Neurogenic low back pain

Spinal nerves are protected in the spinal canal before they penetrate out of the neuroforamen. However, during the degeneration and aging process, the bony structure and soft tissue such as ligaments surrounding the nerves see hypertrophic change which possibly compromises the neuronal structure including nerve roots and their low back branches. Most patients with spondylosis, spinal stenosis and spondylolisthesis with or without radicular symptoms suffer from symptoms of back pain (Singh et al., 2005). Before the radicular symptoms are displayed, the narrowing canals compress the nerves and more or less cause local inflammatory and mechanical mechanism of back pain. Singh et al. divided two groups of patients with spinal stenosis; one group was congenital and young, while the other was degenerated and elderly. Since the pain is possibly from inflammation or mechanical stress directed to the nerves, neuronal modulation or temporary blocks are suitable for those without severe spinal stenosis or not suitable for surgical intervention. The clinical presentation of neurogenic LBP always combined both low back symptoms and neurogenic claudication with or without radicular pain (Singh et al., 2005). Diagnosis of neurogenic LBP depends on clinical presentation and dynamic lateral lumbar X ray imaging. Pedicle width narrowing and neuroforamen compromise can be seen on CT or MRI. The treatment choices for neurogenic pain before shifting to surgery includes epidural neuroforamen steroid injection (Benny & Azari 2011) selective nerve block (Thackeray et al., 2010) and RF for adjacent DRG (Van Boxem et al., 2011).

2.3 Low back pain from facet joint arthritis or degeneration (anatomy and biomechanics, clinical presentation and physical examination, diagnostic imaging)

2.3.1 Anatomy and pathogenesis of facet joints pain

Each vertebra has two sets of facet joints. One pair faces upward (superior articular facet) and one downward (inferior articular facet). The joint surfaces are coated with cartilage allowing joints to move smoothly against each other. These joints allow flexion (bend forward), extension (bend backward), and twisting motion. Since these joints are just like other joints in the whole body, and they share axial loading with IVD, they can be affected by degeneration and arthritis. Facet joints have been implicated as a cause of chronic spinal pain in 15% to 40% of patients with chronic low back pain (Schwarzer et al., 1995a; 1995b). They are innervated by medial branches of dorsal rami from the spinal nerves (Fig. 3) and theoretically, facet joint pain can be treated by denervation of the medial branches of the dorsal rami, which supply the sensory innervation of the joints (Shealy, 1976; Bogduk & Long, 1980; Dreyfuss et al., 2000).

The lumbar facet syndrome was first described by Ghormley in 1933 (Ghormley, 1933). After detailed anatomical study of the lumbar zygapophysial nerve supply by Bogduk and Long in 1979 (Bogduk & Long 1979), several control studies showed initial benefits for pain relief by radiofrequency medial branch neurotomy (Dreyfuss et al., 2000; van Kleef et al., 1999). There are several diseases that contribute to facet joint disease, including degeneration, synovial cysts, ankylosing spondylitis, and trauma. A controlled trial has shown that RF medial branch neurectomy is not a placebo (van Kleef et al., 1999) and an observational study has shown that, provided patients are carefully selected using controlled diagnostic blocks, and provided a correct surgical technique is used, some 60% of patients can expect at least 80% relief of their pain at 12 months, and 80%of patients can expect at least 60% relief (Dreyfuss et al., 2000).

The nerve supply of the facet joint originates from two levels (Fig. 3); one branch of the primary ramus arises from the nerve root at the same level as the joint and another branch from the level above. Therefore therapeutic injection of the facet joint should include the joint above the suspected level (Lynch & Taylor 1986). In the lumbar region, the medial branch of the posterior ramus lies in a groove on the base of the superior articular facet, where it lies in direct contact with the base of the superior surface of the transverse process, passing between the mammillary and accessory processes. The nerve actually passes under the mammilloaccessory ligament, and this is the most reliable site for locating the nerve in

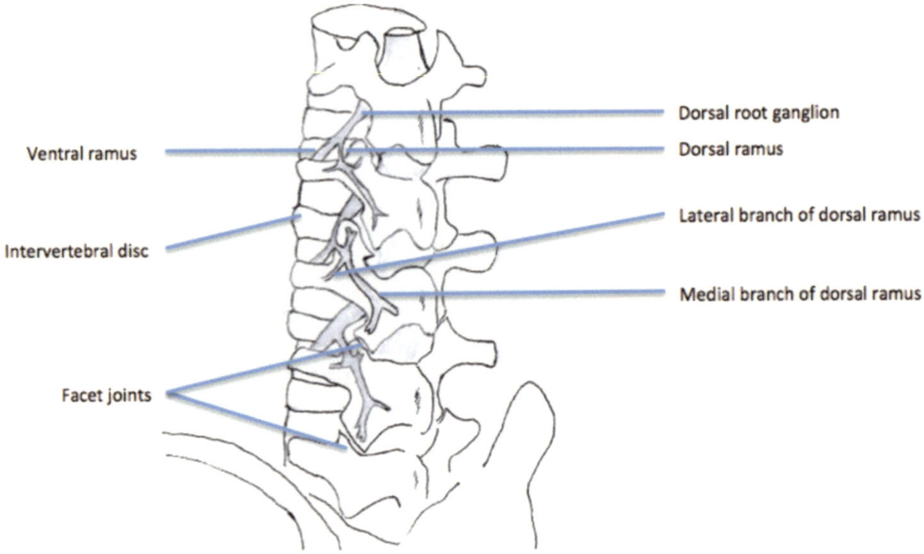

Fig. 3. The branches of lumbar roots

the lumbar spine. Studies support the idea that RF denervation is a treatment of choice for initial pain relief; however, it does not produce permanent pain relief because the nerve eventually regenerates, usually within 12-18 months (North et al., 1994). Lord et al. found the median time of return to 50% of pre-procedure pain was 263 days (Lord et al., 1996). Dreyfuss et al. found that pain relief may last for about 12 months (Dreyfuss et al., 2000). So, repeated treatment may be necessary in some patients.

2.3.2 Clinical presentation and physical examination of lumbar facet syndrome

Like other types of LBP, pain from facet joint arthritis or degeneration is always related to other pathologies. So there is no specific clinical presentation or physical examination finding for lumbar facet syndrome. Because the lumbar facet joints are like other synovial joints, the pain related to arthritis causes local tenderness of affected joints. All of the lumbar facet joints are capable of producing some referred pain. Pain emanating from the upper facet joints tends to extend into the flank, hip, and upper lateral thigh, whereas pain from the lower facet joints is likely to penetrate deeper into the thigh, usually laterally and/or posteriorly. Infrequently, the L4–L5 and L5–S1 facet joints can provoke pain extending into the lower lateral leg and, in rare instances, even the foot (Cohen & Raja 2007).

2.3.3 Image diagnosis and diagnostic block

Lumbar facets hypertrophy, joint interface widening, adjacent neuroforamen and spinal canal narrowing are easily seen in CT and MRI studies. However, the clinical symptoms and image findings often show no correlation (van Kleef et al., 2010). Diagnostic blocks using local anesthetics are performed either in the joint space or medial branch nerve region. However, both are associated with significant false-positive and false-negative rates (Cohen & Raja 2007). Technically, half way between the upper edge of the transverse process and the ligamentum mammilloaccessorium was suggested by Dreyfuss et al. because the infiltration of the anesthetic agent will influence the proximal segmental nerves and caused false-positive results (Dreyfuss et al., 1997). Because the double block causes a high false-negative rate, it was not recommended for use right now (Bogduk & Holmes 2000).

2.4 Sacroiliac joint pain (anatomy and biomechanics, clinical presentation and physical examination, diagnostic imaging, intraarticular diagnostic block)

Sacroiliac joint (SIJ) pain was believed one of the causes of chronic LBP and single anesthetic and steroid injection has proved 35% effectiveness in patients who underwent failed fusion surgery (van Kleef et al., 2010). The prevalence of SIJ pain is from 16-44% depending on different diagnostic tools (Cohen 2005). Since this kind of pathology can be treated as a single block or minimally invasive procedure, data for SIJ pain have accumulated in recent years. RF applying to SIJ is to block the nerve fibers to the SIJ. The treatment efficacy, possible mechanisms and techniques will be discussed later.

2.4.1 Anatomy and biomechanics

The SIJ is one of the origins of chronic low back pain that has always been neglected (Maigne & Planchon 2005). It is the largest axial joint in the body, with an average surface area of 17.5 cm^2 (Cohen 2005). The anterior third of the intersurface between the sacrum and the ilium is the true synovial joint and the rest of the junction is comprised of a strong and complicated ligamentous network. The ligamentous network of women is weaker which allows motility for parturition. Like other true synovial joints, age-related change, degeneration or arthritis developed on the cartilage surface of the SIJ. Each SIJ is composed of the true synovial joint and part of the ligamentous network.

The pain sources from SIJ can be divided into two parts, one is intra-articular and the other is extra-articular. The intra-articular SIJ pain may be due to the degenerative process of the articular cartilage and chronic arthritis, or autoimmune arthritis. Cohen reviewed the SIJ pathology and summarized some risk factors, including leg length discrepancy, gait abnormality, prolonged vigorous exercise, scoliosis, and spinal fusion to sacrum. (Cohen 2005) The other pain origin of SIJ is extra-articular structures such as ligaments weakening and muscle trauma and inflammation and hypermobility of SIJ caused by iatrogenic trauma. The innervation of the SIJ is divided into two parts, anterior and posterior. The posterior nerve supply to the SIJ comes from the lateral branches of the L4-S3 dorsal rami and the anterior innervation of SIJ is from the ventral rami of L4-S2 (Cohen 2005).

2.4.2 Clinical presentation, physical examination and diagnostic imaging

Since SIJ pain could be caused by either intra-articular or extra-articular structures, the diagnosis of SIJ pain is more complicated. Except for the risk factors in the medical history, physical examinations which involve the distraction of these joints can be helpful; for example, Patrick's test and Gaenslen's test. However, inconsistent predictable values were found in both physical examinations and medical history. Radiological studies showed the diagnosis of SIJ pathology is not correlated to the clinical presentation.

2.4.3 Diagnostic block of sacroiliac joint pain

Intraarticular anesthetics injection (diagnostic block) of SIJ can reasonably relieve the pain from SIJs. However, there are limited data supporting the diagnostic tool that strongly reflects the real pathology. Besides, the SIJ is technically difficult to approach even by a good experienced pain physician. The local anesthetics leakage to surrounding structures may cause false-positive results. Even with the CT-guided injection method, there is less than a 30% accomplishment rate for adequate injection (Rosenberg et al., 2000). According to these findings, diagnostic SIJ block is not reliable to diagnose true SIJ pain.

2.5 Low back pain combined with radicular pain (anatomy and biomechanics, clinical presentation and physical examination, diagnostic imaging, epidural nerve block)

Before the development of PRF (non-thermal), RF has been used directly targeting neural tissue, including movement disorders (Carr 1971; Cala et al., 1976), cardiac arrhythmias (Baszko et al., 2002), trigeminal neuralgia, and cervical and lumbar radicular pain, and the treatment target is the reflecting DRG (van Kleef et al., 1996). After PRF was developed, it was used more and more because of a lower complication rate (Chua et al., 2010). Although lacking good randomized control studies, RF or PRF applying to adjacent DRG of low back or radicular pain can still be used for selected patients before moving to surgical treatment.

2.5.1 Anatomy and biomechanics

The mechanisms of lumbosacral radicular (LSR) pain can be mainly divided into two possibilities. The first one is mechanical compromise and the other is chemical irritation of adjacent nerve roots. It is difficult to make a clear distinction between them. The mechanical compromise of nerve roots can cause local inflammatory cytokines up-regulation and the chemical irritation of nerve structure can cause nerve swelling and further compromise (Yang et al., 2011). Sometimes, either decompression or epidural steroid injection alone fails to relieve the clinical symptoms of LBP and LSR pain. Blocking or modulation of nerve transmission after inflammation and mechanical vicious circle calm down seems like an alternative treatment.

2.5.2 Clinical presentation and physical examination

LBP with LSR pain are common combined situations in clinical practice. There are two reasons which could explain this. First, the lumbosacral plexus is complicated and the branches from the surrounding structure of the spinal column often cause multiple segments pathology of the lower lumbar and sacral spine. Second, the mechanical compression of the spinal canal also frequently compromises the adjacent DRG or nerve roots which cause the dual symptoms. The clinical presentation of these patients is not easy to differentiate from other pathologies. Generally speaking, compromise of the neuronal structure causes neurologic dermatomes sensory impairment, motor weakness and neurogenic claudication. Physical examinations should be carefully taken, including neurologic signs.

2.5.3 Epidural steroid injection and diagnostic imaging

X-ray studies should be taken in chronic LBP and LSR pain patients including standard anteroposterior and lateral aspects. Dynamic flexion extension lateral film is required to differentiate the possibilities of spondylolisthesis, congenital lumbar stenosis, scoliosis and some unusual osteolytic lesions. A three dimensional view can be seen in CT and MRI studies. CT has benefits of more clear bony structures, including neuroforamens. MRI provides clear information of the spinal canal, nerve roots and disc herniation as well as unusual neogrowth or infectious diseases.

Selective adjacent epidural steroid injection calms down the inflammation process and nerve swelling in both mechanical and chemical irritation. Benny reviewed the efficacy of epidural injection and concluded that there was strong evidence for transforaminal injections in the treatment of LSR pain for both short term and long term relief (Benny & Azari 2011). Once the epidural injection works for pain relief of these patients, repeated injection was suggested if it recurs. Injection at least three times was suggested before shifting to more invasive procedures. RF or PRF for selective DRG treatment should be considered if the result of epidural injection is only temporary.

3. Procedures

3.1 Radiofrequency for intradiscal thermotherapy (evidence of hypothesis, indication, effect of procedure)

The intradiscal electrothermal therapy (IDET) is based on the mechanisms of AF tear healing and could be enhanced by thermal effect. The intradiscal volume and pressure decrease as well. Biologic study showed there are several types of radiofrequency intradiscal thermoplasty including spinecath (Oratec Interventions, Inc., Menlo Park, CA), disctrode, transdiscal biacuplasty (Fig. 4). The IDET procedures of spinecath and disctrode use a navigable intradiscal catheter with a thermal resistive coil. The procedure was performed under local anesthesia with lidocaine. All catheter placements are under fluoroscopy guidance. Preoperative administration of intravenous antibiotics two hours before the procedure is suggested.

In spinecath, the operator uses a 30-cm catheter with a 5-cm active electrothermal tip inserted anteriorly into the annulus or nucleus via a 17-gauge introducer. The active tip was advanced anterior-laterally inside the nuclear tissue and directed circuitously to return posteriorly, providing an ideal position to heat the entire posterior annulus. Once a satisfactory position was obtained in the anteroposterior, lateral views, the catheter was connected to a lead and passed to an independent technician. In all cases, the catheter tips were within 5 mm of the posterior vertebral margin upon review of saved fluoroscopic films. The disctrode was designed with a different approach which let the catheter directly pass through the posterior part of AF. The temperature during IDET begins at 65°C and was increased incrementally by 1°C every 30 seconds to achieve a final temperature of 90°C. The final temperature was maintained for 4 minutes, giving a total treatment time of 16.5 minutes.

The procedures of transdiscal biacuplasty include two RF electrodes. The symptomatic disc was reached in oblique position after cutaneous-subcutaneous anesthesia using lidocaine 1%. To facilitate the intervention, first both posterolateral parts of the disc were bilaterally accessed by 17 G introducer needle (Baylis Medical Inc., Montreal, Canada). Then, two RF probes (Baylis Medical Inc., Montreal, Canada) specially designed for cooled RF practice, wherein closed circuit sterile water circulates, were fitted into the disc after they were passed through the introducers. To ensure that the probe tip was at optimal depth in the posterior annulus, the location of the probe in the tissue was controlled in the lateral and AP positions, with the radiopaque band at its tip taken as reference. The temperature = 45°C, Ramp Rate = 2.0°C/min, Time = 15 minutes.

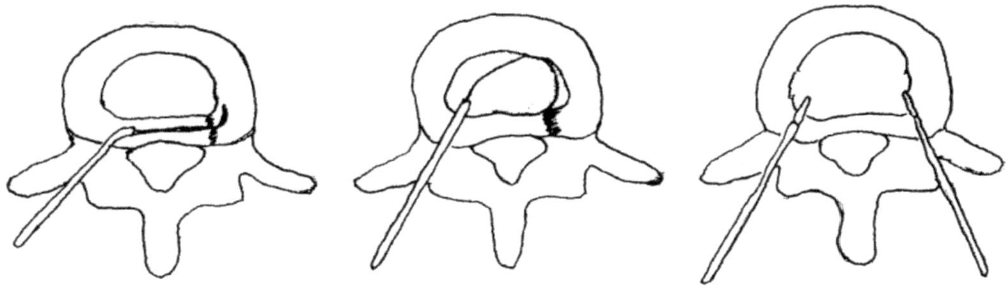

A B C

A: Disctrode; B: Spinecath; C: Transdiscal biacuplasty

Fig. 4. Different approach for RF intradiscal thermotherapy

3.2 Pulsed radiofrequency for L2 dorsal root ganglion for discogenic low back pain and other lumbar dorsal root ganglion for neurogenic low back pain

The PRF application procedure for L2 DRG was carried out with the patient in the prone position. The skin over the operative area was sterilized and then infiltrated with 2%

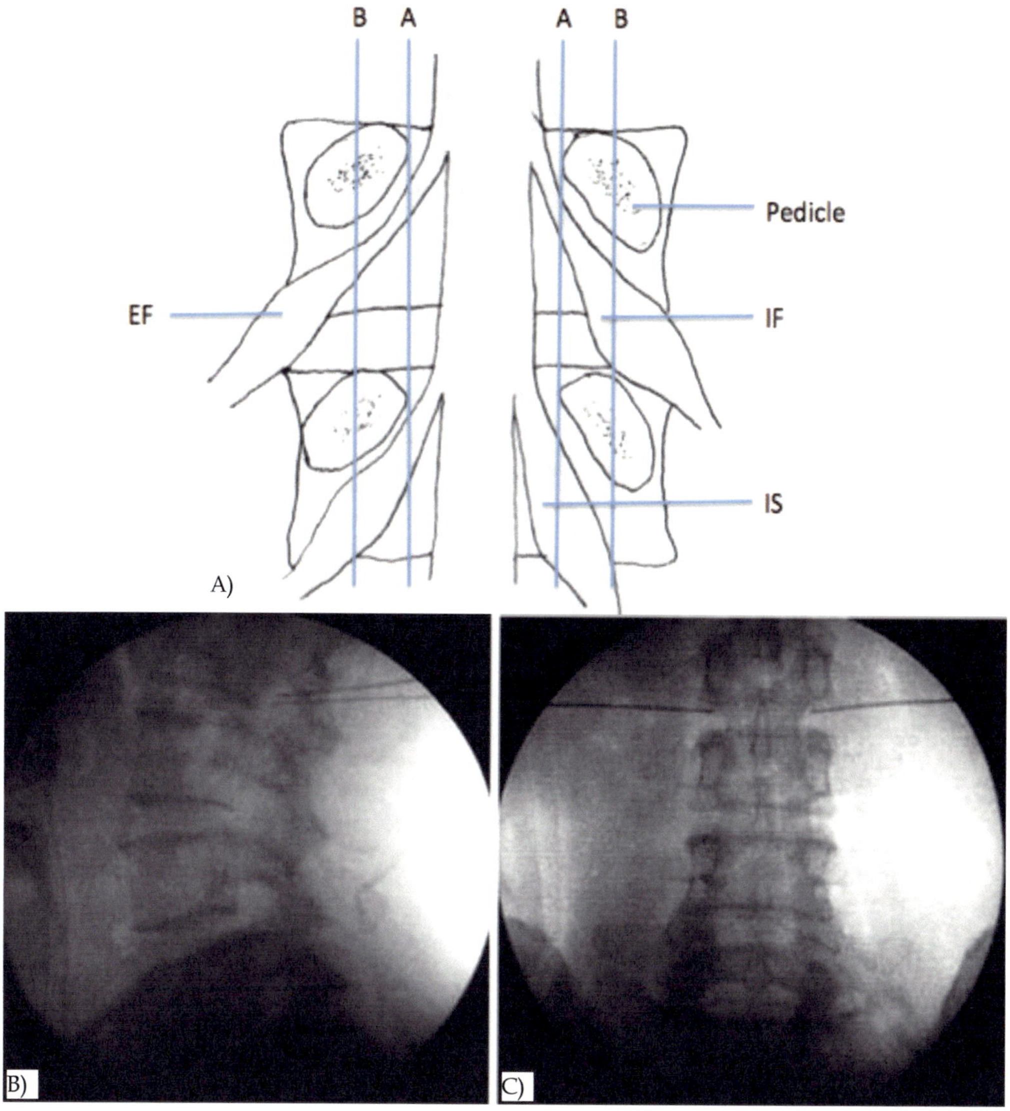

Fig. 5. DRG anatomy and ideal RF needle tips position landmark on fluoroscopy
A: Positions of dorsal root ganglia (DRG) were determined by two schematic lines and classified into three types. Line A: aligning the medial borders of L4 and L5 pedicles, Line B: aligning the centers of L4 and L5 pedicles, Intraspinal type (IS) : DRG located proximal to line A, Intraforaminal type (IF) : DRG located between line A and B, Extraforaminal type (EF) : DRG located distal to line B. B & C: Lateral (B) and anteroposterior (C) radiographs showing ideal RF needle tips position of L2 DRG.

lidocaine solution. Under C-arm fluoroscopic guidance, a 10-cm 22-gauge carved tip cannula with a 1-cm active tip electrode was placed toward the DRG near the intervertebral foramen. A RF generator was used. The RF electrode was positioned when sensory stimulation (50 Hz) reproduced the patient's pain at less than 0.5 V, which indicated the location of the ganglion. The aim was to produce a tingling sensation in the dermatomal distribution of the nerve in question (Tsou et al., 2010).

Because the DRG location has been studied, the authors divided three types of location according to the DRG and pedicle relationship (Moon et al., 2010) (Fig. 5A). The needle was advanced deeper into the intervertebral foramen until the patient felt a tingling sensation (Tsou et al., 2010) (Fig. 5B & 5C). Then 2-Hz PRF waves were applied for 120 seconds at 45 V while making sure that the electrode tip temperature did not exceed 42°C. The DRG near the intervertebral foramen of L2 root was targeted for discogenic LBP and other levels according to the radicular pain or evidence of compression noted on MRI. Motor function testing is usually not necessary with PRF because there is no risk of motor root damage with this procedure.

3.3 Radiofrequency in medial branch block for facet joint pain

The patient is placed in prone position on the radiolucent table. The anatomical landmarks of the spinal structures reflected on the skin are marked under fluoroscopic guidance including midline and facet joints and transverse processes. The skin is then sterilized in the standard fashion. Local anesthesia with 2% lidocaine was injected into the subcutaneous tissue but not extended. Then a 10-cm, 22-gauge cannula with a 5-mm exposed tip was introduced percutaneously under fluoroscopic guidance to the medial branch of the distal portion of the spinal posterior rami nerve. The tip depth and site were adjusted according to the sensation similar to the clinical presentation. The point is the most sensitive area of soreness, numbness, heaviness and distention using the techniques of twirling, rotating the tip around the lesions. Then stimulation at 5 Hz with 0.5 msec pulse duration was used to confirm the nerve position. The temperature of the electrode tip was then raised to 80°C for 90 seconds.

The nerve supply of the facet joint originates from two levels. One branch of the primary ramus arises from the nerve root at the same level as the joint and another branch from the level above. Therefore therapeutic injection of the facet joint should include the joint above the suspected level (Lynch and Taylor, 1986). For example, the facet joint between the L4 and L5 vertebral bodies is innervated by the medial branch nerves from the L3 and L4 nerve roots. In the lumbar region, the medial branch of the posterior ramus lies in a groove on the base of the superior articular facet, where it lies in direct contact with the base of the superior surface of transverse process, passing between the mammillary and accessory processes. The nerve actually passes under mammilloaccessory ligament, and this is the most reliable site for locating the nerve in lumbar spine. The L5-S1 facet joint is innervated by three nerves, L4, L5, and S1.

3.4 Radiofrequency for sacroiliac joint pain

RF for sacroiliac joint (SIJ) pain is to lesion the possible afferent nerve from SIJs. All procedures can be done in an outpatient setting using local anesthesia. Inclusion criteria includes axial low back or buttock pain ≥ 6 months in duration with tenderness overlying the SIJ(s); failure to respond to conservative therapy (e.g. physical therapy and pharmacotherapy), including long-term (>2 months) pain relief with SIJ corticosteroid injections; and ≥ 75% pain relief as calculated from a 6-hour post-block pain diary following a single diagnostic SIJ injection (Cohen & Abdi 2003; Cohen et al., 2008). At each level, placement of the electrode in close proximity to the nerve was confirmed using electrostimulation at 50 Hz, with concordant sensation achieved at ≤ 0.5 V. Prior to lesioning, the absence of leg contractions was verified with stimulation at 2 Hz up to 2 V. After satisfactory electrode placement, 0.5 ml of lidocaine 2% was injected through each cannula to reduce thermal pain and ensure blinding. The RF probe was then reinserted and a 90-second, 80° C lesion was made using a RF generator set to the lowest audible volume to blend in with ambient noise. For S1-3 lateral branch procedures, the RF needle targeting points illustrated on Fig. 6. To ensure that anesthetic spread to adjacent foramina did not impede sensory testing, electrodes were placed and stimulated at contiguous levels before denervation commenced. Once the needles were properly positioned, monopolar electrodes were sequentially inserted into the cannulae and 2.5-minute lesions were made using a water-cooled RF heating system (Pain Management SInergy System, Baylis Medical) and generator (PMG-115-TD, V2.0A, Baylis Medical). Using cooling-probe technology, the tissue temperature immediately adjacent to the cooled electrode was maintained at 60° C, while the target tissue was heated to 75° C, resulting in a lesion diameter ranging between 8 and 10 mm (Fig. 6). For safety reasons, this aggressive lesioning precludes using cooling probe technology for lumbar primary dorsal rami.

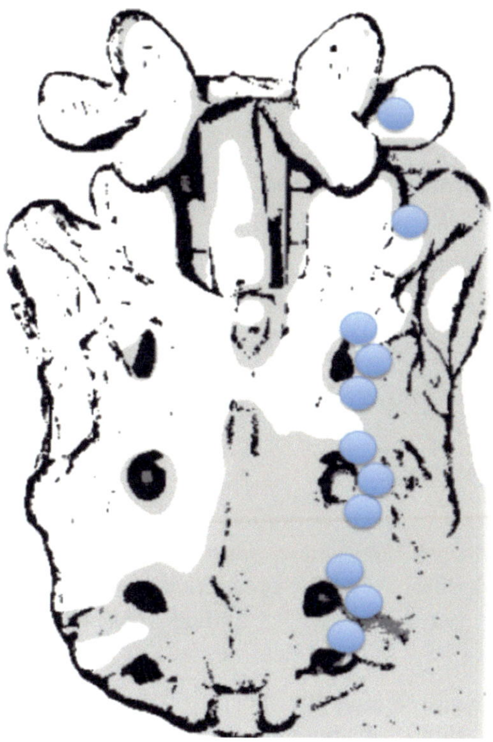

Fig. 6. The RF needle targeting points for sacroiliac joint pain

4. Evidence base medicine of radiofrequency application for low back pain

4.1 Intradiscal radiofrequency annuloplasty for discogenic low back pain

Among three major intradical RF annuloplasty procedures, IDET with spinecath was most commonly used and well-studied (Freeman et al., 2005; Pauza et al., 2004). Although the first results showed 50-70% efficacy, however, different opinions were noted in two randomized control trials and various positive and negative studies. In fact, no one will argue that the outcome is worse in severe degenerative disc disease and multi-segment degeneration. The evidence of transdiscal biacuplasty and disctrode is lacking. Both need prospective control trials to prove their efficacy.

4.2 Medial branches radiofrequency for low back pain from facet joints

Van Boxem et al. reviewed five recent randomized trials on the efficacy of RF facet denervation for chronic LBP. Three of them are positive, one is negative and one is equivocal (van Boxem et al., 2008). Among all the applications of RF and PRF, cervical facet and lumbar facet syndrome are most evidently effective. Although most of the symptom relief period is limited, these minimal invasive and safe procedures are worth using to treat selected patients with LBP from facet origins.

4.3 Radiofrequency for low back pain from sacroiliac joints

The diagnosis of SIJ-related LBP is difficult and there are still limited clinical studies supporting the efficacy of RF procedures in the treatment of SIJ pain. The treatment of SIJ intraarticular steroid injection is promising. Before a good prospective controlled study proves its efficacy, RF procedure applying to SIJ pain should be used as a second line procedure (Cohen 2005).

4.4 Radiofrequency for lumbar dorsal root ganglion in low back pain combined with or without lumbosacral radicular pain

Most prospective controlled trials for RF applying to DRGs are small or limited by inadequate study design or relative short term follow up (Malik & Benzon 2008). However, the best evidence trial of RF for DRG in treatment of LSR pain showed negative results (Geurts et al., 2003) Even if there are some retrospective studies with positive efficacy, a well-designed, randomized controlled trial is necessary.

5. Complication and management

The possible complications of RF denervation include bleeding, infection, nerve damage, broken electrodes and post-denervation neuritis. However, the complication rate is relatively lower than other more invasive procedures. And even though there are some case reports which mention permanent nerve damage (Abbott et al., 2007), the incidence decreases with the use of PRF, which is a less destructive procedure.

There are different considerations of IDET procedures even the complications are infrequent. One retrospective study review the complications of 1675 IDET procedures, six nerve injuries and six post-IDET disc herniation were reported. Other complications includes catheter breakage, temporary bladder dysfunction. Furthermore, the intradiscal heat may cause endplate injury and accelerate disc degeneration (Derby et al., 2008).

6. Conclusion

Chronic low back pain is a complicated situation that influences most members of the population at sometime during their lifetime. There are numerous of treatment modalities developed according to the diagnosis of different pathologies including non-invasive treatment and surgical intervention. Because chronic low back pain diagnosis is difficult, and the cause is multi-factorial, most treatment results for chronic LBP are unexpected. RF is a less invasive procedure that targets the pain transmission route. It can be applied reasonably to all pain problems. The evidence for the efficacy of RF for different kinds of etiologies causing chronic LBP is accumulating. Although there are some negative results and weak evidence of its efficacy, RF is still a treatment of choice because of low risk.

7. References

Abbott, Z., Smuck, M., Haig, A., & Sagher, O. (2007). Irreversible spinal nerve injury from dorsal ramus radiofrequency neurotomy: a case report. *Archives of Physical Medicine and Rehabilitation*, Vol. 88, No. 10, *pp*. 1350-1352. ISSN 0003-9993

Andersson, G. B. (1999). Epidemiological features of chronic low-back pain. *Lancet*, Vol. 354, No. 9178, *pp*. 581-585. ISSN 0410-6736

Andersson, G. B. , Mekhail, N. A., & Block, J. E. (2006). A randomized, double-blind, controlled trial: intradiscal electrothermal therapy versus placebo for the treatment of chronic discogenic low back pain. *Spine*, Vol. 31, No. 14, *pp* 1637-1638. ISSN 1528-1159

Baszko, A., Rinaldi, C. A., Simon, R. D. B., & Gill, J. S. (2002). Atrial fibrillation current and future treatments: radiofrequency ablation and novel pacing techniques. *International Journal of Clinical Practice*, Vol. 56, No 5., *pp*. 370-376. ISSN 1368-5031

Benny, B., & Azari, P. (2011). The efficacy of lumbosacral transforaminal epidural steroid injections: a comprehensive literature review. *Journal of Back and Musculoskeletal Rehabilitation*, Vol. 24, No. 2, *pp*. 67-76. ISSN 1878-6324

Bogduk, N, & Holmes, S. (2000). Controlled zygapophysial joint blocks: the travesty of cost-effectiveness. *Pain Medicine*, Vol. 1, No. 1, *pp*. 24-34. ISSN 1526-2375

Bogduk, N, & Long, D. M. (1979). The anatomy of the so-called "articular nerves" and their relationship to facet denervation in the treatment of low-back pain. *Journal of Neurosurgery*, Vol. 51, No. 2, *pp*. 172-177. ISSN 0022-3085

Bogduk, N, & Long, D. M. (1980). Percutaneous lumbar medial branch neurotomy: a modification of facet denervation. *Spine*, Vol. 5, No. 2, *pp*. 193-200. ISSN 0362-2436

Bogduk, N, Tynan, W., & Wilson, A. S. (1981). The nerve supply to the human lumbar intervertebral discs. *Journal of Anatomy*, Vol. 132, No. 1, 39-56. ISSN 0021-8782

Brown, M. F., Hukkanen, M. V., McCarthy, I. D., Redfern, D. R., Batten, J. J., Crock, H. V., Hughes, S. P. F., Polak, J. M., (1997). Sensory and sympathetic innervation of the vertebral endplate in patients with degenerative disc disease. *The Journal of Bone and Joint Surgery British*, Vol. 79, No.1, *pp*. 147-153. ISSN 0301-620X

Cahana, A., Vutskits, L., & Muller, D. (2003). Acute differential modulation of synaptic transmission and cell survival during exposure to pulsed and continuous radiofrequency energy. *The Journal of Pain*, Vol. 4, No. 4 , *pp*. 197-202. ISSN 1526-5900

Cala, L. A., Mastaglia, F. L., & Vaughan, R. J. (1976). Localisation of stereotactic radiofrequency thalamic lesions by computerised axial tomography. *Lancet*, Vol. 2, No. 7995, *pp*. 1133-1134. ISSN 0140-6736

Carr, E. M. (1971). Chronically implantable radiofrequency electrode for lesion production. Technical note. *Journal of Neurosurgery*, Vol. 35, No. 4, *pp*. 495-497. ISSN 0022-3085

Carragee, E. J. (2005). Clinical practice. Persistent low back pain. *The New England Journal of Medicine*, Vol. 352, No. 18, *pp*. 1891-1898. ISSN 1533-4406

Carragee, E. J., & Hannibal, M. (2004). Diagnostic evaluation of low back pain. *The Orthopedic Clinics of North America*, Vol. 35, No. 1, *pp*. 7-16. ISSN 0030-5898

Chao, S.-C., Lee, H.-T., Kao, T.-H., Yang, M.-Y., Tsuei, Y.-S., Shen, C.-C., & Tsou, H.-K. (2008). Percutaneous pulsed radiofrequency in the treatment of cervical and lumbar radicular pain. *Surgical Neurology*, Vol. 70, No. 1, *pp*. 59-65. ISSN 0090-3019

Chua, N. H. L., Vissers, K. C., & Sluijter, M. E. (2010). Pulsed radiofrequency treatment in interventional pain management: mechanisms and potential indications—a review. *Acta Neurochirurgica*, Vol. 153, No. 4, *pp*. 763-771. ISSN 0942-0940

Cohen, S. P. (2005). Sacroiliac joint pain: a comprehensive review of anatomy, diagnosis, and treatment. *Anesthesia and Analgesia*, Vol. 101, No. 5, *pp*. 1440-1453. ISSN 0003-2999

Cohen, S. P., & Abdi, S. (2003). Lateral branch blocks as a treatment for sacroiliac joint pain: A pilot study. *Regional Anesthesia and Pain Medicine*, Vol. 28, No. 2, *pp*. 113-119. ISSN 1098-7339

Cohen, S. P., & Raja, S. N. (2007). Pathogenesis, diagnosis, and treatment of lumbar zygapophysial (facet) joint pain. *Anesthesiology*, Vol. 106, No. 3, *pp*. 591-614. ISSN 0003-3022

Cohen, S. P., Hurley, R. W., Buckenmaier, C. C., Kurihara, C., Morlando, B., & Dragovich, A. (2008). Randomized Placebo-Controlled Study Evaluating Lateral Branch Radiofrequency Denervation for Sacroiliac Joint Pain. *Anesthesiology*, Vol. 109, No. 2, *pp*. 279-288. ISSN 1528-1175

Coppes, M. H., Marani, E., Thomeer, R. T., & Groen, G. J. (1997). Innervation of "painful" lumbar discs. *Spine*, Vol. 22, No. 20, *pp*. 2342-2349; discussion 2349-2350. ISSN 0362-2436

Davis, T. T., Delamarter, R. B., Sra, P., & Goldstein, T. B. (2004). The IDET procedure for chronic discogenic low back pain. *Spine*, Vol. 29, No. 7, *pp*. 752-756. ISSN 1528-1159

Derby R, Baker RM, Lee CH, Anderson PA.(2008) Evidence-informed management of chronic low back pain with intradiscal electrothermal therapy. *The Spine Journal*, Vol. 8, No.1, *pp*. 80-95. ISSN 1529-9430

Deyo, R. A., & Weinstein, J. N. (2001). Low back pain. *The New England Journal of Medicine*, Vol. 344, No. 5, *pp*. 363-370. ISSN 0028-4793

Dreyfuss, P, Halbrook, B., Pauza, K., Joshi, A., McLarty, J., & Bogduk, N. (2000). Efficacy and validity of radiofrequency neurotomy for chronic lumbar zygapophysial joint pain. *Spine*, Vol. 25, No. 10, *pp*.1270-1277. ISSN 0362-2436

Dreyfuss, P, Schwarzer, A. C., Lau, P., & Bogduk, N. (1997). Specificity of lumbar medial branch and L5 dorsal ramus blocks. A computed tomography study. *Spine*, Vol. 22, No. 8, *pp*. 895-902. ISSN *0362-2436*

Edgar, M. A. (2007). The nerve supply of the lumbar intervertebral disc. *The Journal of Bone and Joint Surgery British* Vol. 89, No. 9, *pp*. 1135-1139. ISSN 0301-620X

Erdine, S., Yucel, A., Cimen, A., Aydin, S., Sav, A., & Bilir, A. (2005). Effects of pulsed versus conventional radiofrequency current on rabbit dorsal root ganglion morphology. *European Journal of Pain*, Vol. 9, No. 3, *pp*. 251-256. ISSN 1090-3801

Freeman, B. J. C., Fraser, R. D., Cain, C. M. J., Hall, D. J., & Chapple, D. C. L. (2005). A randomized, double-blind, controlled trial: intradiscal electrothermal therapy versus placebo for the treatment of chronic discogenic low back pain. *Spine*, Vol. 30, No. 21, *pp*. 2369-2377; discussion 2378. ISSN *1528-1159*

Freemont, A. J. (2009). The cellular pathobiology of the degenerate intervertebral disc and discogenic back pain. *Rheumatology*, Vol. 48, No. 1, *pp*. 5-10. ISSN 1462-0332

Freemont, A. J., Peacock, T. E., Goupille, P., Hoyland, J. A., O'Brien, J., & Jayson, M. I. (1997). Nerve ingrowth into diseased intervertebral disc in chronic back pain. *Lancet*, Vol. 350, No. 9072, *pp*. 178-181. ISSN *0140-6736*

Freemont, A. J., Watkins, A., Le Maitre, C., Baird, P., Jeziorska, M., Knight, M. T. N., Ross, E. R. S., O'Brien, J. P., Hoyland, J. A. (2002). Nerve growth factor expression and innervation of the painful intervertebral disc. *The Journal of Pathology*, Vol. 197, No. 3, *pp*. 286-292. ISSN 0022-3417

Geurts, J. W. M., van Wijk, R. M. A. W., Wynne, H. J., Hammink, E., Buskens, E., Lousberg, R., Knape, J. T. A., Groen, G. J. (2003). Radiofrequency lesioning of dorsal root ganglia for chronic lumbosacral radicular pain: a randomised, double-blind, controlled trial. *Lancet*, Vol. 361, No. 9351, *pp*. 21-26. ISSN 0140-6736

Ghormley, R. K. (1933). Low back pain. With special reference to the articular facets, with presentation of an operative procedure. *Journal of the American Medical Association*, Vol. 101, No. 23, *pp*. 1773 -1777.

Helm, S., Hayek, S. M., Benyamin, R. M., & Manchikanti, L. (2009). Systematic review of the effectiveness of thermal annular procedures in treating discogenic low back pain. *Pain Physician*, Vol. 12, No. 1, *pp*. 207-232. ISSN 1533-3159

Higuchi, Y., Nashold, B. S., Jr, Sluijter, M., Cosman, E., & Pearlstein, R. D. (2002). Exposure of the dorsal root ganglion in rats to pulsed radiofrequency currents activates dorsal horn lamina I and II neurons. *Neurosurgery*, Vol. 50, No. 4, *pp*. 850-855; discussion 856. ISSN 0148-396X

Hurri, H., & Karppinen, J. (2004). Discogenic pain. *Pain*, Vol. 112, No. 3, *pp*. 225-228. ISSN 0304-3959

Kapural, L., & Mekhail, N. (2006). A randomized, double-blind, controlled trial: intradiscal electrothermal therapy versus placebo for the treatment of chronic discogenic low back pain. *Spine*, Vol. 31, No. 14, *pp*. 1636; author reply 1636-1637. ISSN 1528-1159

Karasek, M., & Bogduk, N. (2000). Twelve-month follow-up of a controlled trial of intradiscal thermal anuloplasty for back pain due to internal disc disruption. *Spine*, Vol. 25, No. 20, *pp*. 2601-2607. ISSN 0362-2436

Kallewaard, J. M., Terheggen, M. A. M. B., Groen G. J., Sluijter, M. E., Derby, R., Kapural, L., Mekhail, N., van Kleef, M. (2010). Discogenic low back pain. *Pain Medicine*, Vol. 10, No. 6, *pp*. 560-579. ISSN 1533-2500

Keller, A., Boyle, E., Skog, T. A., Cassidy, J. D., & Bautz-Holter, E. (2011 in press). Are Modic changes prognostic for recovery in a cohort of patients with non-specific low back pain? *European Spine Journal* ISSN 1432-0932

Kvarstein, G., Måwe, L., Indahl, A., Hol, P. K., Tennøe, B., Digernes, R., Stubhaug, A., Tønnessen, T. I., Beivik, H. (2009). A randomized double-blind controlled trial of intra-annular radiofrequency thermal disc therapy--a 12-month follow-up. *Pain*, Vol. 145, No. 3, *pp*. 279-286. ISSN 1872-6623

Lindblom, K. (1951). Technique and results of diagnostic disc puncture and injection (discography) in the lumbar region. *Acta Orthopaedica Scandinavica*, Vol. 20, No. 4, *pp*. 315-326. ISSN 0001-6470

Lord, S. M., Barnsley, L., Wallis, B. J., McDonald, G. J., & Bogduk, N. (1996). Percutaneous radio-frequency neurotomy for chronic cervical zygapophyseal-joint pain. *The New England Journal of Medicine*, Vol. 335, No. 23, *pp*. 1721-1726. ISSN 0028-4793

Lynch, M. C., & Taylor, J. F. (1986). Facet joint injection for low back pain. A clinical study. *The Journal of Bone and Joint Surgery British*, Vol. 68, No. 1, *pp*. 138-141. ISSN 0301-620X

Maigne, J. Y., & Planchon, C. A. (2005). Sacroiliac joint pain after lumbar fusion. A study with anesthetic blocks. *European Spine Journal*, Vol. 14, No. 7, *pp*. 654-658. ISSN 0940-6719

Malik, K., & Benzon, H. T. (2008). Radiofrequency applications to dorsal root ganglia: a literature review. *Anesthesiology*, Vol. 109, No. 3, *pp*. 527-542. ISSN 1528-1175

Maus, T. (2010). Imaging the Back Pain Patient. *Physical Medicine and Rehabilitation Clinics of North America*, Vol. 21, No. 4, *pp*. 725-766. ISSN 1558-1381

Moon, H. S., Kim, Y. D., Song, B. H., Cha, Y. D., Song, J. H., & Lee, M. H. (2010). Position of dorsal root ganglia in the lumbosacral region in patients with radiculopathy. *Korean Journal of Anesthesiology*, Vol. 59, No. 6, *pp*. 398. ISSN 2005-7563

Nagda, J. V., Davis, C. W., Bajwa, Z. H., & Simopoulos, T. T. (2011). Retrospective review of the efficacy and safety of repeated pulsed and continuous radiofrequency lesioning of the dorsal root ganglion/segmental nerve for lumbar radicular pain. *Pain Physician*, Vol. 14, No. 4, *pp*. 371-376. ISSN 2150-1149

Nakamura, S., Takahashi, K., Takahashi, Y., Yamagata, M., & Moriya, H. (1996). The afferent pathways of discogenic low-back pain. Evaluation of L2 spinal nerve infiltration. *The Journal of Bone and Joint Surgery British*, Vol. 78, No. 4, *pp*. 606-612. ISSN 0301-620X

Nakamura, S., Takahashi, K., Takahashi, Y., Morinaga, T., Shimada, Y., & Moriya, H. (1996). Origin of nerves supplying the posterior portion of lumbar intervertebral discs in rats. *Spine*, Vol. 21, No. 8, *pp*. 917-924. ISSN 0362-2436

Navani, A., Mahajan, G., Kreis, P., & Fishman, S. M. (2006). A case of pulsed radiofrequency lesioning for occipital neuralgia. *Pain Medicine*, Vol. 7, No. 5, *pp*. 453-456. ISSN 1526-2375

North, R. B., Han, M., Zahurak, M., & Kidd, D. H. (1994). Radiofrequency lumbar facet denervation: analysis of prognostic factors. *Pain*, Vol. 57, No. 1, *pp*. 77-83. ISSN 0304-3959

Ohnmeiss, D. D., Vanharanta, H., & Ekholm, J. (1999). Relationship of pain drawings to invasive tests assessing intervertebral disc pathology. *European Spine Journal*, Vol. 8, No. 2, *pp*. 126-131. ISSN 0940-6719

Ohnmeiss, D. D., Vanharanta, H., & Ekholm, J. (1999). Relation between pain location and disc pathology: a study of pain drawings and CT/discography. *The Clinical Journal of Pain*, Vol. 15, No. 3, *pp*. 210-217. ISSN 0749-8047

Ohtori, S., Takahashi, K., Chiba, T., Yamagata, M., Sameda, H., & Moriya, H. (2002). Substance P and calcitonin gene-related peptide immunoreactive sensory DRG neurons innervating the lumbar intervertebral discs in rats. *Annals of Anatomy*. Vol. 184, No. 3, *pp*. 235-240. ISSN 0940-9602

Pauza, K. J., Howell, S., Dreyfuss, P., Peloza, J. H., Dawson, K., & Bogduk, N. (2004). A randomized, placebo-controlled trial of intradiscal electrothermal therapy for the treatment of discogenic low back pain. *The Spine Journal*, Vol. 4, No. 1, *pp*. 27-35. ISSN 1529-9430

Peng, B., Chen, J., Kuang, Z., Li, D., Pang, X., & Zhang, X. (2009). Expression and role of connective tissue growth factor in painful disc fibrosis and degeneration. *Spine*, Vol. 34, No. 5, *pp*. E178-182. ISSN 1528-1159

Peng, B., Hao, J., Hou, S., Wu, W., Jiang, D., Fu, X., & Yang, Y. (2006). Possible pathogenesis of painful intervertebral disc degeneration. *Spine*, Vol. 31, No. 5, *pp*. 560-566. ISSN 1528-1159

Racz, G. B., & Ruiz-Lopez, R. (2006). Radiofrequency procedures. *Pain Practice*, Vol. 6, No. 1, *pp*. 46-50. ISSN 1533-2500

Richebé, P., Rathmell, J. P., & Brennan, T. J. (2005). Immediate early genes after pulsed radiofrequency treatment: neurobiology in need of clinical trials. *Anesthesiology*, Vol. 102, No. 1, *pp*. 1-3. ISSN 0003-3022

Rosenberg, J. M., Quint, T. J., & de Rosayro, A. M. (2000). Computerized tomographic localization of clinically-guided sacroiliac joint injections. *The Clinical Journal of Pain*, Vol. 16, No. 1, *pp*. 18-21. ISSN 0749-8047

Saal, J A, & Saal, J. S. (2000). Intradiscal electrothermal treatment for chronic discogenic low back pain: a prospective outcome study with minimum 1-year follow-up. *Spine*, Vol. 25, No. 20, *pp*. 2622-2627. ISSN 0362-2436

Saal, Jeffrey A, & Saal, J. S. (2002). Intradiscal electrothermal treatment for chronic discogenic low back pain: prospective outcome study with a minimum 2-year follow-up. *Spine*, Vol. 27, No. 9, *pp*. 966-973; discussion 973-974. ISSN 1528-1159

Sachs, B. L., Vanharanta, H., Spivey, M. A., Guyer, R. D., Videman, T., Rashbaum, R. F., Johnson, R. G., Hochschuler, S. H., Mooney, V. (1987). Dallas discogram description. A new classification of CT/discography in low-back disorders. *Spine*, Vol. 12, No. 3, *pp*. 287-294. ISSN 0362-2436

Schwarzer, A. C., Aprill, C. N., Derby, R., Fortin, J., Kine, G., & Bogduk, N. (1995a). The prevalence and clinical features of internal disc disruption in patients with chronic low back pain. *Spine*, Vol. 20, No. 17, *pp*. 1878-1883. ISSN 0362-2436

Schwarzer, A. C., Wang, S. C., Bogduk, N., McNaught, P. J., & Laurent, R. (1995b). Prevalence and clinical features of lumbar zygapophysial joint pain: a study in an Australian population with chronic low back pain. *Annals of the Rheumatic Diseases*, Vol. 54, No. 2, *pp*. 100-106. ISSN 0003-4967

Shaheen, N. J., Sharma, P., Overholt, B. F., Wolfsen, H. C., Sampliner, R. E., Wang, K. K., Galanko, J. A., et al. (2009). Radiofrequency ablation in Barrett's esophagus with dysplasia. *The New England Journal of Medicine*, Vol. 360, No. 22, *pp*. 2277-2288. ISSN 1533-4406

Shealy, C. N. (1976). Facet denervation in the management of back and sciatic pain. *Clinical Orthopaedics and Related Research*, Vol. Mar-Apr., No.115, *pp*. 157-164. ISSN 0009-921X

Singh, K., Samartzis, D., Vaccaro, A. R., Nassr, A., Andersson, G. B., Yoon, S. T., Phillips, F. M., et al. (2005). Congenital lumbar spinal stenosis: a prospective, control-matched, cohort radiographic analysis. *The Spine Journal*, Vol. 5, No. 6, *pp*. 615-622. ISSN 1529-9430

Smith, H. P., McWhorter, J. M., & Challa, V. R. (1981). Radiofrequency neurolysis in a clinical model. Neuropathological correlation. *Journal of Neurosurgery*, Vol. 55, No. 2, *pp*. 246-253. ISSN 0022-3085

Thackeray, A., Fritz, J. M., Brennan, G. P., Zaman, F. M., & Willick, S. E. (2010). A pilot study examining the effectiveness of physical therapy as an adjunct to selective nerve root block in the treatment of lumbar radicular pain from disk herniation: a randomized controlled trial. *Physical Therapy*, Vol. 90, No. 12, *pp*. 1717-1729. ISSN 1538-6724

Tsou, H.-K., Chao, S.-C., Wang, C.-J., Chen, H.-T., Shen, C.-C., Lee, H.-T., & Tsuei, Y.-S. (2010). Percutaneous pulsed radiofrequency applied to the L-2 dorsal root ganglion for treatment of chronic low-back pain: 3-year experience. Journal of Neurosurgery. *Spine*, Vol. 12, No. 2, *pp*. 190-196. ISSN 1547-5646

Tsou, H.-K., Chao, S.-C., Kao, T.-H., Yiin, J.-J., Hsu, H.-C., Shen, C.-C., & Chen, H.-T. (2010). Intradiscal electrothermal therapy in the treatment of chronic low back pain: Experience with 93 patients. *Surgical Neurololgy International*. Vol. 1. *pp*. 37. ISSN 2229-5097

Van Boxem, K., van Bilsen, J., de Meij, N., Herrler, A., Kessels, F., Van Zundert, J., & van Kleef, M. (2011). Pulsed Radiofrequency Treatment Adjacent to the Lumbar Dorsal Root Ganglion for the Management of Lumbosacral Radicular Syndrome: A Clinical Audit. *Pain Medicine*. Vol. 12, No. 9, *pp*. 1322-1330. ISSN 1526-4637

Van Boxem, K., van Eerd, M., Brinkhuizen, T., Brinkhuize, T., Patijn, J., van Kleef, M., & van Zundert, J. (2008). Radiofrequency and pulsed radiofrequency treatment of chronic pain syndromes: the available evidence. *Pain Practice*, Vol. 8, No. 5, *pp*. 385-393. ISSN 1533-2500

Van Kleef, M, Barendse, G. A., Kessels, A., Voets, H. M., Weber, W. E., & de Lange, S. (1999). Randomized trial of radiofrequency lumbar facet denervation for chronic low back pain. *Spine*, Vol. 24, No. 18, *pp*. 1937-1942. ISSN 0362-2436

Van Kleef, M, Liem, L., Lousberg, R., Barendse, G., Kessels, F., & Sluijter, M. (1996). Radiofrequency lesion adjacent to the dorsal root ganglion for cervicobrachial pain: a prospective double blind randomized study. *Neurosurgery*, Vol. 38, No. 6, *pp*. 1127-1131; discussion 1131-1132. ISSN 0148-396X

Van Kleef, Maarten, Vanelderen, P., Cohen, S. P., Lataster, A., Van Zundert, J., & Mekhail, N. (2010). Pain originating from the lumbar facet joints. *Pain Practice*, Vol. 10, No. 5, *pp*. 459-469. ISSN 1533-2500

Van Zundert, J., de Louw, A. J. A., Joosten, E. A. J., Kessels, A. G. H., Honig, W., Dederen, P. J. W. C., Veening, J. G., et al. (2005). Pulsed and continuous radiofrequency current adjacent to the cervical dorsal root ganglion of the rat induces late cellular activity in the dorsal horn. *Anesthesiology*, Vol. 102, No. 1, *pp*. 125-131. ISSN 0003-3022

Walsh, T. R., Weinstein, J. N., Spratt, K. F., Lehmann, T. R., Aprill, C., & Sayre, H. (1990). Lumbar discography in normal subjects. A controlled, prospective study. *The Journal of Bone and Joint Surgery, American*, Vol. 72, No. 7, *pp*. 1081-1088. ISSN 0021-9355

Wichman, H. J. (2007). Discography: over 50 years of controversy. *Wisconsin medical journal*, Vol. 106, No. 1, *pp*. 27-29. ISSN 1098-1861

Yang, G., Marras, W. S., & Best, T. M. (2011). The biochemical response to biomechanical tissue loading on the low back during physical work exposure. *Clinical Biomechanics*, Vol. 26, No. 5, *pp*. 431-437. ISSN 1879-1271

Zhang, J., Shi, D.-S., & Wang, R. (2011 in press). Pulsed radiofrequency of the second cervical ganglion (C2) for the treatment of cervicogenic headache. *The Journal of Headache and Pain*. ISSN 1129-2377

Zhou, Y., & Abdi, S. (2006). Diagnosis and minimally invasive treatment of lumbar discogenic pain--a review of the literature. *The Clinical Journal of Pain*, Vol. 22, No. 5, *pp*. 468-481. ISSN 0749-8047

Part 3

Surgical Treatment

9

Nonfusion Techniques for Degenerative Lumbar Diseases Treatment

Leonardo Fonseca Rodrigues, Paula Voloch and Flávio Cavallari
Hospital São Vicente de Paulo/ Hospital Federal do Andaraí, Brazil

1. Introduction

Conservative treatment is the "gold standard" treatment for low back pain, in spine degenerative conditions. However, in cases where there is a failure in conservative measures, surgical treatment becomes an option (Roh et al, 2005). These procedures traditionally included decompression of spinal elements, correction of deformity and arthrodesis of the diseased spinal segment but, in some conditions, they both may be used in a combined manner (Schwarzenbach et al, 2005).

The technique of fusion with the use of only bone graft was first reported by Hibbs and Albee in 1911 (*apud in* Huang et al, 2005), for prevention of progression of Pott disease. Pioneers in using metallic instrumentation for stabilization, associated with bony fusions, were Harrington (1976) for scoliosis surgery, Roy Camille (1979) and Steffe (1986) with screw-plate system, Magerl (1984) with external fixation for frature treatment, and Dick (1985), with the internal fixator (*apud in* Schwarzenbach et al, 2005). Since then, lumbar fusion became the "gold standard" surgical treatment for a wide range of painfull conditions. The primary goal of lumbar stabilization is to treat pain from disc or facet, in the instable spinal unit. In these cases pain emerges apparently under load (Christiansen et al, 2004).

However, no surgical treatment is perfect. Christiansen and coworkers (2004) obtained positive results in approximately 70% of cases of fusion surgery. An important complication, in the medium-term follow-up, is degeneration of the disc, adjacent to a fusion segment (Rham and Hall, 1996), known as adjacent disc degeneration (ADD). In this study, ADD occured in 30% of cases, five years after fusion. Articular hypermobility in the segment above fusion segment was reported by Luk and collaborators (1987) in 50% of cases, of which 30% had also stenosis of the spinal canal.

Another post-operative complication, related to fusion, is pseudoarthrosis, compromising the final result of the surgery (Kornblum et al, 2004). In order to achieve good results in fusion, consolidation of the bony fusion is critical (Butterman et al, 1998). However, a study of Muholand and Sengupta (2002) noted that bony consolidation, with achieved fusion segment, does not represent necessarily a clinical success.

Rham and Hall (1996), in their study, also demonstrated that, in pseudoarthrosis, micromovements in the facet joint preserves hypermobility in the adjacent segments, acting like a "protective factor for the development of the adjacent segment degeneration". This finding was also described in 2004 by Ghiselli and collaborators.

With all these evidences, nonfusion techniques arise, aiming the prevention of ADD, and the fact that this new technology does not require bone graft, since these techniques don´t depend on bony consolidation.

1.1 The lumbar stability

In 1990, White and Panjabi defined instabillity of the spine as "the loss of the spine´s ability to maintain its patterns of displacement under physiologic loads so there is no initial or additional neurologic deficit, no major deformity, and no incapacitating pain".

The importance of lumbar stability was originally established by Kruton (1944). Morgan and King (1957) reported that instability was a primary cause of low back pain. The degenerative process of the lumbar spine was better understood after studies of Kirkaldy-Willis and coworkers (1978), and the development of the disease was described later by Kirkaldy-Willis and Farfan (1982), using a concept of three phases: 1) temporal dysfunction, 2) unstable phase, and 3) restabilization. In the last phases, 2 and 3, patients often have stenosis, or deformities, like degenerative scoliosis, often requiring surgery for stabilization, decompression and/or correction of the deformity. **(Figure 1)**

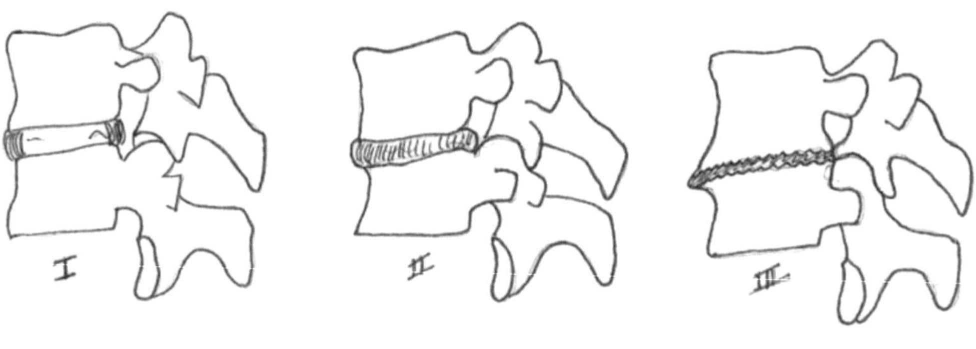

Fig. 1. The degenerative cascade described by Kirkaldy-Willis and Farfan (1982). At the third phase, the disc lost height and facet hypertrophy promotes segment stabilization, but also narrowing the neural foramen and the vertebral canal (stenosis)

The intervertebral disc plays the most important role in spine stabilization (Roh et al, 2005). Disc degeneration is a physiological process with aging. The extracellular matrix structure changes, mainly in proteoglycans concetration at the *nucleus pulposus*, leading to disc dehydration causing, because of that, morphological changes in the disc (Biyany et al, 2004). With these changes, biomechanical function of the disc is altered, and the load in this dysfunctional disc starts to injury other structures, such as the endplates, the facet joints and the fibrous annulus (Bernick et al, 1991). Additionally, these degenerative changes can cause a number of effects in the spine and nerve roots. Protrusion or disc herniation can cause radicular compression, central stenosis and considering that there are nociceptors located there, it will, as well, lead to low back pain (Roh et al, 2005).

The basic functions of the spine are: to provide stability, giving mobility to the body, to protect the spinal cord, and to control neural information in order to move the upper and lower limbs (Harms and Tabasso, 1999). For this reason, this architecture has passive elements (bones, joints and ligaments) and active elements (muscles).

Therefore, the spinal stabilizing system consists of three subsystems: spinal column, muscles surrounding the spine, and motor control unit. The spine carries load, and provides information about position, motion and loads of the spinal column (proprioception). With this information, the control unit turns it into action by the muscles (active elements), which must provide dynamic changes in the spinal column, altering the spinal posture and loads (Panjabi, 1992).

1.2 Biomechanics of the degenerated spine

Biomechanics of the spine is not simple, because it involves complex movements of flexion, lateral inclination and rotation, and the combination of all these movements. As the spine has a huge amount of spinal units, which provide the movements, its center of rotation is not static. As movement changes, the center of rotation changes as well, and so does the loading on the spine structures, having different points of axial load in the same functional unit, with focus in the intervertebral disc and facet joints (Lumsden et al, 1968). This mobility is possible due to the possibility of intervertebral disc deformation, but is limited by the disc architecture, vertebral body, and the structures in the posterior arch (Harms and Tabasso, 1999).

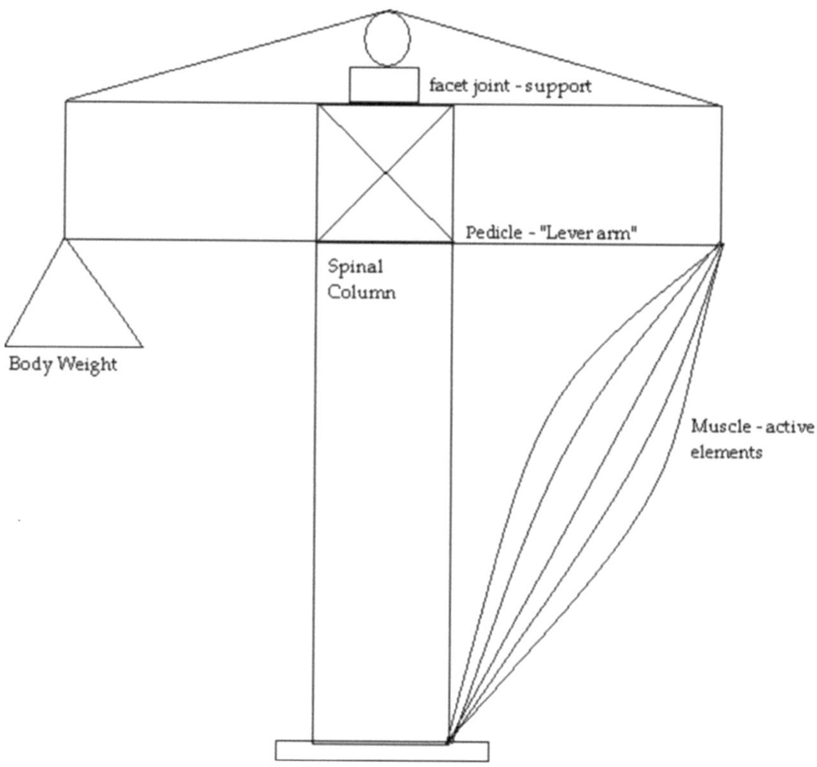

Fig. 2. The "crane", of the lumbar stability. To be stable, all the elements, active and passive, must be intact. (Adapted from Harms and Tabasso, 1999)

For Better understanding, we can imagine the spine as a crane (**Figure 2**). In standing position, the body center of gravity is located anterior to the spine, anterior to the vertebral bodies and intervertebral discs. Thus, an axial load is distributed as an axial compressive load in the anterior column, holding 80% of the axial load, and the remainig 20 % as a shear force in the posterior column (Harms and Tabasso, 1999). So, the anterior column receives loads primarily by compression forces, and the posterior column also resists stretching, torque and tilt. Due to these characteristics, the anterior column acts like a distraction device, and the posterior column as a tension band (Harms and Tabasso, 1999). The tensile forces in the posterior columns are actively made by the muscles, and supported by the facet joints and ligaments. The lever arm of this stabilization system depends on the pedicular sizes, influencing in the effectiveness of the posterior musculature (Harms and Tabasso, 1999).

The function and effectiveness of the posterior elements to provide stability depends on the integrity of the anterior column (Harms and Tabasso, 1999). Kirkaldy-Willis and Farfan described this degenerative cascade (1982), the degeneration of the disc (anterior column) causing an overload in posterior elements, thus inducing a degeneration of muscles and facet joints.

Modic (1984, 1991), using Magnetic Ressonance Imaging (MRI) studies, described degenerative changes in the intervertebral disc, with overload to the endplates (**Figure 3**). Biomechanical failure on the facet joints, and muscular failure, with overload to the endplates, leads to a noceceptive pain (Kusslich et al, 1991), and the progression of the disease leads to cyst formation on the facets, hypertrophy, with narrow disc space, that can cause central or *foraminal stenosis* (Dubois et al, 1999).

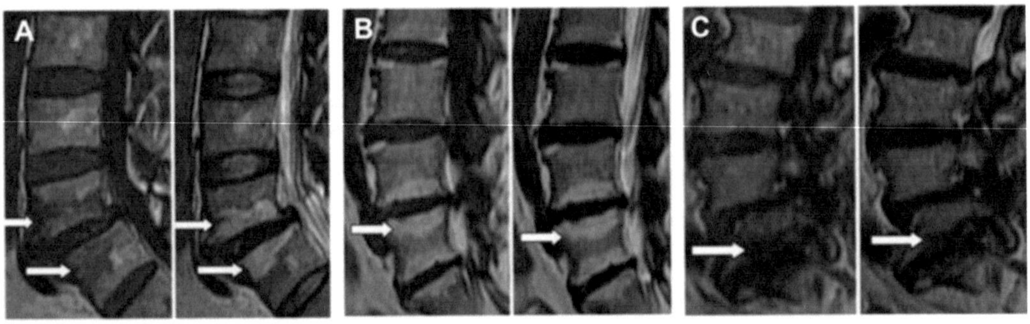

Fig. 3. The overload in the endplates, caused by disc degeneration, induces changes in the MRI. **A)** Modic type 1, the endplates are black in T1 incidence and white in 2 incidence (edema). **B)** The enplates are white in both T1 and T2 incidences (fat). **C)** The endplates are black in both incidences (sclerotic). (Adapted from Zhang et al, 2008)

Albert and Manniche (2007) demonstrated, in a randomized controlled trial with 181 patients, that Modic changes type 1 is more strongly associated with non-specific low back pain than Modic changes type 2. They also suggested in this study, that disc herniation is a strong risk factor for developing Modic changes in the same level, during the following year (Albert and Manniche 2007).

1.3 The adjacent segments

Over the years, the "gold standard" technique to treat severe degenerative lumbar spondylosis has been spinal fusion (Lehman et al 1987, Ko et al 2010). However, since the beginning of this use, the damaging effects of creating rigid segments in the spine, with overload to the adjacent levels (transition syndrome) have been discussed (Fymoyer et al, 1979, Stokes et al, 1981, Aota et al, 1995, Rahm et al, 1996, Christiensen et al, 2004, Fritzel et al, 2003, Cheh et al, 2007, Kumar et al, 2001, Wiltse et al, 1999, Miyakoshi et al, 2000, Lee et al, 1988, Min et al, 2008, Yang et al, 2008, Korovesis et al, 2009).

No surgical technique is perfect, even in this "gold standard" method, patients are subject to a number of short and long-term morbidities. The relative immobility of fused spinal segments transfers stress to adjacent segments, leading to acceleration of adjacent level degeneration, because the sagital alignment of a fused spinal segment is fixed and cannot adapt to variations in posture (Weinhoffer et al, 1995).

A series of studies have shown, in cadavers and in vitro, that fusion increases intradiscal pressures, end plate stresses, and annular stresses at adjacent segments (Lee et al, 1984, Weinhoffer et al, 1995, Cunningham et al, 1997, Rohlman et al, 2001, Eck et al, 2002, Rao et al, 2005, Sudo et al, 2006). The restricted motion in the fused segments, in a active body, having fixed sagittal alignement, increases motion and stress at adjacent levels, in sitting, supine and erect postures (Huang et al, 2005).

This stress doesn't lead to hipermobility in the adjacent levels after fusion since degeneration progresses. Avoidance of hypermobility at the adjacent levels is frequently attributed to nonfusion technology. A few studies already reported about such an effectiveness of dynamic stabilization techniques (Olsewki et al, 1996, Phillips et al, 2002, Shono et al, 1998, Panjabi et al, 2007).

The incidence of adjacent disc degeneration is not clear. But, it has shown clinical evidence. Sears and coworkers (2011), in a retrospective cohort study, associate the risk of a new surgery for adjacent level degeneration with the number of levels fused. They concluded that, although young patients who underwent single-level fusions are at low risk, patients who underwent fusion of three or four levels had a threefold increased risk of further surgery, compared with single-level fusion, and a predicted 10-year prevalence of 40%.

Szpalski and coworkers (2002) published a comprehensive review of nonfusion implants, which comprises posterior dynamic stabilization, interspinous devices, and total lumbar disc replacement. The potential reduction of the adjacent disc disease is mainly attributed to the avoidance of increased stress at the adjacent segments. Such increased stress is anticipated in instrumented fusion procedures, leading to hypermobility at the adjacent segments. Shono and coworkers (1998) demonstrated that hypermobility at the adjacent levels was proportional to the length and rigidity of the instrumented constructs.

2. When is surgery necessary?

Low back pain is the first symptom of disc degeneration. The degenerate intervertebral disc is associated with structural failure, with radial failures, prolapse, endplate damage, annular protrusion, internal disc rupture, and disc space narrowing (Dubois et al, 1999, Schnake et al, 2006). Especially the discs, posterior and capsular ligaments, as well as the vertebral endplates have been found to be the major sources of nociception leading to pain (Kusslich et al, 1991). With the progression of the disease, hydration of the nucleus pulposus decreases, and this composition alters, leading to loss of dic height and reduction of its intradiscal pressure. As described by Kirkaldy-Willis (1982), this cascade evolves, leading to overload the annulus fibrosus and the facet joints. Loading with inadequate nuclear turgor leads to shearing forces in the transitional zone between the nucleus and annulus (Huang et al, 2005). As a result, we have ruptures and radial tears in the annulus fibrosus, and overload of the facet joints. This change leads to instabillity. In addition to disc protrusion toward the spinal canal, disc height decreases, and there is spinal canal or neural foramen stenosis, that leads to radicular pain (Yu et al, 1988, Urban et al, 2003).

Pain from degenerative diseases may arise from stenosis, facet overload, the disc itself, and eccentrically loaded vertebral endplate. In patients with radicular pain, secundary to radicular compression, consequece to disc prolapse, surgical procedures have to be carried out, if the conservative therapy fails. However, operative treatment, like discectomy or nucleotomy, leads to progression of the disc degeneration (Dunlop et al, 1984, Gottfried et al, 1986, Brinckmann et al, 1991). In 2004, Jansson and collaborators published that approximately 10% of all operated discs reherniate and approximately 27% of all operated patients have to undergo a second operation within 10 years.

Nonsurgical management must be considered, in low back pain, especially in patients without radicular compression signs.

Stabilization devices leave the pain-generating disc tissues in situ, but restrict certain types of motion and alter load transfer through the functional spinal unit (Huang et al, 2005). Fusion implants are designed to unload the disc and facets by load sharing.

In 1954, Verbiest described the so called neurogenic intermittent claudication secundary to lumbar spine stenosis. Recent studies show that clinical or nonsurgical tratment have poor results comparing to surgical procedures (Weinstein et al, 2008). Surgical trearment based on decompression alone presented poor results, related to progression of symptoms and deformity (Hanley et al, 1995). At the same time, adding an arthrodesis to the decompression procedure increases the operative time and blood loss, and consequently the complication rate (Di Silvestre et al, 2010).

3. Nonfusion techniques: Advantages and disadvantages

For many years arthrodesis has been acknowledged as the gold standard treatment for a wide variety of spinal pathologies such as deformities, unstable and painful conditions of the lumbar motion segment (Mayer et al, 2002).

Nevertheless, spinal fusion in degenerative disc disease when there is no instability or disturbed curvature, though often performed, is not a consensus among the spinal community (Greenough et al, 1994, 1998, Kozak et al, 1994, Mayer et al, 1998). In most cases, there is indication of arthrodesis when all kinds of conservative therapies fail.

However, the results seem to not always justify these decisions (Mayer et al, 2002). Fritzell and coworkers (2003) observed a 12% 2-year incidence rate of major complications following lumbar arthrodesis, with a reoperation rate of 14.6%. Complications include pseudarthrosis, bone graft donor site pain, instrumentation failure, infection and simple failure to relieve pain (Frelinghuysen et al, 2005, Tropiano et al, 2005). Not to mention the possibility of an adjacent segment degeneration, which have made spinal surgeons think of an alternative method that could avoid such complications (Rham et al, 1996).

It's important to mention that there are increasing numbers of patients who have undergone spinal fusion for degenerative disc disease with images showing adjacent level degenerative changes, but not necessarily with a strong clinical impact. In a long term follow-up of 30 years, there was a significantly higher incidence of radiographic changes at adjacent levels after lumbar fusion, but this was not accompanied by a significant change in the functional outcomes (Kumar et al, 2001).

Non fusion technologies in spine surgery are being developed to address the arthrodesis' disadvantages (Jansen and Marchesi, 2008).

Non fusion implant types include total disc replacements, prosthetic nuclear implants, and posterior stabilization devices.

Potential advantages of nonfusion implants (Mayer et al, 2002, Huang et al, 2005):

1. Elimination of the need for bone graft.
2. Reduction in surgical morbidity
3. Elimination of pseudarthrosis
4. Reduction of adjacent level degeneration

Pseudarthrosis and the need for bone graft are truly eliminated, as well as the above mentioned reduction in surgical morbidity. Nevertheless, it is questionable whether nonfusion technologies significantly decrease the incidence of adjacent level disease, especially if segmental motion is not well maintained. Furthermore, if bone graft substitutes prove to be efficacious and economically viable alternatives to autogenous bone grafting, avoidance of autograft harvest will no longer be a significant advantage of nonfusion implants.

Potential disadvantages of nonfusion implants (Mayer et al, 2002):

1. Mechanical failure and device migration
2. Implant subsidence
3. Same level degeneration

Considering the fact that nonfusion implants are characterized by motion, therefore they are subject to mechanical failure or migration.

It is believed that subsidence is a significant contributor to poor outcomes after total disc replacements and this is probably the most significant challenge to long-term outcomes with these implants. Optimized implant design and end plate coverage may diminish the chances of subsidence happening.

The preservation of segmental motion obtained in nonfusion technology created the concept of symptomatic same level degeneration, as opposed to what is seen in a solid fusion. The possible sources of same level degeneration are the intervertebral disc, the facet joints, and the ligamentum flavum.

In conclusion, the potential pitfalls and benefits of nonfusion implants have to be carefully considered before the selection of this technology. Long-term randomized prospective studies are necessary and are currently unavailable, so non fusion procedures should be reserved for use in a small population of highly selected patients.

4. Total disc arthroplasty

4.1 Overview

Lumbar fusion has been developed for several decades and became the standard surgical treatment for symptomatic lumbar degenerative disc disease (DDD). Disc arthroplasty devices have been designed in an attempt to replace functionally the intervertebral discs, as opposed to the gold standard method (fusion) which could not achieve that (Fekete et al, 2010). Hence, a method of motion preservation would be the best alternative to spinal arthrodesis, considering that it could theoretically prevent adjacent level degeneration. However, the long-term stability, endurance and strength of the prosthesis are unknown for the majority of implants (Freeman et al, 2006).

The most important functions of the intervertebral discs are the transmission of load and the maintenance of motion and disc height. Nevertheless, none of the implants currently available can reproduce totally the kinematics of a healthy intervertebral disc (Fekete et al, 2010).

There are basically two types of disc arthroplasty devices: nucleus or total disc replacement (TDR) devices, the latter being the most frequently used.

The first total disc arthroplasty implant to be widely used was the three component SB-Charité prothesis, a metal-polyethylene-metal construct, devised by Schellnack and Buttner-Janz in 1984. Since then, three different prototypes of the SB-Charité prosthesis have been developed. Of all the other types of total disc arhroplasty, the Pro-disc prosthesis, devised by Marnay, also in the 1980s, has been widely used (Zigler et al, 2004) (**Figure 4**).

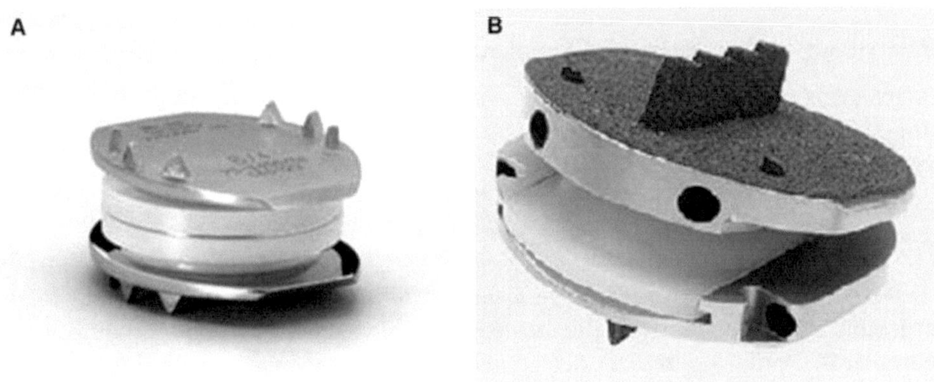

Fig. 4. Total disc replacement in lumbar Spine: **A)** Sb-Charité, **B)** ProDisc prosthesis

Each artificial disc comprehends two or three components including two endplates and an articulating mechanism with either a metal-on-metal or metal-on-polymer surface. In order to keep the disc in place and providing stability within the host vertebral body, devices feature different designs, such as teeth-like components called spikes or fins that are driven into the vertebral bone, a porous coated surface on the endplates, promoting bony in-growth around these structures, or are secured into the recipient vertebral body with screws (Mayer et al, 2005, Jansen and Marchesi, 2008).

Arthroplasty devices can be classified based on their biomechanical properties, as (Errico et al, 2005):

1. Constrained implant: Have mechanical restrictions in motion within the physiological range, providing a fixed center of rotation.
2. Semi-constrained implant: allows motion in the physiological range
3. Non-constrained implant: allows hypermobility in comparison to the physiological range

The healthy tri-joint complex (intervertebral disc and the two facet joints) represents a semi-constrained system that allows physiological motion and prevents abnormal (excessive) motion. This motion unit in its healthy state allows for six potential motion directions: compression, distraction, flexion, extension, lateral bending, and axial rotation (McCullen et al, 2003).

Unlike spinal fusion, artificial disc replacement (ADR) is designed to preserve motion at the target spinal level. As well as possibly providing greater pain relief, this motion preservation may potentially decrease stress on and mobility of the adjacent segment structures, factors that are thought to contribute to adjacent segment disease. ADR can also restore pre-degenerative disc height and spinal

Alignment and the benefit of not depending upon a bone graft. Other theoretical advantages include maintenance of mechanical characteristics, decreased perioperative morbidity compared with fusion, and early return to function (Fritzell et al, 2001).

4.2 Indications

The leading indication for total disc arthroplasty is symptomatic, degenerative, monosegmental instability of the lumbar spine between L2 and S1. The patient must be refractory to all kinds of conservative treatment, having persistent pain (intensity greater than 5 on the visual analog scale) for at least six months. Age should range preferably between 30 and 50 years old. Furthermore, there must be a correlation between imaging studies and symptoms. MRI shows degeneration of the disc, with only mild loss of height of intervertebral space. Provocative discography reproduces the patient's typical pain (Kraemer et al, 2009).

Contraindications to the implantation of disc prosthesis include: osteoporosis, infection, deformities, tumors, malformations, multisegmental degeneration and psychosocial disturbances (Kraemer et al, 2009).

4.3 Surgical method

Insertion of the prosthesis involves an anterior approach and is usually performed by a general surgeon and a spine surgeon. Potential problems associated with ADR may include injury to other structures (vascular, neurologic, intestinal, or urogenital), infection, loosening, polyethylene or metal wear, loss of motion over time, impact on adjacent discs and facet joints, subsidence, implant failure, heterotopic ossification, and device related endplate fracture (Geisler et al, 2004, 2008).

4.4 Clinical results

It is still uncertain, though, whether TDR is really more effective and safer than the gold standard treatment, lumbar fusion. To systematically compare the effectiveness and safety of TDR to that of arthrodesis treating lumbar DDD, Yajun and collaborators performed a meta-analysis, which has been published in 2010. The authors observed that the group of patients submitted to TDR had slightly better functioning and less back or leg pain without clinical significance, and significantly higher satisfaction status in TDR group compared with lumbar fusion group at the 2-year follow-up. Later on, at five years follow-up, these outcomes have not shown significant differences between comparing groups. The complication and reoperation rate of two groups are similar both at two and at five years.

The authors concluded that TDR does not show significant superiority for the treatment of lumbar DDD compared with fusion. The benefits of motion preservation and the long-term complications are still unable to be concluded. More high-quality RCTs with long-term follow-up are imperative to come to new conclusions.

In a systematic review of the literature, Freeman and coworkers (2006) stated that significant facet joint osteoarthritis is a contraindication to TDR, but that could be a difficult situation to identify in its early stages.

Moreover, the future of facet joints following a total disc replacement is obscure and facet joint hypertrophy, which accelerates spinal stenosis, may be a potent long-term complication that kind of implant. Not to mention that revision procedures will unquestionably be technically difficult with a great risk of vascular injury, particularly at the L4/5 level.

Therefore, that review of the literature concluded that the use of TDR may be limited to the treatment of degenerative disc disease in its early stages, with preservation of disc height. That would limit its indications, eliminating its uses in the majority of patients.

Up until now, only few studies have examined the direct effects of disc arthroplasty on adjacent levels. These studies show contradictory conclusions. While some of them support the idea of decreased adjacent-level degeneration, although lacking a clinical significance (Huang et al, 2006) others raise concerns about the high rates of index-level facet joint arthrosis and adjacent-level degeneration, despite motion preservation (van Ooij et al, 2003, Shim et al, 2007, Siepe et al, 2007) A trustworthy analysis of these results is difficult , considering the limitations in study design as well as the differences in the kinematics of the various implants examined.

5. Interspinous implants

5.1 Overview

With population aging, degenerative spine disorders became more common. The degenerative cascade, described by Kirkaldy-Willis and Farfan (1982) leads to disc and articular changes, with disc bulging and facet hypertrophy, causing effects in the spine, such as central and/or foraminal stenosis. Verbiest (1954) described neurogenic intermittent claudication secondary to degenerative lumbar stenosis. Thus, some kind of management has to be proposed, either conservative or surgical, in order to relieve symptoms. Weinstein (2008) suggests, in his study, that surgical procedures achieve better results, compared to conservative management.

Symptoms of spinal stenosis most often occur in patients 50 to 70 years old. These symptoms include low back pain, buttock pain, and/or trochanteric and posterior thight pain (Trautwein et al, 2010). Neurogenic claudication occurs when these symptoms exacerbs with walk, in extend position, and relieves when sitting or flexion of the spine.

A surgical treatment, for decompression and fusion of the segments, has increased operative time and blood loss, increasing complication rate in elderly patients (Carreon, et al, 2003, Deyo et al, 1993, Benz et al, 2001). To prevent complications and relief symptoms, with minimally intervention, new techniques have been developed to manage this condition.

Interspinous devices were first described in 1950, by Dr Fred L Knowles (Bono et al, 2007). But, the results are poor, with high number of devices dislodged, needing to be removed. Sénégas (1988) described an interspinous spacer, made of titanium, with Dacron tapes to fix the devices to spinous process. He had success in the treatment of more than 300 patients. After that, other implants have been developed (Coflex, Wallis, X-stop, Diam), with another material types (Titaniumm, peek) (Sengupta, 2004), but follow-up studies are still running to access its efficiency, and precise its indications (**Figure 5**).

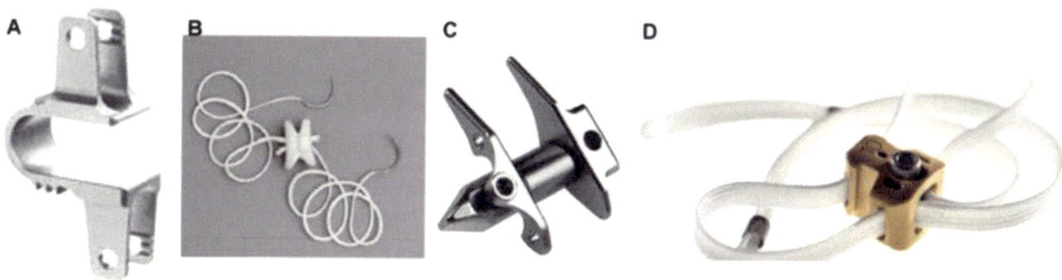

Fig. 5. Interspinous distraction devices: **A)** Coflex, **B)** Diam, **C)** X-stop and **D)** Wallis.

5.2 Indications

The interspinous distraction devices, keep the segment in flexion. In this condition, the device reduces loading to the intervertebral disc, and also reduces spinal and foraminal stenosis (Sengupta et al, 2004). For this reason, this procedure is indicated in patients in whom the symptoms are increased in extension (Gunzburg et al, 2003). For better results, it is indicated for patients aged 50 years or older, with moderately impaired physical function related to neurogenic intermittent claudication, and may be implanted at one or two lumbar levels (Yi et al, 2010). Better results are related to pain relief in lumbar flexion, with or without low back pain, and failure of nonsurgical care (Laurysen et al, 2007). The Wallis mechanism to treat low back pain, caused by degenerative instability, is indicated for: Massive disc herniation, with substantial loss of disc material, reherniated disc with second discectomy, disc herniation in traditional level, like sacralization of L5, adjacent segment degeneration, to previous fusion, and isolated Modic I lesion, that leads to chronic low back pain. (Sengupta et al, 2004). Contraindications include: Disc degeneration grade V of Pfirrmann classification, spondilololisthesys, severe osteoporosis, spinal anatomy that would prevent implantation or cause instability, cauda equine syndrome and active systemic infection or localized infection at the site of implantation (Yi et al, 2010).

5.3 Surgical method

The X-stop device was developed to a minimally invasive approach, with short time surgery, to prevent complications in the elderly patients (Sengupta et al, 2004). For the procedure, patient must be placed in prone or lateral decubitus. They are positioned in flexed position, to keep distraction between the spinous process of the vertebral segment that will be treated. With a 4-5cm midline incision, the spinous processes are approached at the appropriate disc level, which is confirmed radiographically (Zucherman et al, 2004). The supraspinous ligament has to be maintained, to prevent kyphosis and stabilize the implant (Yi, et al, 2010, Zucherman et al, 2007). The distractor is placed through the interspinous ligament, after distraction, to maintain flexion of the segment. The spinal canal is not violated, with no need of laminectomy, laminotomy or foraminotomy.

The Wallis device must be placed with local or general anesthesia. Patient is placed in a prone position, in neutral. A neutral position of physiological lumbar lordosis is best to optimize the effect of the implant. All efforts should be made to avoid subsequent lumbar kyphosis (Sénégas et al, 1988). The supraspinous ligament must be retained, to prevent kyphosis, and stabilize the segment. The procedure itself takes less than 15 minutes (Sénégas et al, 2008).

5.4 Clinical results

Some studies have shown that surgical procedure, using X-stop device for chronic low back pain, substantially superior to conservative treatment, when is related to 1 or 2 level spinal stenosis, in cases where pain is relieved with flexion. These clinical results are based in a 2 years follow-up with claudication Questionnaire criteria (Zucherman, et al, 2004, 2005, Hsu et al, 2006, Anderson et al, 2006).

Comparing patients who received X-stop implants, with patients who underwent laminectomy without fusion (decompression surgery), Kondrashov et al (2007) have shown, in their study, an improvement of 15 points in the Oswestry Disability Index, defining patient success (78% of the X STOP group, versus only 33% of the laminectomy group, had successful outcomes at 4 years follow-up).

Biomechanical studies have shown beneficial effects of X-stop in kinematics of the spine (Kabir et al, 2010), with limitation in flexion/extension movement in the instrumentated level, increase in spinal canal and neural foramen, decrease in intradiscal pressure, decrease facet overload, no degenerative affection in adjacent levels, and no significant changes in biomechanics of the segment. Kutcha et al (2009) indicated, after Oswestry scores and Visual analogic scale evaluation, that X-stop implantation provides short and long term satisfactory clinical outcomes. However, some cases with severe stenosis and claudication have insatisfactory results with these implants.

Barbagallo and coworkers (2009) looked into the complications of X-STOP. Of a total of 69 patients, 8 had complications (1 interoperative and 7 postoperative). These included 4 device dislocations and 4 spinous process fractures, one of this peroperative, in a double-level implant. Of these, 7 patients (10.14%) required revision surgery. Korovessis and coworkers (2009), in a prospective controlled study, concluded that Wallis interspinous implant changed the natural history of adjacent disc degeneration incidence until up to 5 years after surgery. Sénégas and coworkers (2009), in a 13 year follow up study with 107 patients, with canal stenosis or herniated disc, who underwent dymamic stabilization with Wallis, reported that the implants had to be removed in 20 patients, leading to fusion. The other 87 with Wallis had better clinical results in a retrospective evaluation, compared to fusion group. Floman and coworkers (2007), in a retrospective study with 37 patients who underwent primary lumbar disc excision and stabilization with Wallis, have shown 13% of recurrent herniations, suggesting that this implant does not reduce the incidente of recurrent herniations.

Trautwein and coworkers (2010), evaluated Coflex device (interspinous U shaped titanium alloy process), and concluded that fatigue failure of the spinal process and lamina is extremely rare. Kong and coworkers (2007), in a retrospective study, compared clinical results of patients who underwent lumbar decompression surgery, with Coflex placement, with patient who underwent posterior lumbar interbody fusion (PLIF). They assumed that Coflex leads to a lower stress in adjacent levels that PLIF.

6. Pedicular stabilization

6.1 Overview

To reduce pain and disability, spine surgical procedures have three main components: decompression, stabilization and correction of deformity (Schwarzenbach et al, 2005). For many years, spine fusion of the affected segments has been the gold standard procedure for this treatment. However, patients undergoing arthrodesis are subjected to a large number of short and long-term morbidities (Huang et al, 2005).

Considering the concepts of spinal instability, defined by Junghanns (1968), Kirkaldy-Willis and Farfan (1982), and White and Panjabi (1990), and the history of instrumentation in the spine, stabilization methods must diminish pathologic motion, prevent deformity, reduce deformity and compensate iatrogenic destabilization (Schwarzenbach et al, 2005).

Dynamic stabilization with pedicular screws has been developed, as an alternative to fusion, to achieve segmental stabilization, without complications seen in fusion (Di Silvestre et al, 2010).

Henry Graf (1992) first described the system of pedicular screws, surrounded by nonelastic polyester ligament with tension to lock the motion segment in extension. This concept was to lock the facet joints, stopping rotation (Sengupta et al, 2004). This system presented some problems, because the lordosis that the Graf produces results in stenosis of lateral recess, especially if there was any preexisting facet arthropathy or in-folding of the *ligamentum flavum*, and increases load in posterior annulus, which is a feature of painful degeneration of the disc (Grevitt et al, 1995).

In 1994, Dubois proposed the Dynesys system, a pedicular screw-based system, with flexible rods as a dynamic stabilization. Based on Kirkaldy-Willis concept of degeneration, Dynesys attempts to alter the first and second phases, reducing segmental motion to a physiologic level, neutralizing bendig, torsional, and shear forces, thus reducing load on disc (Schwarzenbach et al, 2005)(**Figure 6**).

Strempel and coworkers (2000) introduced a pedicular based stabilization system, with rigid rods and hinged head screws. With this architecture, a division of the load between implant and anterior column is achieved. Screws are made of titanium alloy and, since 2002, are covered by hydroxyapatite for better bone ingrowths (**Figure 6**).

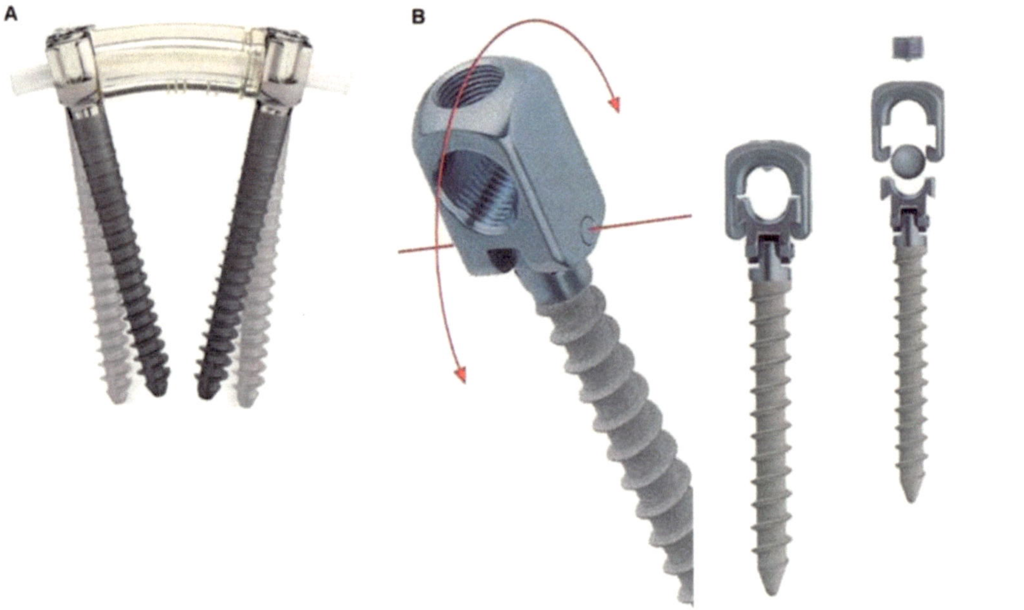

Fig. 6. Pedicular stabilization devices: **A**) Dynesys - dynamic in the flexible rod, **B**) cosmic: Hinged head screws

6.2 Indications

In instability, dynamic devices are indicated in some conditions, based on their design and biomechanical effects (Schwarzenbach et al, 2005). The main goal of Dynesys is to stabilize the degenerated segments in early stages of degeneration, defined by Kirkaldi-Willis (1982). Ko and coworkers (2010) presented a study with patients who underwent dynamic stabilization with Dynesys. These patients had symptomatic low back pain, as result of degenerative spondylolisthesis, radiculopathy, or neurogenic claudication and they failed to respond to conservative treatment.

Strempel and coworkers (2000) said that cosmic system may possibly relieve pain as well as restorate the neurologic function without correction. Fusion is necessary when corrections (mostly in the sagittal plane) are necessary to treat pain. With limitation, this system won't be used when there's a need to treat more than three segments of the spine. Indications for cosmic are: Symptomatic lumbar stenosis, chronically recurring low back pain in the case of discogenic pain and facet syndrome, recurrent disc herniation, combination with a spondylodesis, and extension of an existing spondylodesis in the case of a painful adjacent level degeneration (the last 2 indications are hybrid constructions).

Di Silvestre (2010) consider to use dynamic stabilization system with Dynesys to treat lumbar degenerative scoliosis in the elderly, as an alternative to fusion methods, in order to decrease blood loss (there is no need to decortications of the facets and transverse processes), eliminate need of bone graft, and thus decreasing operative time.

6.3 Surgical method

Patients were treated under general anesthesia, in prone position (Di Silvestre et al, 2010, Ko et al. 2010, Maleci et al, 2011). Medial unique incision must be made, but Wiltse intermuscular plane approach is an option in stabilization procedures without wide decompression (Wiltse et al, 1988, von Strempel, 2000). In cases of stenosis, using Dynesys stabilization, and needing wide laminectomy, patients are positioned with hips flexed in 90 degrees and, after decompression, patients are repositioned to maximum lordosis (Di Silvestre et al 2010).

Pedicular screws are placed, by fluoroscopic control, in a lateral point entry, at the basis of transverse process with convergence angle between 13 and 18º in Dynesys (Di Silvestre et al, 2010), and between 20 and 25º in cosmic (Stremple et al, 2000) horizontal to the sagital plane. The screws must be as long as possible to prevent shear forces. Removing and reinserting of the screws must be avoided, to prevent screw loosening (Di Silvestre et al, 2010, Stremple et al, 2000).

After screw placement, distraction must be taken up to 4mm in cosmic system (Maleci et al, 2011), and 2mm in Dynesys system, to expand the neural foramina. In cosmic system, when decompression is necessary, a transverse stabilizer must be placed (Maleci et al, 2011).

6.4 Clinical results

Nonfusion techniques have been developed to prevent complications seen in spinal fusion, as adjacent segment disease (Cakir et al, 2009). Additionaly, fusion involves longer operative time and blood loss, increasing the complication rate, mainly in elderly (Di Silvestre et al, 2010). Not to mention the need of bone graft, with potencial effects on donor site (Huang et al, 2005).

Graf ligamentoplasty has been often unsatisfactory, not preventing postoperative instability, with high percentage of destatibilzation of the affected segment (Guigui et al, 1994). This system presented some problems, because the lordosis that the Graf produces results in stenosis of lateral recess, especially if there was any preexisting facet arthropathy or in-folding of the ligamentum flavum, and increases load in posterior annulus, which is a feature of painful degeneration of the disc (Grevitt et al, 1995). Kanayama and coworkers (2005) showed poor clinical outcomes with Graf.

Di Silvestre (2010) evaluated clinical results after dynamic stabilization with Dynesys in elderly patients with lumbar degenerative scoliosis, with questionaries (oswestry disability index, Roland Morris, and visual analog scale), and radiologic imaging. In this study, clinical results have shown nonfusion stabilization as a safe procedure in elderly patients, with low complication rate, and statistically significant improvement in clinical outcomes.

Cakir and coworkers (2009), compared patients who underwent surgical treatment with decompression and Dynesys or decompression and fusion, having concluded that, in monosegmental instrumentation, no differences in adjacent level have been found, in a minimum follow-up of 24 months.

In 2006, Schanke and coworkers found signs of degeneration in disc adjacent to Dynesys stabilization in 29% of discs after 2 years. In the same follow-up period, the authors reported complications in 17% of patients, with 4 loosen screws and 1 broken screw out of 96 screws. This proportion was maintained in a continuous follow up, after 2 years, and no progression of instability has been shown (Schaeren et al, 2008). Screw loosening was assessed by Ko and coworkers (2010). Seventy one patients, who underwent decompression and stabilization with Dynesys were evaluated. Loosening of the screws occurred in 19,7% of patients, but this did not affect their clinical improvement. It is interesting to note that such findings had never occurred in the middle vertebras in intermediary level. It´s more likely to occur on marginal segments.

Treatment of the dysfunctional segmental motion was assessed by Cansever and coworkers (2011), using radiologic parameters in postoperative time, in one year follow-up. Their results suggest that decompression with dynamic stabilization were effective for radiologic stability over time.

A recent article described nonfusion method in lumbar spinal fractures (Kim et al, 2011). In a 4 year follow-up (2002 - 2006), their results suggests that this method is one of the most effective to manage thoracolumbar fractures, especially in younger people.

Clinical results published with cosmic system, shows improvement in quality of life after dynamic stabilization, with decrease in visual analog scale of pain (Rodrigues et al. 2010, Strempel et al, 2006, Stremple et al, 2008). As complications, screw loosening was found in 5,2% and 5,03% cases, and just 1 case of adjacent disc degeneration was related. Screw breakage occurred in a low rate, but not all of them were symptomatic.

Rodrigues and coworkers (2010), in a retrospective evaluation of patients submitted to a pedicular dynamic stabilization with cosmic, showed an improvement in quality of life of these patients during the 29,5 months follow-up period. The SF-36 score ranging from 33.15% preoperatively, to 75.99% in the postoperative, was statistically significant using the *student t test* ($p < 0{,}0001$). Maleci and coworkers (2011), using cosmic system, in a 2 year follow up period, showed good results, with a low complication rate. In this article, they emphasized advantages, such as reduction is surgical trauma, avoidance complication in graft donor site, and preservation of intervertebral cartilage. No spontaneous fusion has been observed in the follow-up, but a fibrous rigidity has been present.

7. Conclusion

Nonfusion techniques are new, compared to fusion, as an option in the surgical treatment for low back pain. As new techniques, long-term prospective studies must be designed to achieve their effectiveness.

The effects of fusion are well known in a long term analisys, with a large number of complications. Adjacent disc degeneration, donor site pain, pseudoarthrosys and high blood loss are aspects that must be avoided with nonfusion technologies.

The right indication is the most important key to the success of the surgical treatment. Up to now, good results have been shown with nonfusion surgeries, and these technologies are improving, to avoid complications, and preserve the physiological motion of the spine.

Long-term Follow-up studies must be taken, to a better understanding of these procedures, and indications in a large scale. But, the results obtained up to now are encouraging and hopeful.

8. Acknowledgment

The authors thank Tania Spohr, PhD, for her suggestions regarding the chapther.

In memorian to Jorge Luiz Santana Rodrigues.

9. References

Albert H B, Manniche C. (2007). Modic changes following lumbar disc herniation. *Eur Spine J*, vol 16, No7 (mar 2007), pp.(977-982), ISSN: 0940-6719

Anderson PA, Tribus CB, Kitchel SH. (2006). Treatment of neurogenic claudication by interspinous decompression: application of the X STOP device in patients with lumbar degenerative spondylolisthesis. *J Neurosurg Spine*, vol 4, No 6 (Jun 2006), pp. (463–471), ISSN 1547-5654

Aota Y, Kumano K, Hirabayashi S. (1995). Postfusion instability at the adjacent segments after rigid pedicle screw fixation for degenerative lumbar spinal disorders. *J Spinal Disord*, vol 8, No 6 (Dec 1995), pp. (464–473), ISSN: 1536-0652

Barbagallo GM, Olindo G, Corbino L, et al. (2009). Analysis of complications in patients treated with the X-Stop Interspinous Process Decompression System: proposal for a novel anatomic scoring system for patient selection andreview of the literature. *Neurosurger, v*ol 65, No 1 (Jul 2009), pp. (111–119), ISSN: 0148-396X

Benz RJ, Ibrahim ZG, Afshar P, et al. (2001). Predicting complications in elderly patients undergoing lumbar decompression. *Clin Orthop Relat Res*, vol 384 (Mar 2001), pp. (116-121), ISSN 0009-921X

Bernick S, Walker JM, Paule WJ. (1991). Age changes to the anulus fibrosus in human intervertebral disca. *Spine*, vol 16, No 5 (May 1991), pp. (520-524) ISSN: 0362-2436

Biyani A, Amderssoon GB. (2004). Low Back Pain: Pathophysiology and management. *J Am Acad Orthop Surg*. vol 12, (2004), pp. (106-115). ISSN: 1067-151X

Bonno CM, Vaccaro AR. (2007). Interspinous Process devices in the lumbar spine. *J Spine Disord Tech*, vol 20, No 3 (May 2007), pp. (255-261), ISSN: 15360652

Brinckmann P, Grootenboer H (1991) Change of disc height radial, disc bulge, and intradiscal pressure from discectomy. An in vitro investigation on human lumbar discs. *Spine*, vol 16, No 6 (Jun 1991), pp. (641-646), ISSN 0362-2436

Buttermann GR, Garvey TA, Hunt AF, et al. (1998) Lumbar fusion results related to diagnosis. *Spine*, vol 23, No 1 (Jan 1998), pp. (116-127), ISSN: 0362-2436

Buttner-Janz K, Schellnack K, Zippel H (1987). An alternative treatment strategy in lumbar intervertebral disk damage using an SB Charite modular type intervertebral disk endoprosthesis. Z Orthoplhre Grenzgeb vol 125, No 1 (jan-feb 1987), pp. (1-6), ISSN: 0044-3220

Cakir B, Carazzo C, Schmidt R et al. (2009) Adjacent segment mobility after rigid and semirigid instrumentation of the lumbar spine. *Spine,* vol. 34, No 12 (May 2009), pp.(1287-1291), ISSN 0362-2436

Cansever T, Civelek E, Kabatas S, et al. (2011) Dysfunctional segmental motion treated with dynamic stabilization in the lumbar spine. *World Neurosurg,* vol. 75, No 5 (May-Jun 2011), pp. (743-749), ISSN 1878-8750

Carreon LY, Puno RM, Dimar JR, et al. (2003). Perioperative complications of posterior lumbar decompression and arthrodesis in older adults. *J Bone Joint Surg Am,* vol .85A, No11 (Nov 2003), pp. (2089-2092), ISSN 0021-9355

Cheh G, Bridwell K, Lenke L, et al. (2007). Adjacent segment disease following lumbar/thoracolumbar fusion with pedicle screw instrumentation: a minimum 5-year follow-up. *Spine, vol*32, No 20 (Sep 2007), pp. (2253–2257), ISSN: 0362-2436

Christensen FB. (2004). Lumbar spinal fusion. Outcome in relation to surgical methods, choice of implants and postoperative rehabilitation. *Acta Orthop Scand,*vol 75, No 313 (Oct 2004), pp. (2-43), ISSN:0001-6470

Cunningham B, Kotani Y, McNulty P, et al. (1997). The effect of spinal destabilization and instrumentation on lumbar intradiscal pressure: an in vitro biomechanical analysis. *Spine,* vol22, No 22 (Nov 1997), pp. (2655–2663), ISSN: 0362-2436

Deyo RA, Ciol MA, Cherkin DC, et al. (1993). Lumbar spinal fusion. A cohort study of complications, reoperations, and resource use in the Medicare population. *Spine,* vol 18, No 11 (Sep 1993), pp.(1463-1470), ISSN 0362-2436

Di Silvestre M, Lolli F, Bakaloudis G, Parisini P. (2010). Dynamic stabilization for degenerative lumbar scoliosis in elderly patients. *Spine,* vol 35, No 2 (Jan 2010), pp. (227–234), ISSN: 0362-2436

Dubois G, de Germay B, Schaerer N S & Fennema P (1999). Dynamic Neutralization: A New Concept for Restabilizarion of the Spine. *Lumbar Segmental instability.* Lipincott Marek Szpalski, Robert Gunzbrug and Malcolm H Pope. (23) pp (233-240), Lippincott Williams & Wilkins Healthcare, ISBN 0-7817-1906-2, Philadelphia, USA

Dunlop RB, Adams MA, Hutton WC. (1984). Disc space narrowing and the lumbar facet joints. *J Bone Joint Surg Br,* vol 66, No 5 (Nov 1984), pp.(706-710), ISSN 0301-620X

Eck JC, Humphreys SC, Lim TH, et al. (2002). Biomechanical study on the effect of cervical spine fusion on adjacent level intradiscal pressure and segmental motion. *Spine,* vol 27, No 22 (Nov 2002), pp. (2431–2434), ISSN: 0362-2436

Errico TJ. (2005) Lumbar disc arthroplasty. *Clin Orthop Relat Res,* vol 435 (Jun 2005), pp. (106-117), ISSN 0009-921X

Fekete TF & Porchet F (2010). Overview of disc arthroplasty – past, present and future. *Acta Neurochir,* vol 152, No 3 (Mar 2010), pp.(393-404), ISSN: 0942-0940

Floman Y, Millgram MA, Smorgick Y, et al. (2007). Failure of the Wallis interspinous implant to lower the incidence of recurrent lumbar disc herniations in patients undergoing primary disc excision. *J Spinal Disord Tech,* vol 20, No 5 (Jul 2007), pp. (337–341), ISSN: 1536-0652

France JC, Yaszemski MJ, Lauerman WC, et al. (1999). A randomized prospective study of posterolateral lumbar fusion. Outcomes with and without pedicle screw instrumentation. *Spine,* vol 24, No 6 (Mar 1999), pp. 553-560), ISSN 0362-2436

Freeman BJC & Daverport J (2006). Total disc replacement in the lumbar spine: a systematic review of the literature. *Eur Spine J,* vol 15 (Aug 2006) Suppl 3: (S439-447), ISSN 0940-6719

Frelinghuysen P, Huang RC, Girardi FP, et al (2005) Lumbar total disc replacement part I: rationale, biomechanics, and implant types. *Orthop Clin North Am,* vol 36, No 3 (Jul 2005), pp. (293–299), ISSN 0030-5898

Fritzell P, Hägg O, Nordwall A. (2003). Complications in lumbar fusion surgery for chronic low back pain: comparison of three surgical techniques used in a prospective randomized study. A report from the Swedish Lumbar Spine Study Group. *Eur Spine J,* vol 12, No 2(Apr 2003), pp. (178–189), ISSN 0940-6719

Fritzell P, Hagg O, Wessberg P, Nordwall A (2001) Volvo Award Winner in Clinical Studies: lumbar fusion versus nonsurgical treatment for chronic low back pain: a multicenter randomized controlled trial from the Swedish Lumbar Spine Study Group. *Spine,* vol 26, No 23 (Dec 2001), pp(2521–2532), ISSN: 0362-2436

Frymoyer J, Hanley E, Howe J, et al. (1979). A comparison of radiographic findings in fusion and nonfusion patients ten or more years following lumbar disc surgery. *Spine,* vol 4, No 5 (Sep-Oct 1979), pp. (435–440), ISSN: 0362-2436

Geisler FH, Blumenthal SL, Guyer RD, et al. (2004) Neurological complications of lumbar artificial disc replacement and comparison of clinical results with those related to lumbar arthrodesis in the literature: results of a multicenter, prospective, randomized investigational device exemption study of Charite intervertebral disc. Invited submission from the Joint Section Meeting on Disorders of the Spine and Peripheral Nerves. *J Neurosurg Spine,* vol 1, No 2 (Sep 2004), pp. (143–154), ISSN 1547-5654

Geisler FH, Guyer RD, Blumenthal SL, et al. (2008) Effect of previous surgery on clinical outcome following 1-level lumbar arthroplasty. *J Neurosurg Spine,* vol 8, No 2 (Feb 2008), pp(108–114), ISSN 1547-5654

Ghiselli G, Wang JC, Bhatia NN, et al. (2004). Adjacent segment degeneration in the lumbar spine. *J Bone Joint Surg Am,* vol86, No 7 (Jul 2004), pp.(1497-1503), ISSN: 1535-1386

Gottfried Y, Bradford DS, Oegema TR. (1986). Facet joint changes after chemonucleosis-induced disc space narrowing. *Spine,* vol 11, No 9 (Nov 1986), pp.(944-950), ISSN 0362-2436

Graf H. (1992). Lumbar instability: surgical treatment without fusion. *Rachis,* vol 412, pp.(123–137)

Greenough CG, Peterson MD, Hadlow S, et al. (1998) Instrumented posterolateral lumbar fusion. Results and comparison with anterior interbody fusion. *Spine,* vol 23, No 4 (Feb 1998), pp.(479–486), ISSN: 0362-2436

Greenough CG, Taylor LJ, Fraser RD (1994) Anterior lumbar fusion: results, assessment techniques and prognostic factors. *Eur Spine J*, vol 3, No 4 (1994), pp. (225–223), ISSN 0940-6719

Grevitt MP, Gardner AD, Spilsbury J, et al. (1995) The Graf stabilisation system: early results in 50 patients. *Eur Spine J*, vol 4, no 3 (1995), pp.(169-175), ISSN: 0940-6719

Guigui P, Chopin D. (1994). Assessment of the use of the Graf ligamentoplasty in the surgical treatment of lumbar spinal stenosis: a propos of a series of 26 patients. *Rev Chir Orthop Reparatrice Appar Mot*, vol 80, No 8 (1994), pp.(681-688), ISSN 0035-1040

Gunzburg R, Szpalski M. (2003). The conservative surgical treatment of lumbar spinal stenosis in the elderly. *Eur Spine J*, vol 12, No 2 (Oct 2003), pp. (S176-180), ISSN0940-6719

Hanley EN Jr. (1995). The indications for lumbar spinal fusion with and without instrumentation. *Spine*, vol 20, No 24 (Dec 1995), pp. (143S-153S), ISSN 0362-2436

Harms, J. and Tabasso, G (1999) *Instrumented Spinal Surgery*. by Georg Thieme Verlag, ISBN 85-86703-05-2. Germany

Hsu KY, Zucherman JF, Hartjen CA, et al. (2006). Quality of life of lumbar stenosis- treated patients in whom the X STOP interspinous device was implanted. *J Neurosurg Spine*, vol 5, No 6 (Dec 2006), pp. (500–507), ISSN 1547-5654

Huang RC, Girardi FP, Lim MR, et al. (2005). Advantages and disadvantages of Nonfusion Technology in Spine Surgery. *Orthop Clin North Am*, vol 36, No 3 (Jul 2005), pp. (263–269), ISSN 0030-5898

Huang RC, Tropiano P, Marnay T, et al. (2006) Range of motion and adjacent level degeneration after lumbar total disc replacement. *Spine J*, vol 6, No 3 (May-Jun 2006), pp. (242–247) ISSN: 1529-9430

Jansen ME, Marchesi DG. AO (2008) Evidence-based spine surgery. Special edition total disc replacement. *Lumbar spine*.vol.4 No .3.pp (6-12)

Jansson KA, Nemeth G, Granath F, et al. (2004). Surgery for herniation of a lumbar disc in Sweden between 1987 and 1999 an analysis of 27,576 operations. *J Bone Joint Surg Br*, vol 86, No 6 (Aug 2004), pp. (841-847),ISSN 0301-620X

Junghanns H, Schmoral G. (1968). Die gesunde und die kranke Wirbelsaule in Rontgenbild und Klinik. Thieme, 556 pages

Kabir SM, Gupta SR, Casey AT. (2010). Lumbar interspinous spacers: a systematic review of clinical and biomechanical evidence. *Spine*, vol 35, No 25 (Dec 2010), pp. (E1499-506), ISSN 0362-2436

Kanayama M, Hashimoto T, Shigenobu K, et al. (2005). Non-fusion surgery for degenerative spondylolisthesis using artificial ligament stabilization: surgical indication and clinical results. *Spine*, vol 30, No 5 (Mar 2005), pp. (588-592), ISSN 0362-2436

Kim YM, Kim DS, Choi ES, et al. (2011). Nonfusion method in thoracolumbar and lumbar spinal fractures. *Spine*, vol 36, No 2 (Jan 2011), pp. (170-176), ISSN 0362-2436

Kirkaldy-Willis WH, Farfan HF (1982). Instability of the lumbar spine. *Clin Orthop Relat Res*, vol 165, pp(110-123), ISSN 0009-921X

Kirkaldy-Willis WH, Wedge JH, Yong-Hing K, et al. (1978). Pathology and pathogenesis of lumbar spondylosis and stenosis. *Spine*, vol 3, No 4 (Dec 1978), pp.(319-328), ISSN: 0362-2436

Ko C. C. Tsai HW, Huang WC, et al. (2010) Screw loosening in the Dynesys stabilization system: radiographic evidence and effect on outcomes. *Neurosurg Focus*, vol 28, No 6 (June 2010), pp. (E10), ISSN 1092-0684

Kondrashov DG, Hannibal M, Hsu KY, et al (2007). X STOP versus decompression for neurogenic claudication: Economic and clinical analysis. *The Internet J Minimally Invasive Spinal Technology* . Vol 1. No 2. (2007) ISSN: 1937-8254

Kong DS, Kim ES, Eoh W. (2007). One-year outcome evaluation after interspinous implantation for degenerative spinal stenosis with segmental instability. *J Korean Med Sci*, vol 22 No 2 (Apr 2007), pp. (330–335), ISSN 1011-8934

Kornblum MB, Fischgrund JS, Herkowitz HN, et al. (2004) Degenerative lumbar spondylolisthesis with spinal stenosis: a prospective long-term study comparing fusion and pseudarthrosis. *Spine*, vol 29, No 7 (Apr 2004), pp. (726-733), ISSN: 0362-2436

Korovessis P, Repantis T, Zacharatos S, et al.(2009). Does Wallis implant reduce adjacent segment degeneration above lumbosacral instrumented fusion? *Eur Spine J*, vol 18, No 6 (Jun 2009), pp. (830–840), ISSN 0940-6719

Kozak JA, Heilman AE, O'Brien JP (1994) Anterior lumbar fusion options. *Clin Orthop Relat Res*, vol 300, (Mar 1994), p. (45–51), ISSN: 0009-921X

Kraemer J (2009). Intervertebral Disk Prostheses. In: Intervertebral disk diseases: causes, diagnosis, treatment and prophylaxis 3rd edition. Thieme,Stuttgart, New York, pp 297-303.

Kruton F (1944) The instability associated with disc degeneration in the lumbar spine. *Acta-Radio*, vol 25 (1944), pp. (593-609)

Kuchta J, Sobottke R, Eysel P, Simons P. (2009). Two-year results of interspinous spacer (X-Stop) implantation in 175 patients with neurologic intermittent claudication due to lumbar spinal stenosis. *Eur Spine J*, vol 18, No 6 (Jun 2009), pp. (823-829), ISSN 0940-6719

Kumar M, Jacquot F, Hall H. (2001). Long-term follow-up of functional outcomes and radiographic changes at adjacent levels following lumbar spine fusion for degenerative disc disease. *Eur Spine J*, vol 10, No 4 (Aug 2001), pp. (309–313), ISSN 0940-6719

Kusslich D, Ulstrom C, Michael C. (1991). The tissue origin of low back pain and sciatica: a report of pain response to tissue etimulation during operations on the lumbar spine using local anesthesia. *Orthop Clin North Am*, vol 22, No 2 (Apr 1991), pp. (283-301), ISSN: 0030-5898

Lauryssen C. Appropriate selection of patients with lumbar spinal stenosis for interspinous process decompression with the X STOP device. *Neurosurg Focus*. 2007 Dec 15;22(1):E5. ISSN 1092-0684

Lee C, Langrana N. (1984). Lumbosacral spinal fusion. A biomechanical study. *Spine*, vol 9, No 6 (Sep 1984), pp. (574-581), ISSN 0362-2436

Lee CK. (1988). Accelerated degeneration of the segment adjacent to a lumbar fusion. *Spine*, vol 13, No 3 (Mar 1988), pp. (375–377), ISSN: 0362-2436

Lehmann TR, Spratt KF, Tozzi JE, et al. (1987) Long-term follow-up of lower lumbar fusion patients. *Spine*, vol 12, No 2 (Mar 1987), p. (97–104), ISSN: 0362-2436

Luk KD, Lee FB, Leong JC, Hsu LC. (1987) The effect on the lumbosacral spine of long spinal fusion for idiopathic scoliosis. A minimum 10-year follow-up. *Spine*, vol 12, No 10 (Dec 1987), pp.(996-1000), ISSN: 0362-2436

Lumsden RM, Morris JM. (1968) An in vivo study of axial rotation and immobilization at lumbosacral joint. *J Bone Joint Surg*, vol 50, No 8 (Dec 1968), pp. (1591-1602), ISSN: 1535-1386

Maleci A, Sambale RD, Schiavone M, et al. (2011). Nonfusion stabilization of the degenerative lumbar spine. *J Neurosurg Spine*, vol 15, No 2 (Aug 2011), pp. (151-158) ISSN 1547-5654

Mayer HM (1998) Microsurgical anterior approaches for anterior interbody fusion of the lumbar spine. In: McCulloch JA, Young PH (eds) Essentials of spinal microsurgery. Lippincott-Raven,Philadelphia, pp 633–649

Mayer HM (2005) Total lumbar disc replacement. *J Bone Joint SurgBr*, vol 87, No 8 (Aug 2005), pp. (1029–1037), ISSN 0301-620X

Mayer HM , Korge A; (2002) Non-fusion technology in degenerative lumbar spinal disorders: facts, questions, challenges. *Eur Spine J*, vol 11, No.2 (Oct 2002), pp. (S85–S91), ISSN 0940-6719

McCullen GM, Yuan HA (2003) Artificial disc: current developments in artificial disc replacement. *Curr Opin Orthop;* vol 14, no 3 (Jun 2003), pp. (138–1430), ISSN: 10419918

Min J, Jang J, Jung B, et al. (2008). The clinical characteristics and risk factors for the adjacent segment degeneration in instrumented lumbar fusion. *J Spinal Disord Tech*, vol 21, No 5 (Jul 2008), pp. (305-309), ISSN 1536-0652

Minns RJ, Walsh WK. (1997). Preliminary design and experimental studies of a novel soft implant for correcting sagittal plane instability in the lumbar spine. *Spine*, vol 22, No 16 (Aug 1997), pp. (1819-1825), ISSN 0362-2436

Miyakoshi N, Abe E, Shimada Y, et al. (2000). Outcome of one-level posterior lumbar interbody fusion for spondylolisthesis and postoperative intervertebral disc degeneration adjacent to the fusion. *Spine*, vol 25, No 14 (Jul 2000), pp.(1837-1842). ISSN 0362-2436

Modic M, Pavlicek W, Weinstein M, et al. (1984). Magnetic ressonance imaging of the intervertebral disc disease. Clinical and pulse sequence considerations. *Radiology*, vol 152, No 1 (Jul 1984), p(103-111), ISSN 0033-8419

Modic M, Ross J. (1991). Magnetic ressonance imaging in the evaluation of low back pain. *Orthop Clin North Am*, vol 22, No 2 (Apr 1991), pp. (283-301), ISSN 0030-5898

Morgan FP, King T. (1957) Primary instability of lumbar vertebrae as common cause of low back pain. *J Bone Joint Surg*, vol 39, No 1 (1957), pp. (6-22), ISSN: 0301-620X

Mulholand RC, Sengupta DK. (2002). Rationale, principles and experimental evaluation of the concept of soft stabilization. *Eur Spine J*, vol 11, No 2 (Oct 2002), pp. (S198-205), ISSN 0940-6719

Olsewski JM, Schendel MJ, Wallace LJ, et al. (1996). Magnetic resonance imaging and biological changes in injured intervertebral discs under normal and increased mechanical demands. *Spine*, vol 21, No 17 (Sep 1996), pp.(1945–1951), ISSN: 0362-2436

Panjabi M, Henderson G, Abjornson C, et al. (2007). Multidirectional testing of oneand two-level ProDisc-L versus simulated fusions. *Spine*, vol 32, No 12 (May 2007), pp. (1311-1319), ISSN: 0362-2436

Panjabi M, Malcolmson G, Teng E, et al. (2007). Hybrid testing of lumbar CHARITE discs versus fusions. *Spine*, vol 32, No 9 (Apr 2007), pp. (959-966), ISSN: 0362-2436

Panjabi MM. (1992). The stabilizing system of the spine. Part II. Neutral zone and instability hypothesis. *J Spinal Disord*, vol 5, No 4 (Dec 1992), pp. (390-396), ISSN 0895-0385

Phillips FM, Reuben J, Wetzel FT. (2002). Intervertebral disc degeneration adjacent to a lumbar fusion. An experimental rabbit model. *J Bone Joint Surg Br.*, vol 84, No 2 (Mar 2002), pp.(289-294), ISSN 0301-620X

Putzier M, Schneider SV, Funk JF, et al. (2005). The surgical treatment of the lumbar disc prolapse: nucleotomy with additional transpedicular dynamic stabilization versus nucleotomy alone. *Spine*, vol 30, No 5 (Mar 2005), pp. (E109-114), ISSN 0362-2436

Rahm MD, Hall BB. (1996). Adjacent segment degeneration after lumbar fusion with instrumentation: a retrospective study. *J Spinal Dis*, vol 9, No 5 (Oct 1996), pp. (392-400), ISSN: 0895-0385

Rao R, David K, Wang M.(2005). Biomechanical changes at adjacent segments following anterior lumbar interbody fusion using tapered cages. *Spine*, vol 30, No 24 (Dec 2005), pp. (2772-2776), ISSN: 0362-2436

Rodrigues LF, Voloch P, Gurgel S, Cavallari F. (2010). Retrospective evaluation by SF-36 questionnaires of patients submitted to pedicular dynamic stabilization for treatment of degenerative lumbar diseases. *Coluna/Columna*, Vol.9. No2. (Abr/jun 2010), pp.(104-112), ISSN: 1808-1851

Roh JS, Teng AL, Yoo JU, Davis J, Furey C & Bohlman HH (2005). Degenerative Disorders of the Lumbar and Cervical Spine. *Othop Clin N Am*, vol 36, No 3 (Jul 2005), pp. (255-262), ISSN: 0030-5898

Rohlmann A, Neller S, Bergmann G, et al. (2001). Effect of an internal fixator and a bone graft on intersegmental spinal motion and intradiscal pressure in the adjacent regions. *Eur Spine J.*, vol10, No 4 (Aug 2001), pp. (301-308), ISSN 0940-6719

Schaeren S, Broger I, Jeanneret B. (2008). Minimum four-year follow-up of spinal stenosis with degenerative spondylolisthesis treated with decompression and dynamic stabilization. *Spine*. Aug 15;33(18):E636-42. ISSN 0362-2436

Schnake KJ, Schaeren S, Jeanneret B. (2006). Dynamic stabilization in addition to decompression for lumbar spinal stenosis with degenerative spondylolisthesis. *Spine*, vol 31, No 4 (Feb 2006), pp. (442-449), ISSN 0362-2436

Schwarzenbach O, Berlemann U, Stoll T M, et al (2005). Posterior Dynamic Stabilization System: DYNESYS. *Orthop Clin N Am*, vol 36, No 3 (Jul 2005), pp. (363-372), ISSN 0030-5898

Sears W R Sergides IG, Kazemi N, et al (2011). Incidence and prevalence of surgery at segments adjacent to a previous posterior lumbar arthrodesis. *Spine J*, vol 11, No 1 (Jan 2011), pp.(11-20), ISSN 1529-9430

Sénégas J, Etchevers JP, Baulny D, et al (1988) Widening of the lumbar vertebral canal as an alternative to laminectomy, in the treatment of lumbar stenosis. *Fr J Orthop Surg* No 2. pp. (93-99). ISSN: 1749-799X

Sénégas J, Vital JM, Pointillart V, et al. (2009). Clinical evaluation of a lumbar interspinous dynamic stabilization device (the Wallis system) with a 13-year mean follow-up. *Neurosurg Rev*, vol 32, No 3 (Jul 2009), pp. (335–341), ISSN 0344-5607

Sénégas J. (2002). Mechanical supplementation by non-rigid fixation in degenerative intervertebral lumbar segments: the Wallis system. *Eur Spine J*, vol 11, No 2 (Oct 2002), pp.(S164-169), ISSN 0940-6719

Sénégas J. (2008). Dynamic lumbar stabilization with the Wallis interspinous implant. *Interactive Surgery*. vol 3, No 4 (2008), pp. (221-228), ISSN 1778-3968

Sengupta DK. (2004). Dynamic stabilization devices in the treatment of low back pain. *Orthop Clin North Am*, vol 35, No 1 (Jan 2004), pp. (43-56), ISSN: 0030-5898

Shim CS, Lee SH, Shin HD, et al (2007) CHARITE versus ProDisc: a comparative study of a minimum 3-year follow-up. *Spine*, vol 32, No 9 (Apr 2007), pp. (1012–1018), ISSN 0362-2436

Shono Y, Kaneda K, Abumi K, et al. (1998). Stability of posterior spinal instrumentation and its effects on adjacent motion segments in the lumbosacral spine. *Spine*, vol 23, No 14 (Jul 1998), pp. (1550–1558), ISSN 0362-2436

Siepe CJ, Mayer HM, Heinz-Leisenheimer M, et al (2007) Total lumbar disc replacement: different results for different levels. *Spine*, vol 32, No 7 (Apr 2007), pp. (782–790), ISSN 0362-2436

Stokes I, Wilder D, Frymoyer J, Pope M. (1981). 1980 Volvo award in clinical sciences. Assessment of patients with low-back pain by biplanar radiographic measurement of intervertebral motion. *Spine*, vol 6, No 3 (May-Jun 1981), pp. (233-240), ISSN 0362-2436

Strempel A, Moosmann D, Stoos C, Martin A. (2007). Posterior nonfusion stabilization of the degenerated lumbar spine with cosmic. In: Szpalski M, Gunzburg R, Le Huec JC, Brayada-Bruno M, eds. Nonfusion techniques in spine surgery. Philadelphia: Lippincott Williams & Wilkins; 2000:410-430 ISBN: 0781769728.

Strempel,A., Moosmann, D., Stoss, C., et al. (2006). Stabilization of the degenerated lumbar spine in the non fusion technique with Cosmic posterior dynamic system *World Spine J*, vol. 1, No 1 (2006), pp.(40-47)

Stremple, A. (2008). Dynamic stabilisation: cosmic system. *Interact Surg*, Vol 3, No 4(2008),pp.(229–236). ISSN: 1778-3968

Sudo H, Oda I, Abumi K, et al. (2006). Biomechanical study on the effect of five different lumbar reconstruction techniques on adjacent-level intradiscal pressure and lamina strain. *J Neurosurg Spine*, vol 5, No 2 (Aug 2006) pp. (150–155), ISSN 1547-5654

Szpalski M, Gunzburg R, Mayer M. (2002). Spine arthroplasty: a historical review. *Eur Spine J*, vol 11, No 2 (Oct 2002), pp.(S65-84), ISSN 0940-6719

Trautwein FT, Lowery GL, Wharton ND, et al. (2010). Determination of the in vivo posterior loading environment of the Coflex interlaminar-interspinous implant. *Spine J*, vol 10, No 3 (Mar 2010) pp. (244–251), ISSN 1529-9430

Tropiano P, Huang RC, Girardi FP, et al (2005) Lumbar total disc replacement. Seven to eleven-year follow-up. *J Bone Joint Surg Am*, vol 87, No 3 (Mar 2005), pp. (490–496), ISSN 0021-9355

Turner JA, Ersek M, Herron L, et al. (1992). Surgery for lumbar spinal stenosis: attempted meta-analysis of the literature. *Spine*, vol 17, No 1 (Jan 1992), pp. (1-8), ISSN 0362-2436

Urban JPG, Roberts S. (2003). Degeneration of the intervertebral disc. *Arthritis Res Ther*, vol 5, No 3 (2003), pp. (120-130), ISSN 1478-6354

van Ooij A, Oner FC, Verbout AJ (2003) Complications of artificial disc replacement: a report of 27 patients with the SB Charite disc. *J Spinal Disord Tech*, vol 16, No 4 (Aug 2003), pp.(369–383), ISSN 1536-0652

Verbiest H (1954). A radicular syndrome from developmental narrowing of the lumbar vertebral canal. *J Bone Joint Surg Br*, vol 36-B, No 2 (May 1954), pp. (230-237), ISSN: 0301-620X

Weinhoffer S, Guyer R, Herbert M, et al. (1995). Intradiscal pressure measurements above an instrumented fusion. A cadaveric study. *Spine*, vol 20, No 5 (Mar 1995), pp.(526–531), ISSN 0362-2436

Weinhoffer SL, Guyer RD, Herbert M, et al. (1995) Intradiscal pressure measurements above an instrumented fusion. A cadaveric study. *Spine*, vol 20, No 5 (Mar 1995), pp.(526-531), ISSN 0362-2436

Weinstein JN, Tosteson TD, Lurie JD, et al. (2008). SPORT Investigators: Surgical versus nonsurgical therapy for lumbar spinal stenosis. *N Engl J Med*, vol 358, No 8 (Feb 2008), pp.(794-810), ISSN: 0028-4793

White A.A., Panjabi M.M (Eds.), Clinical biomechanics of the spine, 2nd ed, JB Lippincott, Philadelphia, PA, 1990

Wiltse L, Radecki S, Biel H, et al.(1999) Comparative study of the incidence and severity of degenerative change in the transition zones after instrumented versus noninstrumented fusions of the lumbar spine. *J Spinal Disord*, vol 12, No 1 (Feb 1999), pp.(27–33), ISSN 0895-0385

Wiltse LL, Spencer CW. (1988). New uses and refinements of the paraspinal approach to the lumbar spine. *Spine*, vol 13, No 6 (Jun 1988), pp. (696-706) ISSN 0362-2436

Wippermann BW, Schratt HE, Steeg S, et al. (1997). Complications of spongiosa harvesting of the ilial crest. A retrospective analysis of 1,191 cases. *Chirurg*, vol 68, No 12 (Dec 1997), pp. (1286-1291), ISSN 0009-4722

Yajun W, Yue Z, Xiuxin H, Cui C . A meta-analysis of artificial total disc replacement versus fusion for lumbar degenerative disc disease. *Eur Spine J*, vol 19, No 8 (Aug 2010), pp. (1250–1261), ISSN 0940-6719

Yang J, Lee J, Song H. (2008). The impact of adjacent segment degeneration on the clinical outcome after lumbar spinal fusion. *Spine*, vol 33, No 5 (Mar 2008) pp. (503–507), ISSN 0362-2436

Yi X, McPherson B. (2010). Application of X Stop device in the treatment of lumbar spinal stenosis. *Pain Physician*, vol 13, No 5 (Sep-Oct 2010), pp.(E327-326), ISSN 1533-3159

Yu S, Haughton VM, Ho PS, Sether LA Wagner M, Ho KC.(1988). Progressive and regressive changes in the nucleus pulposus II. The adult. *Radiology*, vol 169, No 1 (Oct 1988), pp. (93-97) ISSN 0033-8419

Zhang YH, Zhao CQ, Jiang LS, et al (2008). Modic changes: a systematic review of the literature. *Eur Spine J*, vol 17, No 10 (Oct 2008), pp. (1289-1299), ISSN 0940-6719

Zigler JE. (2004) Spinal Jn Lumbar spine arthroplasty using the ProDisc II. *Spine J*, vol 4, No 6 (Nov-Dec 2004), pp.(260S-267S), ISSN 1529-9430

Zucherman J, Hsu K, Hartjen C, et al. (2004). A prospective randomized multicenter study for the treatment of lumbar spinal stenosis with the X STOP interspinous implant: 1-year results. *Eur Spine J*, vol 13, No 1 (Feb 2004), pp.(22–31), ISSN 0940-6719

Zucherman JF, Hsu KY, Hartjen CA, et al. (2005). A multicenter, prospective, randomized trial evaluating the X STOP interspinous process decompression system for the treatment of neurogenic intermittent claudication. Two year follow-up results. *Spine*, vol30, No 12 (Jun 2005),. pp.(1351–1358), ISSN 0362-2436

10

Posterior Dynamic Stabilization: The Interspinous Spacer

Antoine Nachanakian,
Antonios El Helou and Moussa Alaywan
Division of Neurosurgery, Saint George Hopsital University Medical Center, Balamand Univerity, Lebanon

1. Introduction

Neurogenic claudication was first related to lumbar stenosis (Verbiest H, 1954). Since, decompressive surgery was indicated in patients who failed to respond to conservative therapy. It is followed by posterior fusion in cases where the motion segment showed instability or in case where a secondary instability is expected to develop after surgery. However, while many patients have benefited from fusion procedures, successful fusion has not always been accompanied by clinical improvement (Adelt D et.al, 2007)

Evidence is growing that fusion may in fact have undesirable long-term effects on the remainder of the spine, particularly on the immediately adjacent motion segments(Kong D-S et.al, 2007; Etebar S, Cahil DW, 1999). This adjacent-level degeneration is typically seen rostral to a fused segment but may also occur caudal to a fusion, especially when the fusion occurs at the L4–5 level. The phenomenon is thought to be due to the altered biomechanics of the fused spine, where in abnormal forces acting upon the intervertebral discs and facet joints adjacent to the fused segment precipitate the accelerated failure of these stabilizing elements (Kanayama M et.al, 2001). From this evidence for adjacent-segment degeneration emerged the concept of "dynamic" or nonfusion stabilization of the lumbar spine.

2. Rational

Posterior dynamic stabilization, in which pedicle screw fixation is coupled with a flexible longitudinal connecting system, presumably allows for the normalization of intersegmental motion (Kaech DL et.al, 2001). This stands in contrast to traditional fusion surgery, in which the goal is complete and immediate elimination of motion and, ultimately, arthrodesis (Kaech DL et.al, 2000). While both strategies seek to address the underlying pathology of microinstability, the dynamic stabilization approach promises to do so in a more physiological manner. By "restoring" normal motion, mobility is theoretically preserved rather than eliminated, and the forces acting above and below the construct are altered to a lesser extent, reducing the potential undesirable effects of fusion (Kaech DL, Jinkins JR, 2002). Recently, new concepts, such as soft stabilization, dynamic stabilization, and motion preservation, have been explored as alternative treatment options to lumbar fusion.

Interspinous process spacers have been introduced as a possible alternative to spinal decompression and fusion for the treatment of neurogenic intermittent claudication (NIC) and discogenic lower back pain (Bowers C et.al, 2010). The interspinous devices distract the Neural Foramen, unload the intervertebral disc, and limit spinal extension, improving central canal and foraminal stenosis. Interspinous Distracter (ISD) is designed to stabilize the motion segment after neural elements decompression in lumbar stenosis, tolerating flexion and extension in this segment thus preserving the adjacent segment from deterioration.

The first interspinous device, the Wallis system (Abbott Spine), was developed in 1986 and used in patients with recurrent disc herniation. It was found to improve outcome in patients who underwent a second discectomy incorporating the Wallis device (Mariottini A, 2005). The second generation of the Wallis implant, made with elastic polyetheretherketone (PEEK), has been shown to reduce pain severity in patients with mild to moderate disc degeneration, lateral recess, central spinal stenosis, and significant lower back pain when used in combination with other surgical interventions. Other interspinous spacers used in Europe but not approved for use in the US include the DIAM (Medtronic Sofamor Danek) and the Coflex (Paradigm Spine) (Mariottini A, 2005; Sénégas J, 2002). The X-Stop device (St. Francis Medical Technologies) was approved by the US FDA in November 2005 and has been shown to be superior to nonoperative therapy in patients with NIC (Zucherman JF, 2004).

While the time course and prevalence of adjacent-segment disease are not fully known, there is increasing evidence in the spine literature that its effects may be seen soon after fusion surgery and in as many as 30% of patients(Christie SD et.al, 2005). In a recently published large retrospective analysis (Chen et.al, 2001) reported rate of clinical adjacent-segment disease was 30.3% and showed that patients in whom adjacent-level disease developed had significantly worse Oswestry Disability Index scores than those without adjacent-level disease. They further identified age > 50 years at time of surgery, increasing length of fusion, and extension of the fusion to L1–3 as significant risk factors for the development of adjacent level disease. No significant difference was identified between posterior and circumferential fusion.

Our experience is based on 87 cases performed between September 2008 and January 2011 with different lumbar spine pathologies (Table1). The ages of our patient were between 45 and 70 years, with a mean age of 55 years. All patients were treated with Interspinous Distracter (ISD).

Number of cases	Pathology	Male / female ratio
33	Bi-Foraminal stenosis	24/9
21	Lig. flavum hypertrophy	6/15
18	Suspended vertebrae	12/6
9	Facet syndrome	All females
6	Adjacent syndrome	3/3

Table 1. Pathology of the patients at the time of presentation.

3. Decision making

3.1 Diagnostic criteria

Preoperative patient evaluation included plain lumbar film, lumbar MRI, lumbar CT and lumbar osteodensitometry.

MRI assesses the canal and foramina stenosis and the joint synovium. In the latter, weakness or absence of the intracapsular (intra-articular) white signal on T2 weighted sequences signals is characteristic of degenerative disease of the joint.

Lumbar CT assesses the lumbar facets.

All patients with confirmed osteoporosis by osteodensitometry were excluded from this type of treatment.

3.2 Indications

Patients are eligible for enrolment if they have:

- Degenerative disk disease and subsequent bilateral foraminal stenosis (Figure 1).

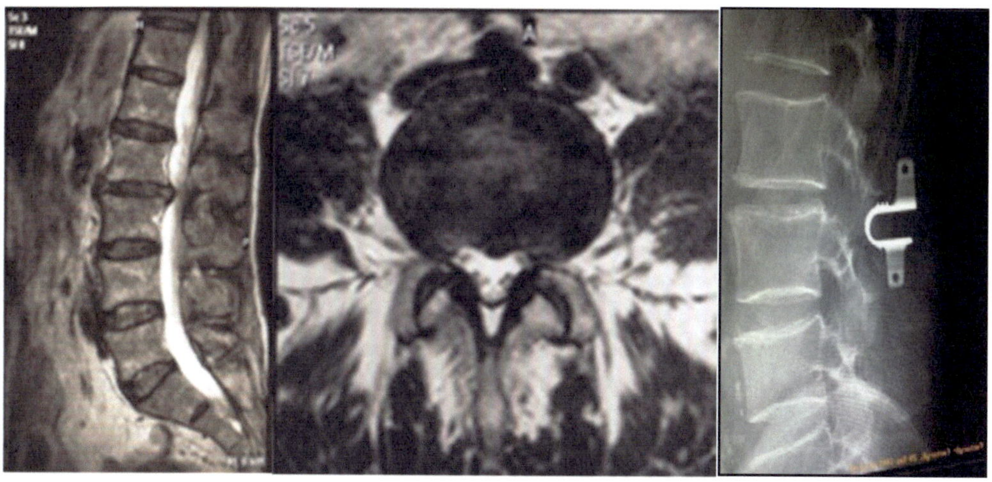

Fig. 1. Biforaminal stenosis at the level of L2-L3

- Foramino-canalar stenosis, due to ligamentum flavum hypertrophy, declare symptoms consisting of bilateral lower limb paresthesia upon walking.
- Low back pain irradiating to both lower limbs due to suspended vertebra was shown mainly due to facet degenerative disease (Figure 2).
- Facet joint syndrome (Figure 3).
- Adjacent segment syndrome which refers to degenerative changes that occurs in the mobile segment next to spinal fusion (Schlegel JD, et.al 1996). It's exact mechanism remains uncertain, but fusion technique specifically shifts the center of rotation leading to increase stress on the facets and/or disc of the adjacent mobile segment. It increases mobility of the adjacent segment, and the intradiscal pressure immediately neighboring a fused segment. And so, it can lead to disc degeneration. Finally, posterior dynamic

stabilization is done to decrease and/or avoid the harmful effects of rigid fusion like, listhesis, instability, hypertrophic facet joint arthritis, herniated nucleus pulposus, and stenosis.

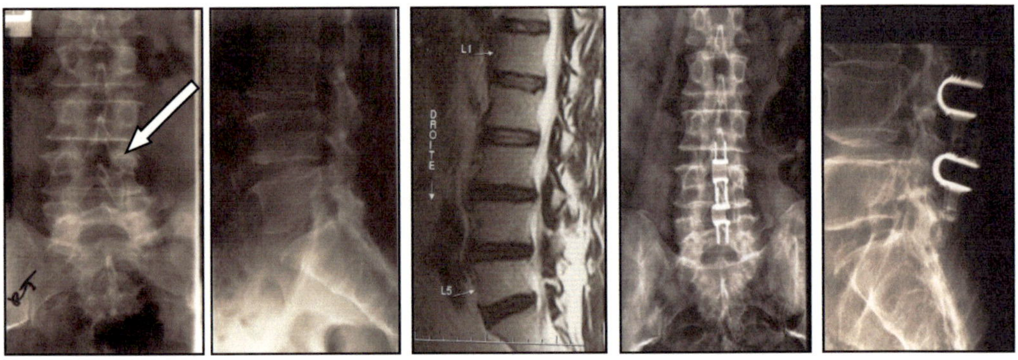

Fig. 2. L4 **Suspended vertebra** treated with 2 interspinous spacers L3-L4 and L4-L5, fixing the L4.

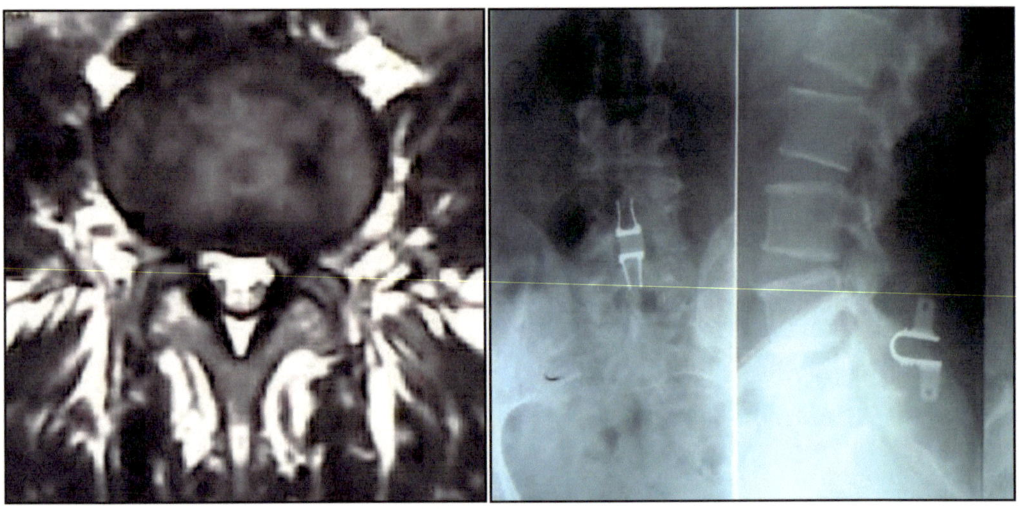

Fig. 3. Facet syndrome: lack of synovial fluid at the level of L4-L5 joint; a sign of degeneration.

3.3 Contra indications

- Presence of lumbar stenosis of more than 2 adjacent levels
- The level of stenosis above L1-L2 level or below L4-L5 level
- Have a fracture of the spinous process of the stenotic level
- Operated previously by a laminectomy with removal of the spinous process

- Degenerative and congenital spondylolisthesis
- Osteoporosis defined by the WHO as thinning of bone tissue and loss of bone density over time

3.4 Hospitalization and follow up

In our institution, patients are admitted for 24 hours before the day of the procedure. In pre-op, the patient is asked to hold anti-platelets and other anti-coagulation for 5 days. Only LMWH can be tolerated up to 12 hours before the procedure.

Post op, spinal cord X-ray of the operated region is done. The patient is ambulated with abdominal belt for 1 month.

Usually, the patients had regular follow up at 1, 3 and 12 months.

3.5 Biomechanics of interspinous spacer:

Several models of interspinous distracter (ISD) have been proposed to stabilize the spine tolerating in the same time a certain degree of mobility of the concerned motion segment and preserving the adjacent segment from later damage. Biomechanical studies show that those devices offer a non-rigid fixation and can return a destabilized specimen back to the intact condition in terms of motion in flexion/extension and axial rotation (Samani J, 2002; Tsai KJ et.al, 2006) . It is a biomechanical alternative to a total laminectomy with pedicle screw and rod fixation (Lee CK, 1988).

Furthermore the implant does not significantly change the intradiscal pressures at the adjacent levels, yet it significantly unloads the intervertebral disc at the instrumented level in the neutral and extended positions (Vena P et.al, 2005).

Thus the characteristics of those devices meet the profile needed for cases where minor to moderate instability is expected in the treatment of lumbar stenosis preventing as well a future deterioration of the adjacent motion segment.

3.6 Surgery

3.6.1 Operative technique

3.6.1.1 Preparation

The procedure is done under general anesthesia. All patients were operated in a prone position, flexed on a Wilson surgical frame with the thoracolumbar spine segment in neutral to a slightly kyphotic position, avoiding hyperlordosis for a better interspinous distraction.

3.6.1.2 Product used

Different interspinous spacers types, the DIAM (Medtronic Sofamor Danek), the Coflex (Paradigm Spine) and The X-Stop device (St. Francis Medical Technologies), are used in our institution (Figure 4).

3.6.1.3 The instrument used

A set of lumbar laminectomy is used. In addition, a set of interspinous spacer measurer is utilized to define the depth and width of the spacer to be used.

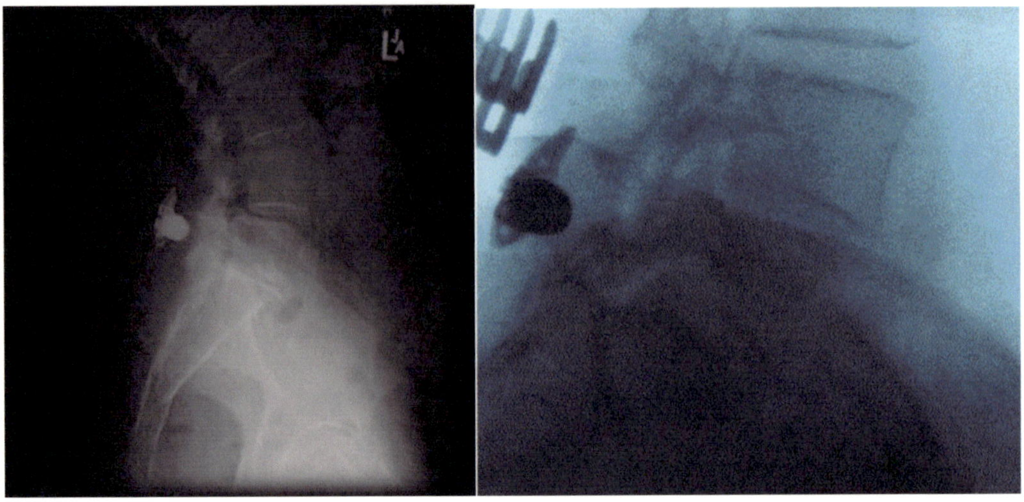

Fig. 4. Right case of DIAM spacer. Left case of X-Stop spacer.

3.6.2 Operative procedure

The level of the procedure is localized under fluoroscopy after positioning. Middline vertical skin incision is done. Dissection of the subcutaneous layer and the paraspinal muscles until identification of the articular facets.

The decompression of the neural elements for stenosis is made through surgical interlaminar fenestrations with flavectomy and opening of the lateral recess, and not by the old-fashioned laminectomy. For the insertion of the ISD the interspinous ligament is resected with temporary disinsertion and retraction of the supraspinous ligament.

Adequate preparation of the interspinous space; removal of all soft tissues and flattening of the bony walls to a straight parallel nidus were ended with an adequate insertion.

Proper depth of the incorporation of ISD was determined following direct spacing of 3-4 mm between the deepest point of the device and the dural sac placing through that space a midsize hook.

One or two interspinous spaces were treated according to the preoperative plan. ISD were inserted, the laterally retracted supraspinous ligament was always stitched to its initial location at the top of the spinous processes (Figure 5).

Discectomy is performed in cases where the protruded/ herniated disc is still compressing the root(s) despite the ligaments resection and the bone recalibrations are done. In cases where the disc is protruded/ herniated, medially dissectomies was not done.

Regular closure of layers and placing of deep hemovac drain ended the surgery.

3.6.3 Post-operative care

The patient is out of bed the day after surgery and discharged on day 3 after surgery, or on day 2 when drain was not inserted.

Fig. 5. Interspinous distracter implanted between spinous processes L3-L4 and L4-L5, the desinserted and retracted laterally supraspinous ligament, will be sutured to its initial place.

Control lumbo-sacral x-ray is done in 2 views to evaluate the created distraction. All patients were put in a lumbar brace, for a period of one month during their daily activities (Figure 6).

3.7 Long term results

Overall improvement was noted in ISD-treated patients, with considerable satisfaction in 75 % of patients on average. The patient at first reported an improvement of their radicular pain with a mean reduction of 3/10 on visual analog scale (VAS) (scale for 0: absent pain to 10 severe intolerable pain necessitating Intra venous treatment). In the pre-operative, radicular pain when existent had a mean score of 8.3/10 on VAS. Whereas, in the immediate post op period, the pain was at 4.9/10 on VAS.(Table 2).

Postoperative walking distance progressively increased during the next 3 months. Patients achieved maximum improvement after an average period of 6 months, with a mean score of 2.1/10 on VAS, and up to 45% of patients were pain free.

The prominent characteristic of this surgery is a low level of postoperative pain.

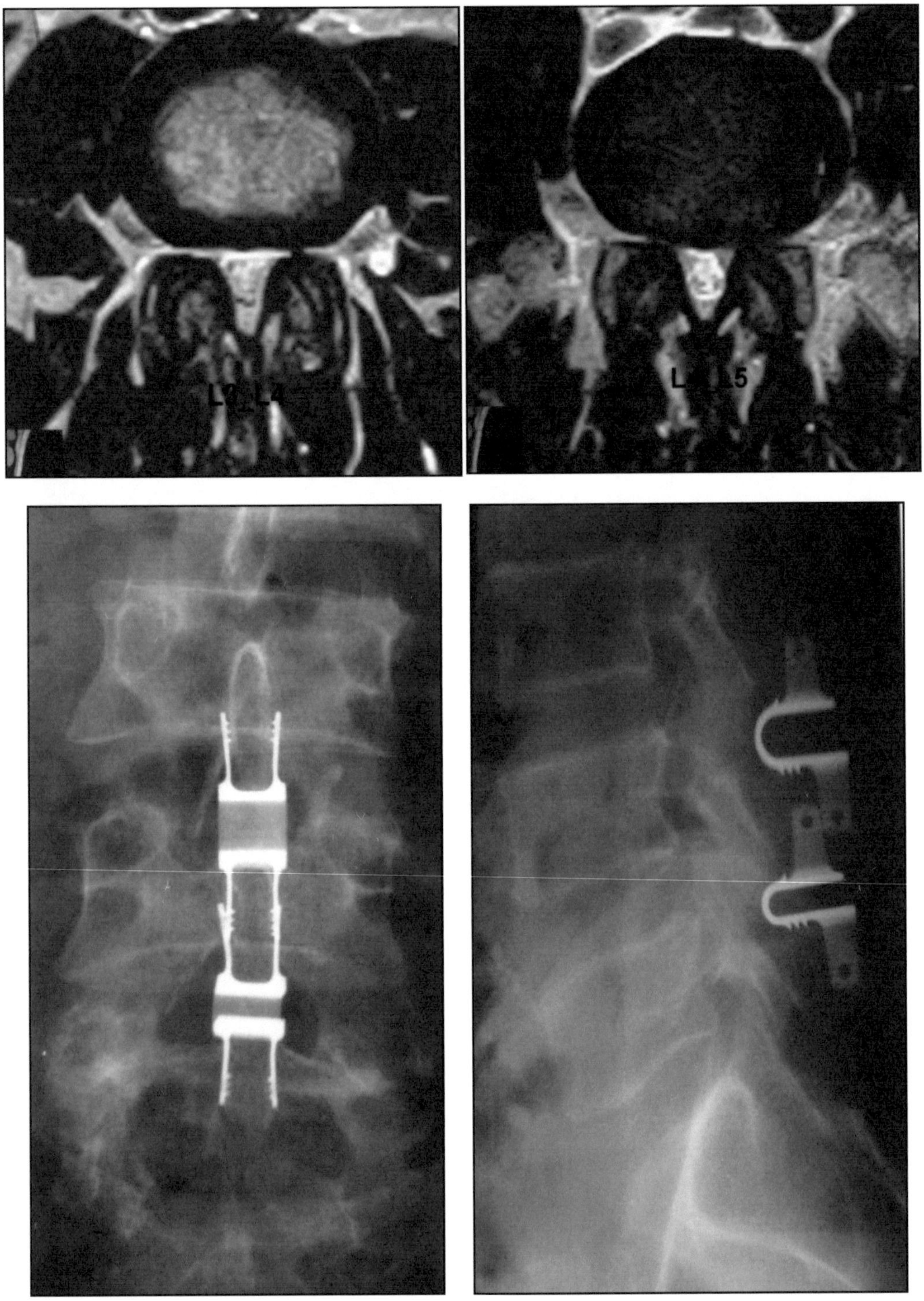

Fig. 6. Top: Bi foraminal stenosis more so at the level L4-L5 then at L3-L4 levels. Bottom: X-ray of adjacent interspinous spacer.

Cases	Improvement after treatment with ISD
Bi-Foraminal stenosis	73 %
Lig. flavum hypertrophy	83 %
Suspended vertebrae	67 %
Facetar syndrome	66 %
Adjacent syndrome	96 %

Table 2. Percentage of improvement in each case.

In postoperative scanning follow-up of the patients examined, a mineralization of the spinous process in contact with the implant was found, in particular at its base which appears to absorb high stresses due to lordosis (Sawnson KE et. al, 2003).

Patients are improved in all their clinical aspects: low back pain, radicular pain and walking distance. Moderate to severe low back pain improved in 75% of patients, leg pain and claudication were improved in 87% and walking distance improved in 74% of the patients (Park S-C et.al, 2009). Patient satisfaction is 89%. These results were achieved by 1 year and did not deteriorate over the long-term.

As described above the procedure has minimal post-operative back pain. And so, after the decompression done by removal of ligamentum flavum and the re-establishment of normal dynamics of the spine play a major role in the resolution of back pain. Restoration of the height of the intervertebral disc relieve the pressure on the sino-vertebral nerve which plays a major role in decreasing paraspinal muscles spasm and though the back pain. In general, back pain evaluated by the VAS with a mean score of 8.2/10 preoperatively resolves in 78 % of patients immediately.

3.8 Complications

In general, material are well tolerated. The rate of complications is between 1 and 10 % over all. Two sets of complications exists; the early and the delayed.

Early complications include device dislocation/malposition, spinous process fractures, erosion of the spinous process, infection, hematoma, and neurological sequelae.

One case of migration was observed in one series (Dieter A et.al, 2007). There were no broken or permanently deformed implants in all series.

3.8.1 How to avoid complications

3.8.1.1 Fracture of the spinous process

A potential complication of placement of an ISP device is fracture of the spinous processes, particularly related to osteopenic patients.

In our experience, we do osteodensitometry for all patients to assess bone density in pre-op. During operation, we should avoid bone erosions of the adjacent spinous processes.

Intraoperative spinous process fracture occurred in < 1% of all cases.

Patient with the delayed spinous process fracture at more than a year had what has recently been referred to as a "sandwich phenomenon" fracture of the middle spinous process in adjacent double-level ISD placement.

3.8.1.2 Neurological manifestations and reoperation

Recurrent symptoms required reoperation, microsurgical decompression and posterolateral fusion in 1.2% of cases (Figure 7).

To avoid this type of complications, a complete posterior decompression through ligamentum flavum excision and discectomy in the presence of herniated disc should be done.

Selection of patient without spondylolysthesis is mandatory to avoid postero-lateral fusion later on.

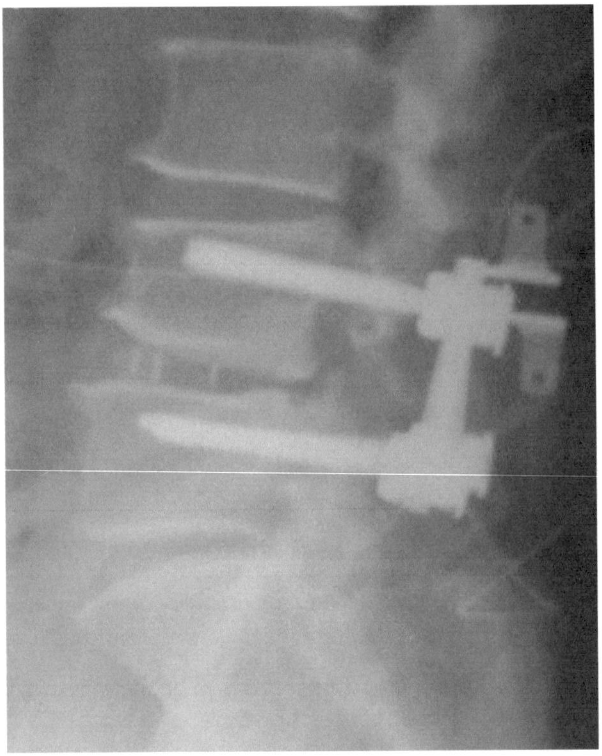

Fig. 7. Adjacent segment syndrome treated by interspinous distracter

4. Conclusion

Interspinous spacer after surgical decompression for spinal stenosis by excision of Ligamentum flavum demonstrates excellent short term and long term results for improvement in back pain, neurogenic claudication and patient satisfaction. It provides restoration of disc height and reduction of vertebral slip. It helps in restoring normal lumbar mobility and decrease nerve roots compression. It offers an alternative to rigid stabilization for lumbar stenosis with mild to moderate instability. If indicated, its use is favored compared to rigid stabilization because it preserve motion and has less harmful effect compared to the risk of failed back syndrome.

5. References

Chen CS, Cheng CK, Liu CL, Lo WH: Stress analysis of the disc adjacent to interbody fusion in lumbar spine. *Med Eng Phys* 23:483–491, 2001

Christian Bowers, M.D., Amin Amini, M.D., M.Sc., Andrew T. Dailey, M.D. and Meic H. Schmidt, M.D.Dynamic interspinous process stabilization: review of complications associated with the X-Stop device. *Neurosurg Focus* 28 (6):E8, 2010

Christie SD, Song KK, Fessler, RG. "Dynamic Interspinous Process Technology." *Spine* 30-2005(16S):S73-78

Dieter Adelt, MD, Jacques Samani, MD, Woo-Kyung Kim, MD, PhD, Marcus Eif, MD, Gary L.Lowery, MD, Phd and Robert J. Chomiak, MS. CoflexTM Interspinous Stabilisation: Clinical and Radiographic results from an international multicenter retrospective study. *Paradigm spine journal*.1-2007:1

Doo-Sik Kong MD, Eun-Sang Kim MD, PhD, Whan Eoh, MD, PhD. One year outcome evaluation after interspinous implantation for degenerative spinal stenosis with segmental instability. *Paradigm spinal journal*.2-2007:1

Etebar S, Cahil DW. Risk factor for adjacent-segmental failure following lumbar fixation with rigid instrumentation for degenerative instability. *J Neurosurg* 1999; 90:163-169.

Kanayama M, Hashimoto T, Shigenobu K, Harada M, Oha F, Ohkoshi Y, Tada H, Yamamoto K, Yamane S. Adjacent-segment morbidity after graft ligamentoplasty compared with posterolateral lumbar fusion. *J Neurosurg* 2001; 95:5-10

Kaech DL, Fernandez C, Lombardi-Weber D. "The Interspinous U: A New Restabilization Device for the Lumbar Spine." *Spinal Restabilization Procedures*, 2000;30:355-362

Kaech DL, Fernandez C, Haninec P. "Preliminary Experience with the Interspinous U Device." *Rachis* 2001; 13:303-304

Kaech DL, Jinkins JR. The interspinous 'U': a new restabilization device for the lumbar spine Spinal Restabilization Procedures. *Elsevier Science B.V.*; 2002:355–362

Lee CK. Accelerated degeneration of the segment adjacent to a lumbar fusion. *Spine* 1988; 13:375-377

Mariottini A, Pieri S, Giachi S, Carangelo B, Zalaffi A, Muzii FV, et al: Preliminary results of a soft novel lumbar intervertebral prothesis (DIAM) in the degenerative spinal pathology. *Acta Neurochir Suppl* 92:129–131, 2005

Sénégas J: Mechanical supplementation by non-rigid fixation in degenerative intervertebral lumbar segments: the Wallis system. *Eur Spine J* 11 (Suppl 2):S164–S169, 2002

Samani, J. "Study of a Semi-Rigid Interspinous U Fixation System." Spinal Surgery, *Child Orthopaedics*, 2002; 1707

Schlegel JD, Smith JA, Schleusener RL. Lumbar motion segment pathology adjacent to thoracolumbar, lumbar, and lumbosacral fusions. Spine 1996;21:970–8

Seong-cheol Park, Sang Hoon Yoon, Yong-Pyo Hong, Ki-jeong Kim, M, Sang-Ki Chung, Hyun-Jib Kim. Minimum 2-Year Follow-Up Result of Degenerative Spinal Stenosis Treated with Interspinous U (CoflexTM). *J Korean Neurosurg Soc* 46 : 292-299, 2009

Swanson KE, Lindsey DP, Hsu KY, Zucherman JF, Yerby SA. The effects of an interspinous implant on intervertebral disc pressures. *Spine*. 2003; 28:26–32.

Tsai KJ, Murakami H, Lowery GL, Hutton WC. A biomechanical evaluation of the stabilization effects of an interspinous device (CoflexTM™). *J Surg Orthop Adv* 2006;15(3):167-72.

Vena P, Franzoso G, Gastaldi D, Contro R and Dallolio V. "A finite element model of the L4–L5 spinal motion segment: biomechanical compatibility of an interspinous device." *Comput Methods Biomech Biomed Engin.* 2005 Feb;8(1):7-16.

Verbiest H, A radicular syndrome from developmental narrowing of the lumbar vertevral canal. *J Bone Joint Surg* 36B-1954:230-237

Zucherman JF, Hsu KY, Hartjen CA, Mehalic TF, Implicito DA, Martin MJ, et al: A prospective randomized multi-center study for the treatment of lumbar spinal stenosis with the X STOP interspinous implant: 1-year results. *Eur Spine J* 13: 22–31, 2004

11

Development and Clinical Evaluation of Bioactive Implant for Interbody Fusion in the Treatment of Degenerative Lumbar Spine Disease

Michal Filip[1], Petr Linzer[1] and Jakub Strnad[2]
[1]*Neurosurgical Dept. KNTB Zlín,*
[2]*LASAK Ltd., Praque,*
Czech Republic

1. Introduction

Due to new information about the pathophysiology and biomechanics of degenerative lumbar spine disease, the surgical treatment of this disease has undergone a significant increase over the past forty years. Novel diagnostic approaches and the development of new materials provided the impetus to produce new types of instrumentation, and these instruments have led to the modernization of interbody fusion including PLIF, TLIF and ALIF methods. These interventions are currently performed in either an open mini-invasive or endoscopic manner. The open interventions are indicated in cases where the spinal canal stenosis is caused by severe degenerative lesions affecting the motion of intervertebral discs, joints, ligaments, or vertebral arch. Despite the development of other surgical techniques (e.g., functional disc substitutes, dynamic stabilization), the posterior interbody fusion represents a powerful approach in the surgical treatment of degenerative stenosis of the spinal canal.

The PLIF method was first applied in the 1940s by Briggs and Milligan who inserted crushed bone grafts into the intervertebral space, and the bone grafts insertion technique was further developed by Cloward (Cloward, 1953). Due to complications associated with autografts (i.e., pain at the sampling site, procedure prolongation, etc.), the PLIF surgical technique was improved in the 1980s, and new implants constructed of various materials were developed (Bienik and Swiecki, 1991; Brantigan et al, 1994; Khoo et al, 2002; Šrámek et al, 2010). Likewise, novel diagnostic tools have been developed including MRI, 3D CT, SPECT-CT (Crock, 1976, Modic et al, 1988; Blumenthal et al, 1988), and new materials (e.g., ceramic, titanium, PEEK) have yielded new types of implants leading to the modernization of the interbody fusion via PLIF techniques (Alexander et al, 2002; Bessho et al, 1997; Brayan et al, 2002; Ciappetta et al, 1997; Kokubo, 1990; Yamamuro, 1995; Hashimoto et al, 2002; Thalgott et al, 2002; Sandhu, 2003). Currently, the majority of implants for PLIF consist of two separate components, including the solid cage shape and osseoconductive material (i.e., TCP, BMP) that ensures osteoblastic activity and the interbody fusion formation. To date, no material with both suitable mechanic properties and high grade bioactivity is currently available. For instance, solid materials (e.g., medical steel, titanium, PEEK) lack bioactivity that is able to support the osseoconduction (Carlson et al, 1988; Williams and McNamara, 1987; Zdeblick and Philips, 2003). Likewise, bioactive or resorbable materials (e.g., glass-ceramic, hydroxyapatite, polysacharides) do not meet the mechanical requirements for fusion implants of intervertebral discs (Hench el al, 1971; Filip et al, 1996; Sobale et al, 1990; McAfee, 1986).

Currently, the majority of implants consist of cages that form various shapes. The perimeter is constructed from a solid material that ensures the structural strength. The centre of the cage is hollow and is filled by bone grafts or osseoconductive material (e.g. bi- or tri-calcium phosphate) to promote bone fusion in this part of the implant. The optimal implant for an interbody fusion should imitate the properties of the bone tissue by combining sufficient mechanical strength as well as bioactive surface. Therefore, the mechanical strength and the shape of the implant should ensure the primary stability of the segment of the lumbar spine following the operation. Furthermore, the bioactive surface should allow stimulation of osteoblast proliferation at the interface of the implant and bone, and should promote activation of their migration along the implant surface. The bioactive surface should also act as a conductor for osteoblast migration to the fixed vertebral bodies to form the fusion. This quality would prevent the requirement for additional filling of the implant by osseoconductive material. The aim of our work was to create an implant with optimal strength and bioactivity in an attempt to replace the use of autografts and two-compartment implants for PLIF.

2. Research

2.1 Advantages and disadvantages of PLIF surgical technique using autografts

When the conservative treatment fails, patients from categories LS syndrome and FBS syndrome (Failed Back Surgery syndrome) are often referred to surgical management via posterior interbody fusion (Benzel et al, 2003; Cloward, 1953; Crock, 1976; Daniaux, 1986; Dove, 1990; Gurr et al, 1999; Cho et al, 2002). The indication for this surgery is based on neurological finding. In general, the patient predominantly either suffers from back pain associated with progression of root lesion in a lower extremity or with neurogenic claudications in the lower limbs, and shows no reaction to the full conservative therapy algorhytm (Anderson, 2000; Brinnckman et al, 1989; Brodke et al, 1997; Cloward, 1953, Hrabálek et al, 2009; Paleček et al, 1994; Fischgrund et al, 1997). Furthermore, the disease is supported by graphic images of compression of the neural structures caused by degenerative lesions (Knudson, 1944; Crock, 1978; Modick et al, 1983; Sonntag and Theodore, 2000). The desired clinical effect can be achieved by the decompression of neural structures together with spondylodesis of the affected spine segment using PLIF (Steffee, 1988; Hashimoto el al, 2002; Dick, 1987; Wang et al, 2005). When surgical treatment is necessary, no acceptable scientific long-term evidence of efficacy exists for any type of surgical treatment of the degenerative lumbar spine disease (Brodke et al, 1997; Benzel, 2003; Sonntag and Theodore, 2000; Paleček et al, 1994).

We performed PLIF using autografts that were developed in the 1980s. An autograft (mostly iliac crest bone grafts) stripped of connective tissue was inserted under compression into the intervertebral space. The best stability was achieved by transpedicular fixation of the operated segment necessary for osteointegration of the grafts via their remodeling and PLIF formation (Bauer and Muschler, 2000). The advantage of a recently collected autolog bone graft has been the presence of live bone cells with mineralized extracellular matrix. The biological activity, structure and proteins of bone morphogenesis are important prerequisites of the fusion. In addition, clinical experiences from the first half of the 20th century have proved better surgical outcomes with autolog grafts in comparison to simple decompression (Cloward, 1953; Dawson et al, 1981; Dick, 1987; Carlson et al, 1988). Autolog grafts in this form have been the gold standard for PLIF in the majority of spondylosurgical clinics through the end of the 20th century. Despite improving surgical outcomes with a growing number of operated patients, new complications still exist regarding this otherwise successful surgical technique (Kurz et al, 1989).

The most common complications associated with this surgery include problems with bone graft sampling in that limitations are present in bone size and structure that may be safely collected from a live patient in cases of extensive intervention. Furthermore, patients can suffer from unpleasant reactions including debilitating postoperative pain, infection, seroma, cosmetic defects, nerve injury, hip fractures, vessel injury and blood loss. These adverse reactions can occur in 10 to 39% of cases (Arrington et al, 1996; Banwart el al, 1994; Banwart et al, 1995). Therefore, these reactions and other problems have led to search for artificial materials for PLIF. The optimal material for PLIF substituting bone grafts should ideally have the following characteristics. First, the material should show solid structural support (load resistance immediately after implantation). Second, the material should display osseoconductivity and bioactivity or the ability to bind with a bone, fusion support without any other additional material (e.g., bone, BCP etc.). Third, the material should provide the possibility for a radiographical assessment of the bone fusion process. Finally, the material should show biomechanic properties (elasticity modulus similar to bone).

2.2 Development of a new implant for PLIF

As described in chapter 2.1, we considered using an implant made from a synthetic material for PLIF in the early 1990s to eliminate the disadvantages of autografts (Madawi et al, 1996). The most available implants were constructed of medical steel (Bagby, 1988). However, these implants did not meet our notion of sufficient strength accompanied by bioactivity. Spondylosurgeons in Charkov (Professor Gruntovskij) have successfully used corundum implants in combination with hydroxyapatite for PLIF in the surgical treatment of degenerative lumbar spine disease in the 1980s. According to results of this clinic, the success of this implant resulted from its prism shape with projections firmly anchored in the intervertebral space that helped the implant to fixate the segment with or without transpedicular fixation following operation. Due to its bioinertion, hydroxyapatite was added, and this soft material was placed around the corundum (Rowlings, 1993; Gogolewski et al, 1993). Therefore, this implant stimulated formation of osteoblasts, and served as a conductor for their migration between adjacent surfaces of adjoining vertebral bodies. In the early 1990s, another type of prosthesis produced from bioactive glass-ceramic was developed by Electric Nippon Glass, and was used by Japanese orthopedists for PLIF (Yamamuro, 1995, Kokubo, 1990). While transpedicular fixation was added to PLIF due to its fragility, the bioactivity of the implant surface allowed fusion due to migration of bone cells along its surface without addition of any supporting material (e.g., bone, hydroxyapatite) (Sobale, 1990; Yamamuro, 1995). Based on these experiences, we began searching for a material for PLIF implant that would combine the advantages of both the shape and the strength of corundum and the bioactivity of glass-ceramic used in the early 1990s. Thereby, the combination of these two properties would allow strong anchoring of this material in the intervertebral space, the restoration of anatomy in the operated segment, the stabilization of unstable segment, and the formation of interbody fusion associated with osseoconductive properties without addition of another material and without the risk of migration.

2.3 Experimental development of glass-ceramic implant (BAS-O)

Unlike bioinert or biotolerant materials, bioactive glass-ceramic material BAS-O, forms a strong chemical bond with live bone tissue (Fatley et al, 1979; Urban, 1992). Material BAS-O is prepared by progressive steps, such as sintering, controlled crystallization and others. The controlled crystallization allows control of processes that determine the bioactive ability of the final material including material transformation, the control of chemical structure, and the structure of the glass phase (Strnad, 1992). The ability of this material to form a strong bond with bone tissue results from the formation of an apatite layer on the material surface resulting in the connection of the bioactive material with body fluid. Crystallographic chemical characteristic of apatite released on the material surface is similar to the organic part of the bone tissue. Thereby, the stability of the operated segment without micromovements and the tight contact of the material without microgaps are necessary for perfect chemical bond BAS-O / live tissue. Otherwise, a risk of connective tissue penetration exists that can prevent the chemical bond on the bone / implant interface (Kokubo, 1990; Urban, 1992).

The most important finding for the planned use of the lumbar implant necessitated that the biochemical and mechanical properties of the glass-ceramic BAS-O mimic the cortical bone tissue. According to the Young model, the shape of their implant exceeded twice the strength of the vertical load, and was close to its flexural strength. Therefore, we based our implant shape on our previous experiences and according to the models that we observed during our study visits. Together with size and shape development, we also created application instrumentation used for the intervertebral space as well as the operation procedure. At this time, the fragility of the ceramic in the contact with steel represented our only disadvantage in that this fragility could cause problems with insertion using metal application instrumentation. The application instrumentation was coated by Teflon in order to prevent damage to the implant. A rectangular prism-shaped implant (25 mm long, 8 mm high and 10 mm wide) was progressively developed after repeated experiments with cadavers from 1991 to 1993 (Filip et al, 1995). "Winglets" have been placed on the opposite sides of the prism (Figure 1).

The winglets cut into the adjacent vertebral bodies after its rotation by 90 degrees, and the implant was firmly attached within the space without a risk of migration into the spinal canal. Due to its bioactivity, the implant should stimulate migration of bone cells along its surface to form interbody fusion. The application technique for the glass-ceramic implant was the same as with other implants for PLIF. During experimental application in cadavers, the implants were well-anchored in the vertebral bodies without compression of dural sac in the spinal canal, and this placement was confirmed by imaging techniques (X-ray and CT).

We also planned to apply such PLIF methods alone in both low-grade instabilities and to add transpedicular fixation in high-grade instabilities.

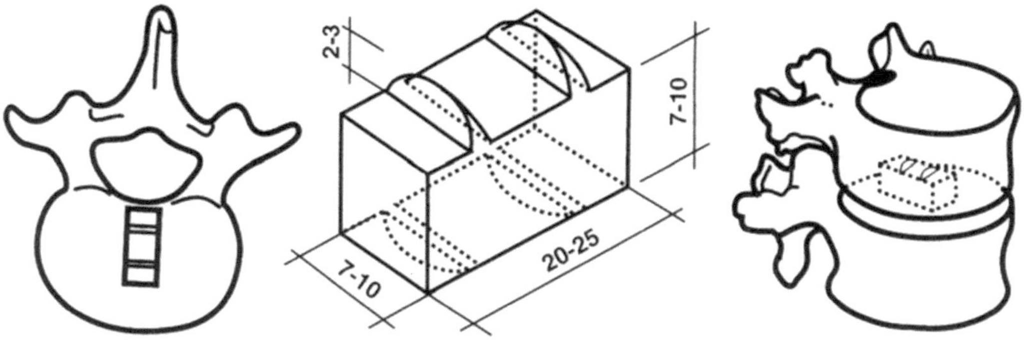

Fig. 1. Outline of the implant and its position in the interbody space (1992).

The implant was inserted using a specially developed instrumentation into the interbody space (Figure 2).

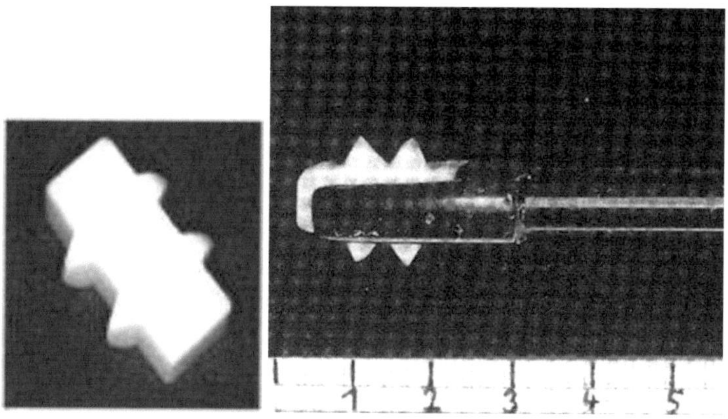

Fig. 2. Glass-ceramic implant plus application fork (1992).

2.4 Implant made of material BAS-0 in clinical practice

Implant BAS-O was registered by the Ministry of Health of the Czech Republic with registration number 89/492/98-IIB in 1992. After registration of the implant and based on experimental results, we introduced the implant into clinical practice in 1994. Based on the advantages from the experimental studies (e.g., stability in the operated space, restoration of anatomy, elimination of the risk of bone grafts sampling, etc.), we expected that these results would be confirmed. From 1994 to 1999, we used this technique in 65 patients observing the indication criteria and the surgical procedure described in the previous chapters. We assessed clinic and graphic postoperative findings in 25 patients out of this population during follow-ups conducted three, six, and twelve months after the intervention. The average age of the patients was 52 years. In 22 patients, the operation represented the first on the spinal segment affected by degenerative instability of various types. We applied the implant BAS-O according to the experimental study. Additionally, PLIF was performed using a pair of implants by the stand alone technique in ten patients (Figure 3), and PLIF was conducted using one or two implants with additional transpedicular fixation by various companies in fifteen patients (Synthes, Stryker etc.).

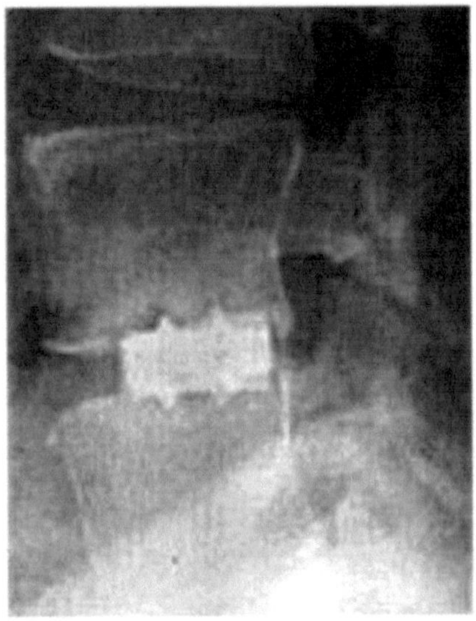

Fig. 3. Fixation of L4/5 instability using a pair of glass-ceramic implants by stand alone technique (1995).

We assessed our results three, six and twelve months after the operation using the ODI score (Oswestry Disability Index; see Table 1). The Oswestry Disability Index (ODI) has become one of the principal condition-specific outcome measures used in the management of spinal disorders. We also used imaging techniques that were available at the time (i.e., X-ray, CT, rarely MRI), and we assessed the change of the implant position in the operated space (i.e., damage, dislocation) using the postoperative imaging techniques (X-ray; see Table 2).

ODI score of our population [%]	**Mean [n] 25**	**Primary instability [n] 8**	**Degenerative listhesis grade I-II [n] 11**	**Isthmic listhesis up to grade II [n] 6**
Before operation	60	55	67	58
3 months after operation	39	36	44	38
6 months after operation	40	39	42	38
12 months after operation	42	40	46	41

Table 1. Clinical assessment according to ODI score.

BAS-O implant position on X-ray	Assessment Month 3	Assessment Month 6	Assessment Month 12
No damage, no migration	25	23	20
Damage, no migration	2	3	5 (20%)

Table 2. Assessment of position change using X-ray.

Our results indicated that we achieved a mean improvement of I grade in the aforementioned population according to ODI assessment during twelve months [60% (severe invalidity) and 42% (moderate invalidity) with mild progression in long-term; 38% at month three and 42% at month twelve]. Using X-ray, we diagnosed implant damage without fragment(s) migration towards the spinal canal or in the prevertebral direction in five patients (20%) after twelve months. Initially, we utilized the sole interbody fixation (stand alone technique) mainly in patients with low grade instability. We found that the clinical condition stabilized in these patients, and the postoperative imaging investigation showed good fixation of the operated segment without prosthetic damage and with adequate postoperative changes around the nerve structures (Figure 4).

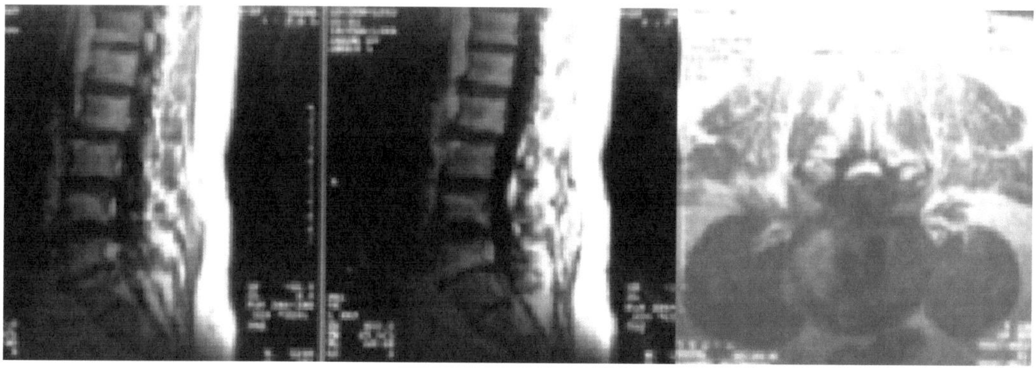

Fig. 4. MRI performed three years after operation of segment L4/5 using a pair of glass-ceramic implants (1998).

In case of 1st or 2nd grade translation as previously defined by Meyerding, we added posterior transpedicular fixation of the whole segment to the implant application. The assessments of the population showed advantages and disadvantages associated with the glass-ceramic implant. For example, the operation time was shortened, and firm anchoring in the interbody space was confirmed due to the shape and elimination of risk associated with bone graft sampling. However, the limiting factor for the universal use, in particular for application stand alone technique, was the mechanical resistance to bending at the ultimate load as well as the probable discongruence of the implant and the bone bed, especially related to shorter implants (under 10 mm). This finding was reflected by implant damage on X-ray in the clinical practice (see Table 2). Due to the relatively good results previously observed (Filip et al, 1996), we extended the stand alone technique to higher grades of translation. However, the lower mechanical resistance of the glass-ceramic to bending was observed, and damage of the implant was detected several months after the operation. This problem affected five patients following operation using this technique during twelve months. Despite implant damage, no migration towards the spinal canal or across vertebral bodies occurred resulting from the construction with retention winglets and the chemical bond (Figure 5).

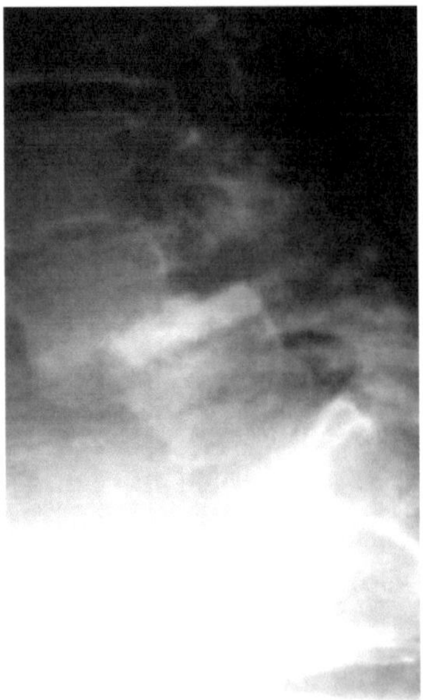

Fig. 5. Slide progression at L4/5 during overload of glass-ceramic implants in 2nd grade instability according to Meyerding.

We also added transpedicular fixation of the affected segment in case a patient experienced clinical impairment(s) due to implant damage and instability progression. We did not observe implant damage in the fixed segments in this study; however, we stopped using the stand alone technique with the glass-ceramic implant for PLIF after this experience. Unfortunately, we failed to directly demonstrate osseoconductive properties of the glass-ceramic implant BAS-0 for PLIF that was associated with fusion of the adjacent vertebral bodies by migrating bone tissue along the glass-ceramic body from 1994 to 1999.

2.5 Experimental development of bioactive titanium in forms by LASAK

The LASAK Company developed bioactive titanium with original surface modification at the end of the 1990s. Due to the limitations of the glass-ceramic implant mentioned above, we have been developing a new type of implant combining bioactive properties and higher mechanical resistance in cooperation with LASAK Company since 1998. Characteristics of this material (higher strength, bioactivity) have provided optimal implant characteristics for PLIF (Yan, 1997; Strnad, 2010). The material used for this implant is technically pure titanium (grade 3) which is dedicated for surgical implants (Regulation ISO 5832-2:1993(E): Implants for surgery, ISO 5835-2). To ensure bioactivity of this material, the implant surface is chemically modified by LASAK technology (Adjudication on Permission to Use a Medical Device No. 82/125/00-IIB by State Institute for Drugs Control of the Ministry of Health of the Czech Republic). Mechanical properties of this material are identical to pure titanium, and its strength and fracture persistence are several times better than characteristics of the bone tissue and the glass-ceramic material (see Table 3).

	Titanium	Bone	Glass-ceramic
Compressive strength (MPa)		100-230	1080
Tensile strength, flexural strength* (MPa)	240–680	200*	170–218*
Elasticity modulus (GPa)	100–120	25	220
Fracture persistence ($MPa \cdot m^{-1/2}$)	~40–100	2-12	2

Table 3. Comparison of mechanical properties of titanium, bone and glass-ceramic BAS-0.

The mechanically and chemically modified surface of the bioactive titanium by LASAK technology is able to induce the production of calcium-phosphate (apatite), and this compound arises from the interaction between the surface of the material and body fluid within hours to days. The chemical and crystallographical properties of this mineral are nearly identical with the bone apatite. Experimental studies with bioactive and bioinert titanium demonstrated that titanium with a bioactive surface better tolerates unfavourable conditions for osseointegration, as gaps between the implant and the bone (Strnad et al, 2003). Bioinert titanium allows penetration of fibrous tissue into the interface implant/bone, and promotes instability or migration of implants towards the spinal canal in conditions requiring spondylosurgery. However, a firm interaction between the calcium-phosphate layer of the implant and the surrounding bone forms immediately after application if bioactive titanium with technological modification according to LASAK is used, which ultimately eliminates this risk.

Strength parameters and bioactivity would be expected to improve conditions for osseointegration in the intervertebral space, as compared to implants generated from bioinert materials and glass-ceramic. Therefore, this type of material appears to be optimal for the development of a new implant for PLIF. Based on our experiences with the glass-ceramic BAS-O, we designed a new implant model constructed from this material. Due to different properties of these two materials (glass-ceramic/biotitanium), we modified the shape of the implant, and we designed new application instrumentation. The basic model was the shape of skewed prism (4°) (20 mm long, 8 mm wide, with graduated high 6, 8 and 10 mm). The implant was equipped with two pairs of projections or winglets (2 mm high) on the opposite sides of the prism. The compression and bending load of our original model for PLIF was virtually mathematically tested using computer technology in cooperation with ČVUT Prague. These tests showed that the shape of the skewed prism with winglets can theoretically ensure the restoration of the anatomy of the operated segment of lumbar spine without a risk of a plunge into the adjacent vertebral bodies both during compression and flexion and without a risk of its damage (Figure 6).

Based on these mathematical analyses, we maintained the basic shape of the implant with the above mentioned parameters. The higher strength of the material allowed us to design simplified application instrumentation. We used a thread in the implant body instead of the Teflon-coated fork used in the implant BAS-O. Due to its strength, no opposite space dilatation was necessary before the application as a result of the bioactive titanium implant, and no risk of damage of the implant shape by metal loaders was detected. Therefore, the handling of the implant during an intervention is easy and safe. The shape of the implant ensured good restoration of the anatomy of the operated area (restoration of the interbody space and its stability) with minimal risk of implant plunge into the adjacent vertebral bodies, as demonstrated by imaging investigations. Other benefits of the new implant included higher strength and shape variability.

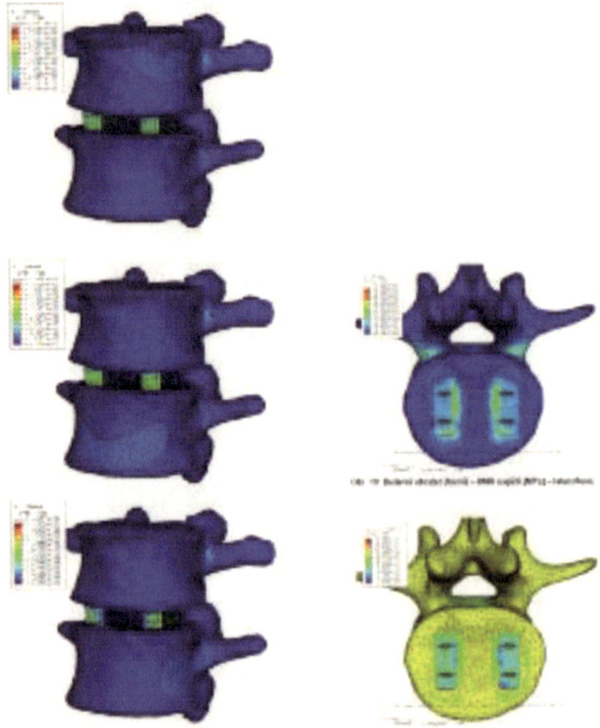

Fig. 6. Illustration of the mathematical testing of the implant model.

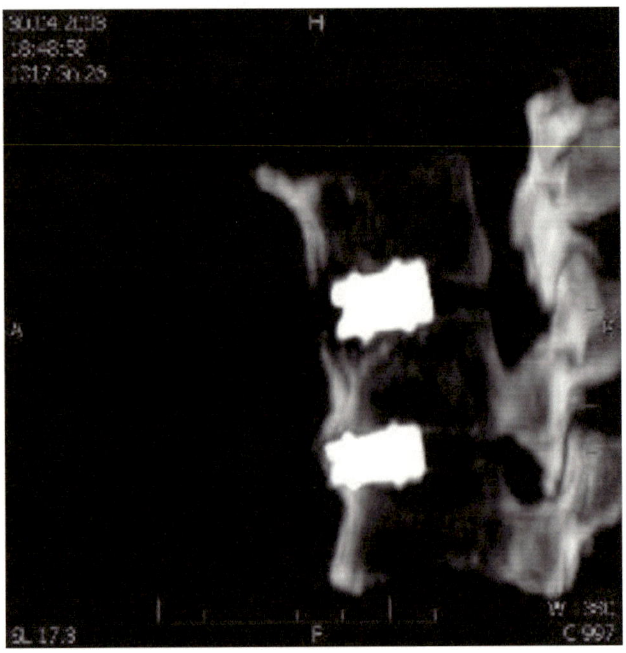

Fig. 7. Implaspin in the intersomatic space of a cadaver by CT (2001).

We removed the whole motion segment with implants from cadavers after experimental operations, and we assessed their localization and the degree of their damage by X-ray and CT scans (Filip et al, 2001). Both investigations showed proper localization of the implants in the intervertebral spaces without any contact with the spinal canal or perforation of the winglets into the adjacent vertebral bodies (figure 7).

Additionally, their shape and surface were not damaged by the new type of instrumentation. Therefore, we assumed that these findings would transfer from experimental studies into clinical practice. However, we were not able to verify the osseoconductive properties of the implant surface in the cadavers. A perfect contact was observed between the surrounding bone tissue resulting from the simple application in cadavers, which was a good precondition for supporting osseointegration in the interbody space via osteoblasts´ migration along its surface. Thus, we verified the osseoconductive properties of the BIO surface of the implant in an animal model (Strnad, 2008). The implant surface in the direct contact with newly produced bone tissue yielded the following values [BIC (%) = 48,5 ± 2,9, 66,0 ± 7,4 and 90,6± 7,0, respectively, two, five and twelve weeks after implantation].

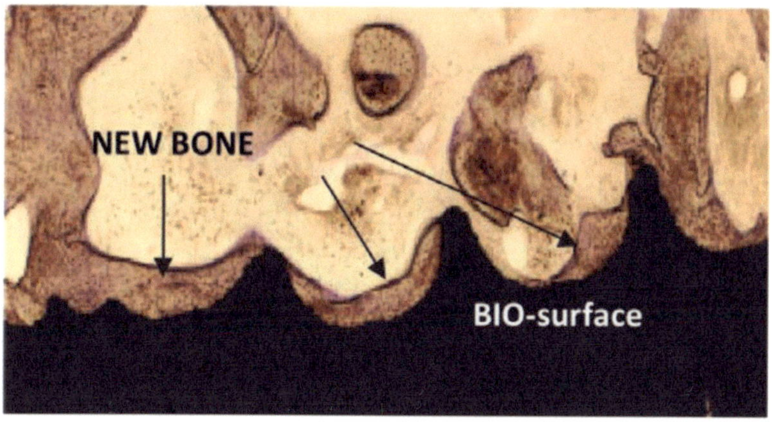

Fig. 8. Histological section of the interface of newly formed bone tissue on the BIO surface of the titanium implant twelve weeks after implantation. This figure illustrates the osseoconductive properties of the surface (optical microscope, toluidine blue staining, original magnification - 200x).

2.6 Implaspin in clinical practice

Encouraged by these experimental results, we began to use this type of implant in clinical practice in indications for PLIF instead of the glass-ceramic implant since 2002 (Figure 9).

The operation technique PLIF was identical to the operation technique used in cadavers (Filip et al, 2010). For example, we decompressed the nervous structures through posterior median line approach, and we then radically removed the degenerated intervertebral disc under the control of the operation microscope. Afterwards, we removed the surfaces of the adjacent vertebral bodies, and we then inserted the bioactive titanium implant using the innovated instrumentation (Figure 10). Finally, we added transpedicular fixation of the whole segment (Synthes, Signus, Easy spine, etc.) (Figure 11).

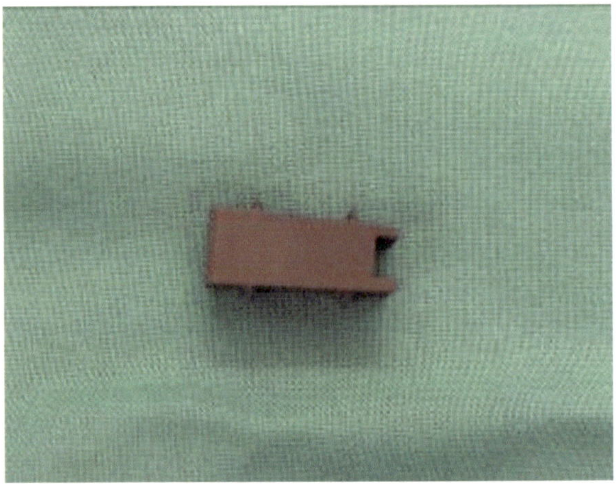

Fig. 9. Implant from bioactive titanium – Implaspin (2002).

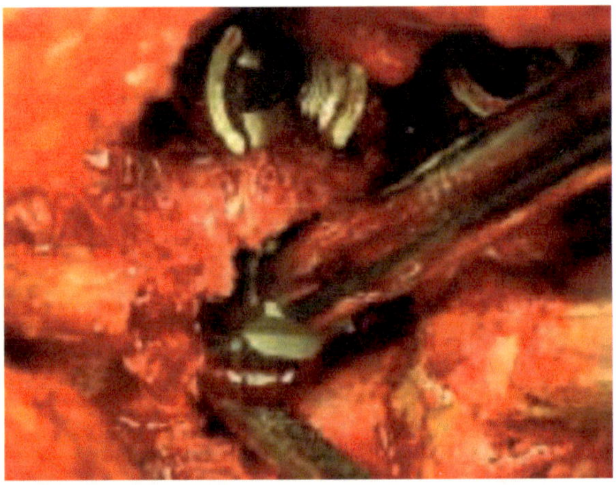

Fig. 10. Insertion of Implaspin into the interbody space.

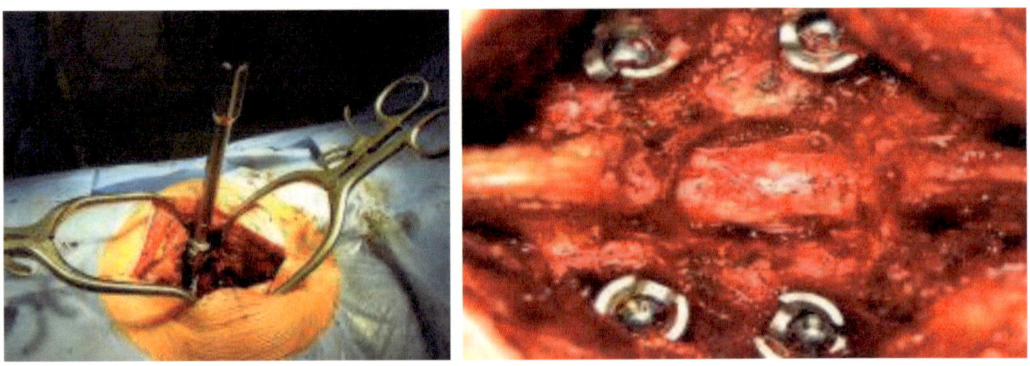

Fig. 11. Transpedicular application of screws (SIGNUS)

To date, we have not observed any complications associated with the implant application into the interbody space. According to the postoperative scans, the implant was always placed in the correct position with winglet penetration into the spongious tissue of the adjacent vertebral bodies. We have selected the size empirically according to the extent of osteochondrosis of the affected disc and the degenerative lesions of the surrounding tissues on scans (X-ray, CT, MRI) during the intervention. In the majority of cases, we used implants (8 or 10 mm high) with angle 4% to maintain lordosis in the lumbar area (Figure 12).

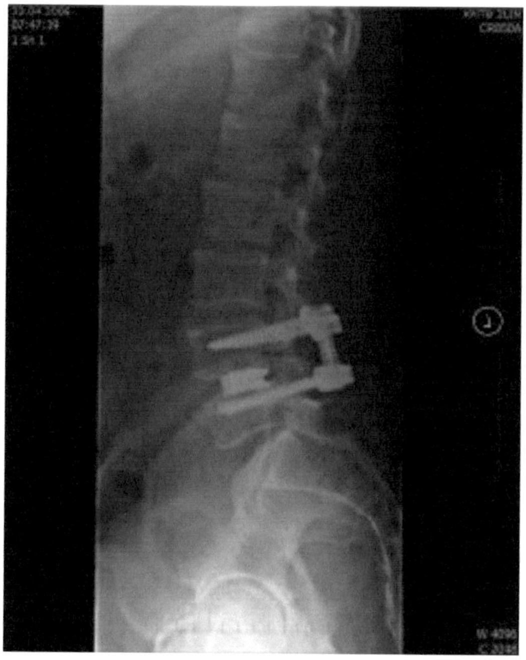

Fig. 12. Fixation L4/5 (Implaspin plus transpedicular screws Signus).

According to the experimental studies, tight contact with the surrounding bone tissue was necessary to activate the bioactivity of the surface. This contact was ensured by the shape of the implant and the winglets that penetrated into the spongious bone tissue of the adjacent vertebral bodies, and was the precondition for migration of the osteoblasts along the implant body resulting in the formation of a junction of the adjacent vertebral bodies by bone tissue without the need to sample bone grafts or to add supporting synthetic materials inside or around the implants.

In 2002 to 2007, operations were performed on 57 patients using the bioactive implant Implaspin in the Neurosurgery Clinic of the Faculty Hospital in Ostrava and in the Neurosurgery Department of Tomáš Baťa´s Regional Hospital in Zlín. We assessed a population of 25 patients with follow-up examinations conducted two or more years following surgery, according to the clinical condition. The follow-ups were also based on the generally used score system ODI and imaging methods (X-ray, CT, MRI) that occurred three, six, twelve and 24 months after surgery. During the follow-ups, we examined the patients for signs of implant damage, instability of the operated segment, and signs of supposed osteoblastic activity of the bioactive surface of the implant on the scans. Results of the ODI questionnaire showed that with Implaspin, our success rate improved by 1 degree (59%–40%), or we stabilized the clinical condition of the majority of the patients long-term (2 and more years), which corresponds to results of other clinics using other implant types (Bessho et al, 1997; Brantigan et al 1993; Brayan et al, 2002; Bienik and Swiecki, 1991; Ciappetta et al, 1997; see Table 4).

ODI score in our population [%]	Mean [n] 25	FBSS [n] 9	IS [n] 6	DI [n] 10
Before surgery	59	65	55	57
3 months after surgery	42	46	40	40
6 months after surgery	40	45	37	38
12 months after surgery	41	45	39	39
24 months after surgery	40	47	35	38

Table 4. Mean Oswestry score values before surgery and at regular visits (FBSS - failed back surgery syndrome; IS - isthmic spondylolisthesis; DI - degenerative instability).

The assessment of the implant position on scans (X-ray, CT, MRI) at postoperative visits demonstrated no signs of implant damage or implant migration out of the intersomatic space. These investigations have not yet shown any signs of instability of the operated segment (i.e., formation of new osteophytes, progression of hypertrophy of the articular facets, and migration of the implant at the site of application). We observed one severe complication in the population which was caused by an inaccurate application of the transpedicular screws. The wound healed in this patient, and the neurological findings stabilized after removal of the screws. The stabilization of the condition may be supported by the implant shape and the winglets which prevented instability even after the removal of the transpedicular screws. This finding was confirmed by the imaging investigations. Based on the clinical condition and the absence of instability signs on imaging investigations, we concluded that the formation of bone fusion was due to osteopblasts´ migration along the bioactive of Implaspin surface.

2.7 Assessment of bioactivity of the implant using SPECT-CT

During our investigations, we attempted to demonstrate the migration of bone cells along the surfaces of the glass-ceramic or biotitanium implants using imaging investigations. Unfortunately, standard CT or MRI were not able to provide this precise information. The CT scans were limited by screw artefacts, and the MRI scans were generally unable to detect changes in bone. In an attempt to resolve these problems, we utilized SPECT-CT, a method that provides up-to-date computed tomography (CT) and gamma camera (SPECT), to detect the activity of the osteoblasts on the body of the titanium implant applied into the interbody space. The computed tomography (CT) can precisely display the anatomic structure of the investigated tissue, and the gamma camera investigation (SPECT) can yield a functional view of the metabolic process in the patient´s body, but without its precise localization or other anatomical details. Thereby, the combination of these investigations provided more complete information on the precise place of the metabolic process as well as its dynamics. In our study, the metabolic process included the activity of the osteoblasts on the surface of the bioactive implant, as applied by the PLIF method.

In 2009, we performed this type of investigation in four patients after surgery for the primary instability of the lower lumbar spine segment using the PLIF operation technique with Implaspin. The study was conducted before the surgery as well as two and six months after the intervention, and we assessed the anatomical changes and metabolic activity at the location where the implants were applied by using the combined scans. The investigation provided preoperative signs of instability localized to the affected space in the area of the disc in all four patients. We detected a hyperintense signal at the operated segment two to three months after the surgery, which was a sign of osteoblast activity on the surface of the implant. We also observed a decrease of this activity (hypointense signal) six months after the surgery as well as a change on the surface of the implant using the combined CT scans.

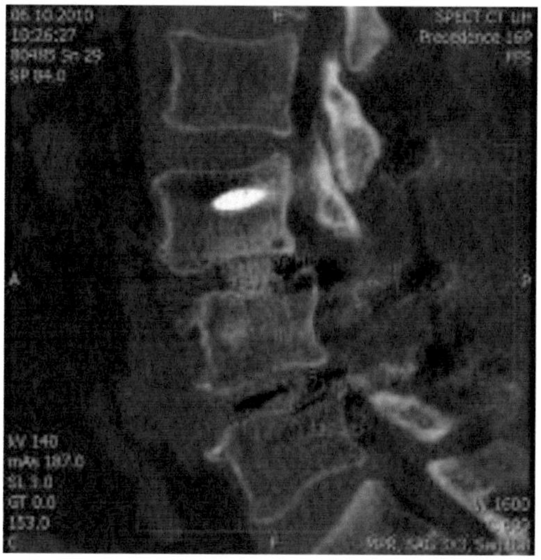

Fig. 13. Implants applied into the spaces L3/4 and L4/5 on SPECT CT. The figure shows the surface of the implant with the bone tissue (grey-black colour) and the titanium screw in the body of the L3 (white colour).

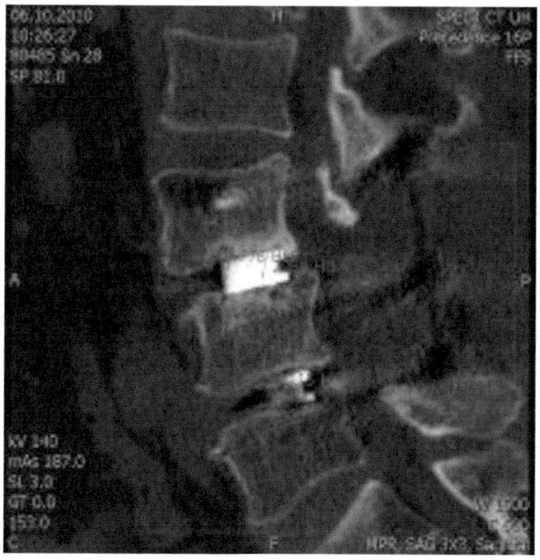

Fig. 14. Implants on a SPECT-CT scan. The bone growth at the border of bone tissue (grey) and Implaspin (white) is visible in the space L4/5.

According to our method, hypointensity signified the completion of the osteoblastic activity. The changes on the CT scans were completed by conducting a measurement using Haunsfield´s units (metal - about 2000 HU; bone tissue 100–300 HU), which provided evidence that the implant was overgrown by bone tissue. This kind of image detects the primary successful binding of the implant via activation of the osteoblasts by its specially adapted bioactive surface (figure 13 and 14). Using this combined imaging technique in all four patients, we demonstrated the migration of bone cells along Implaspin wall and the formation of fusion without the addition of another material, such as autografts or TCP, six months after the surgery. Therefore, the successful fusion was indirectly confirmed using the SPECT-CT improving the postoperative clinical findings.

3. Conclusion

The development of both the material and the shapes of implants continues to progress. Currently, the primary focus of this development is to produce an implant that forms a firm fusion as soon as possible and to ensure the formation of new bone due to its material composition. The current implants for PLIF combine two separate components, including a solid cage shape and osseoconductive material (i.e., TCP, BMP) that ensures the activity of osteoblasts and the formation of the interbody fusion. To date, none of the materials for PLIF available on the market optimally meet both characteristics (see Table 5).

Optimal parameters of the implant	Metal (Titanium, steel)	PEEK or +PEEK carbon fibres	Glass-ceramic	Resorbable implants - polylactides	Bioactive titanium (LASAK Ltd.)
1. Firm structural support (load resistance immediately after implantation)	+	+	+/-	+/-	+
2. Osseoconductivity, bioactivity - ability to bind with a bone, support of fusion without addition of other material (bone, TCP, etc.)	-	-	+	-	+
3. Possibility of radiographic assessment of the bone fusion progression	+	+	+	+	+
4. Biomechanical properties (elasticity modulus similar to bone)	-	+	-	+/-	-

Table 5. Parameters of the implant according to the type of material.

During this investigation, our goal was to develop an implant that would combine both of these components in one unit, ultimately maintaining the strength and bioactive properties present in two-component implants. At the end of the 1990s, we were close to the development of such material due to the implant BAS-0. However, the resistance of the glass-ceramic at the ultimate load provided a limitation that negatively influenced the shape and the application process, as described in chapters 2.3 and 2.4. However, due to these experiences, we and other technicians successfully designed an implant that meets our original conception. This implant is currently used in clinical practice, and experimental studies have confirmed its supposed properties. The combination of the implant's strength and shape with bioactivity enables the smooth application and restoration of anatomy, thereby providing a perfect fixation of the operated segment and stimulating growth of osteoblasts and their migration along its surface. Our original implant Implaspin combines the osteoconductive and osteoplastic properties of the glass-ceramic with the strength of titanium, which was the aim of our research. Thanks to these properties, this implant represents a quality alternative to implants constructed from other materials dedicated to PLIF (see Table 5).

4. Acknowledgement

This study was supported by the Ministry of Industry and Trade of the Czech Republic under the project number: FR-TI3/587

5. References

[1] Anderson, V. C.: A rationale for treatment algorithm of failed back surgery syndrome. Curr Rev Pain 2000; 4: 395-406

[2] Alexander, J., Branch, C., Subach, B., et al.: Applications of a resorbable interbody spacer in posterior lumbar interbody fusion. J Neurosurg (Spine). 2002;97:468-472

[3] Arrington, E., Smith, W., Chanbers, H., et al.: Complications of iliac crest bone graft harvesting. Clin Orthop Rel Res. 1996;329:300-309

[4] Bagby, G.: Arthrodesis by the distraction-compression method using a stainless steel implant. Orthopedics. 1988;11:931-944

[5] Banwart, J., Asher, M., Hassanein, R.: Iliac crest bone harvest donor site morbidity; a statistical evaluation. Spine. 1994;20:1055-1060

[6] Banwart, J., Asher, M., Hassanein, R.: Iliac crest bone graft harvest donor site morbidity. Spine. 1995; 20:1055-1060.Sandhu H, Hs G, Parvatanemi H. Bone grafting for spinal fusions. Orthop Clin North Am. 1999;30:685-698

[7] Bauer, T., Muschler, G.: Bone graft materials: an overview of the basic science. Clin Orthop. 2000;371:10-27

[8] Benzel, E.: Best indications for DDD: Fusion or Non Fusion. In International symposium-Non Fusion techniques in Spinal Surgery. Bordeaux France 2003, Book of Abstracts: 140-1

[9] Bessho, K., Iizuka, T., Murakami, K.: A bioabsorbable polt-L-lactide miniplate and screw system for osteosynthesis in oral and maxillofacial surgery. J Oral Maxillofac Surg. 1997;55:941-945

[10] Bienik J, Swiecki Z.Porous and porous-compact ceramics in orthopedics.Clin Orthop 1991;27:88-94

[11] Blumenthal, S.L., Baker, J., Dosset, A., Selby, D.K.: The role of anterior lumbar vision for internal disc disruption. Spine 1988; 13: 566-9

[12] Brantigan, J., McAfee, P., Cunningham, B., et al.: Interbody fusion using a carbon fiber implant versus allograft bone: a investigational study in the Spanish goat. Spine. 1994;19:1436-1444

[13] Brantigan, J., Steffee, A.: A carbon fiber implant to aid interbody lumbar fusion. Spine. 1993;18:2106-2117

[14] Brinckmann, P, Bigemann, M., Hilweg, D.: Prediction of the Compressive Strength of Human Lumbar Vertebrae. Clin. Biomech., 1989, č. 4, s. S1-S27.

[15] Brodke, D., Dick, J.C., Kunz, D.N., McCabe, R., Zdeblick, T.A.: Posterior Lumbar Interbody Fusion: A Biomechanical comparsion,Including a New Threaded Cage. Spine 1997; 22(1): 26-31

[16] Bryan, B., Gerald, E., Regis, W.H., Brian, R.S., Mark, R.M.: Allograft Implants for Posterior Lumbar Interbody Vision: Results Comparing Cylindrical Dosela and Impacted Wedges. Neurosurgery 2002; 51(5): 1191

[17] Carlson, L., Rostland, T., Albrektsson, B.: Implant fixation improved by close fit. Acta Orthop. Scand., 59: 272-275, 1988.

[18] Cloward R.: Gas-sterilized cadaver bone grafts for spinal fusion operations. Spine. 1980;5:4

[19] Ciappetta, P., Boriani, S., Fava, G.P.: A Carbon Fiber Reinforced Polymer cage for Vertebral Body Replacement: Technical Note. Neurosurgery, 41: 1203-1206, 1997

[20] Clemens, J.A.M., Klein, C.P.A.T., Sakkers, R.J.B.: Healing of gaps around calcium phosphate-coated implants in trabecular bone of the goat. J. Biomed. Mater. Res., 36:55-64, 1997

[21] Cloward, R.: The Treatment of Ruptured Lumbar Intervertebral Disc by Vertebral Body Fusion. 1. Neurosurg., 10, 1953, s. 154-168. Cockin, J.: Autologous bone grafting complications at the donor site. J Bone Joint Surg Br. 1971;53:153

[22] Crock, H. V.: Isolated Lumbat Disc Resotption as a Cause of Nerve Root Canal Stenosis. Clin. Orthop., 1 15, 1976, s. 109-1 15.

[23] Daniaux, H.: Transpedikuläre Reposition und Spongiosplatik bei Wirbelkörperbrüchen der unteren Brust-und Lendenwirbelsäule. Unfalldionogie, 89, 1986, 197-213

[24] Dick, W.: The Fixateno Interne as a Versatile Implant for Spine Surgery. Spine, 12, 1987, 882-900

[25] Dove, J.: Instrumentation postérieure di rachis: Le systéme de Hartshill. Rachis, 2, 1990, s. 487-492.

[26] Fatley T. J., Lynch, K. L., Bensen, M. D.: Tissue Response to Implants of Calcium Phosphate Ceramic in Rabbit Spine. Clin. Orthop., 1979, 1983, s. 246.

[27] Filip, M., Veselský, P., STRNAD, Z.,LANÍK, P.: Replacement of an Intervertebral Disc,using a ceramic prosthesis in the treatment of degenerative diseases of the spine. Acta Chirurgicae Orthopedicae et Traumatologiae Czechoslovakia 62, 1995,p. 226 - 231.

[28] Filip, M., Veselský, P.: Treatment of instability of the lumbar spine by Glass-ceramic Prostheses of the intervertebral disc. Acta Chirurgiae Orthopaedicae et Traumatologiae Czechoslovakia.. Acta Chir Orthop Traum Čech 1996; 63: 83-7

[29] Filip, M., Linzer, P.,Veselský, P., Paleček, T.: Surgical technique verification by the Help of New bioactive titan cage in treatment of degenerative disease of the lumbr spine-experimental study. Acta Chir Orthop Traum Čech 2001; 68: 369-373

[30] Filip M.,Linzer P.,Šámal F.,Kremr J.,Mrůzek M.,Paleček T., Strnad J. : Bioactive cage implaspin in operation treatment of LBP. Bolest No.1, 25- 30, r.2010

[31] Fischgrund, J., Mackay, M., Herkowitz, H., et al.: Degenerative lumbar spondylolisthesis with spinal stenosis: a prospective, randomized study comparing decompressive laminectomy and arthrodesis with and without spinal instrumentation. Spine. 1997;22:2807-2812

[32] Gogolewski, S., Jovanovich, M., Dillon, J., et al.: Tissue response and in vivo degradation of selected polyhydroxyacids: polylactides (PLA), poly(3-hydroxybutyrate) (PHB), and poly(3-hydroxybutyrate-co-3-hydroxyvalerate) (PHB/VA). J Biomed Mater Res. 1993;27:1135-1148

[33] Gurr, K. R., McAfee, P. C., Shih, C.-M.: Biochemical Analysis of Posterior Instrumentation Systems after Decompressive Laminectomy. An Unstable Calf-spine Model. J. Bone Jt Surg., 70-A, June 1988, 680-691.Hamer, A., Stockley, I., Elson, R.: Changes in allograft bone irradiated at different temperatures. J Bone Joint Surg Br. 1999;81:342-344

[34] Hashimoto, T., Shigenobu, K., Kanayama, M., Harada, M., Oha, F., Ohkoshi, Y., et al.: Clinical Results of Single-Level Posterior Lumbar Interbody vision Using the Brantigan I/F Carbon Cage Filled With a Mixture of local Morselized Bone and Bioactive Ceramic Granules. Spine 2002; 27(3): 258-262

[35] Hench, L.L., Splinter, R.J., Allen, W.C.,Greenlee, T.K.: Bonding mechanisms at the interface of ceramic prosthetic materials. J. Biomed. Mater. Res. Symp., 2: 179-149, 1971

[36] Hrabálek, L., Rešková, I., Bučil, J., Vaverka, M., Houdek, M.: Použití titanových a PEEKových implantátů při ALIF stand-alone u degenerativního onemocnění lumbosakrální páteře - prospektivní studie. Česká a slovenská neurologie a neurochirurgie, 72/105 (1), 2009, 38-44

[37] Chaloupka, R., Krbec, M., Ciencala, J., Tichý, V.: Comparison of Surgical Load in Case of 360-degree Fusion in Lumbar Vertebra Spondylolistheses Treated via Rear Instrumentation and the PLIF or ALIF Technique. Acta spondylologica 2003; 2: 69-72

[38] Cho, D., Liau, W., Lee, W., et al.: Preliminary experience using a polyetheretherketone (PEEK) cage in the treatment of cervical disk disease. Neurosurgery. 2002;51:1343-1350

[39] Khoo, LT., Palmer, S., Laich, DT., Fessler, RG.: Minimally invasive percutaneous posterior lumbar interbody fusion. Neurosurgery 2002; 51 (5 Suppl): 166-1

[40] Knudson, R.: The Instability Associated with Disc Degeneration in the Lumbar Spine. Acta Radiol., 25, 1944, s. 594-609.

[41] Kokubo, T.: Bonding Mechanism of Bioactive Glass-ceramics A-W to Living Bone. In: CRC. Handbook od Bioactive Ceramics, Volume I. Bioactive Glasses and Glass-Ceramics, Yamamuro, T., Hench, L. L., Wilson, J. (eds). Boca Raton, CRC Press, 1990, 41 - 49.

[42] Kokubo, T., Shigematsu, M., Naghashima, Y.: Apatite wollastonite-containing glass-ceramic for prosthetic application. Bull. Inst. Chem. Res., Kyoto Univ., 60: 260-268, 1982

[43] Kurz, L., Garfin, S., Booth, R.: Harvesting autogenous iliac bone: a review of complications and techniques. Spine. 1989;14:1324-1331

[44] Madawi. A.A., Powell, M., Crockard, H.A.: Biocompatible osteoconductive polymer versus iliac graft:. Spine. 21:2123-2130, 1996

[45] McAfee, P. C.: Biochemical Approach to Instrumentation of Thoracolumbar Spine: A Review Article, Adv. Orthop. Surg., 8, 1985, 313 - 327.

[46] McAfee, P.C., Bohlmann, H.H., Ducker, T., Elsmont, F.J.: Failure of Stabilization of the Spine with Methylmethal - crylate. A Retrospective Analysis of Twenty-four Cases. J. Bone Jt Surg., 68: 1145-1151, 1986

[47] Modic, M. T., Wienstein, M. A., Pavlicek, W. et al.: Nuclear Magnetic Resonance Imaging of the Spine. Radiology, 148, 1983, s. 757-762.

[48] Modic, M.T., Masaryk, T.J., Ross, J.S., Carter, J.R.: Imaging degenerative disc diseasse. Radiology 1988; 168(1): 177-186

[49] Paleček, T., Wolný E., Veselský, P., Filip M., Drábek, P.: Náš současný přístup k chirurgické léčbě degenerativního onemocnění bederní meziobratlové ploténky. čes. a Slov. Nettrol. Neurochic, 57/90, 1994, č. 5, s. 198-201.

[50] Rowlings, CH.E.III.: Modern Bone Substitutes with emphasis on calcium Phosphate ceramics and osteoinductors. Neurosurgery, 4: 933-936,1993

[51] Russell, G., Hu, R., Raso, J.: Bone banking in Canada: a review. Can J Surg. 1989;32:231-236

[52] Sandhu, HS.: Bone Morphogenetic Proteins and Spinal Surgery. Spine 2003; 28: 64-73

[53] Sobale, K., Hansen, E.S., Rasmussen, H.: Hydroxyapatite coating enhance fixation of porous coated implants. Acta Orthop. Scand., 61: 299-306, 1990

[54] Sonntag, V. K.H., Theodore, N.: Spinal Surgery: The Past Century and the Next. Neurosurgery, 46: 767-778, 2000

[55] Steffee, A.D., Sitkowski, D.J.: Posterior lumbar interbody fusion and plates. Clin Orthop Relat Res 1988 ; 227:99-102

[56] Strnad Z., Role of the glass phase in bioactive glass-ceramics. Biomaterials Vol.13, No. 5 1992, p.317-321

[57] Strnad J., Urban K., Povysil C., Strnad Z.: Secondary Stability Assessment of Titanium Implants with Alkali-Etched Surface: A Resonance Frequency Analysis Study in Beagle Dogs; The International Journal of Oral & Maxillofacial Implants, Volume 23, Number 3, 2008

[58] Strnad J.: Způsob úpravy povrchu titanových implantátů" Pat.č.291685 (2003)

[59] Strnad J., Helebrant A.: Kinetics of bone-like apatite formation in simulated body fluid In: Proc. 5th ESG Conference (eds. Helebrant A., Kasa S., Maryška M.), pp. B2 9-16, Czech Glass Soc., Praha 1999

[60] Thalgott, J., Giuffre, J., Klezl, Z., et al.: Anterior lumbar interbody fusion with titanium mesh cages, coralline hydroxyapatite, and demineralized bone matrix as part of a circumferential fusion. Spine J. 2002;2:63-69

[61] Tullberg, T.: Failure of a carbon fiber implant: a case report. Spine. 1998;23:1804-1806

[62] Urban, K., Strnad, J.: Clinical Application of the Bioactive Glass-Ceramics BAS-O in Orthopaedics, Lékařské zprávy LF UK, Hradec Králové 37, 1992. Vaccaro, A.R., Madigan, L.: Spinal applications of bioabsorbable implants. J Neurosurg. 97[Suppl 4]:407- 412, 2002

[63] Wang, J.C., Mummaneni, P.V., Haid, R.W.: Current treatment strategies for the painful lumbar motion segment: Posterolateral fusion versus interbody fusion. Spine 2005;30:S33-S43

[64] Williams, A., McNamara, R.: Potential of polyetheretherketone and carbon fiber-reinforced PEEK in medical applications. J Mater Sci Lett. 1987;6:188-190

[65] Yamamuro, T.: AW Glass-Ceramic in Spinal Repair. Bioceramics, Vol.8,p.123, Ed.: J. Wilson, L. Henche, D. Greenspan, Ponte Verda, Florida, USA, November 1995

[66] Yan, W.Q., Nakamura, T., Kobayashi, M., Kim, H.M., Miyaji, F., Kokubo, T.: Bonding of chemically treated titanium implant to bone. J. Biomed Mater. Res., 37: 265-275, 1997

[67] Zdeblick, T.A., Philips, F.M.: Interbody Cage Device. Spine 2003; 28 (15 Suppl): S2-S7

12

Surgical Management of Low Back Pain and Degenerative Spinal Disorders

Vu H. Le and S. Samuel Bederman
University of California, Irvine, USA

1. Introduction

Low back pain is the second most common reason to seek a physician in the United States, third most common reason for a surgical procedure, and fifth most common cause for hospitalization (Andersson, 1997). The lifetime prevalence of low back pain is predicted to range from 60 to 80 percent (Hart, 1995; Van Tulder, 2002). The annual prevalence is estimated to be between 15 to 45 percent, with a point prevalence of 30 percent. Low back pain is the most common and most expensive cause of work-related disability in the United States (Atlas, 2000). Between 2002 and 2004, the estimated annual medical costs for all spine related conditions were approximately 193 billion dollars, with about 14 billion dollars in lost wages due to spine disorders (Bone and Joint Decade, 2005). It is a burden to both the individual and society, in terms psychosomatic impairment and socioeconomic impact. In fact, the presence of comorbidity adds to the burden and negatively impacts the patient's functional status (Fanuele et al, 2000).

Fortunately, the majority of these patients recover within 3 months. With conservative care, it has been estimated that about 60 percent recover in 6 weeks, and 80 to 90 percent recover within 3 months (Andersson, 1999). Therefore, only a minority warrants further workup and care that can potentially include surgery. Non-surgical treatment consists of medications, cognitive training, physical therapy, and local injections. Some studies have shown that intensive, structural cognitive behavior therapy including encouragement and daily physical therapy can produce equivalent results compared to fusion in non-specific chronic low back pain (Brox et al, 2006; Fairbank et al, 2005).

In contrast, Fritzell et al (2001) showed that non-intensive, non-structural therapy yielded less optimal results than surgery. What comprises a structural or non-structural non-operative therapy regimen still remains unanswered. Additionally, only a few non-surgical interventional therapies have been shown to be effective, while prolotherapy, facet joint injection, intradiscal steroid injection, and percutaneous intradiscal radiofrequency thermocoagulation have been proven to be ineffective (Chou et al, 2009). The relative merits of non-surgical treatment for these conditions are beyond the scope of this chapter.

The literature on surgical management for low back pain similarly elicits uncertainties due to non-specific diagnosis, but Glassman et al (2009) showed that with diagnostic specificity and stratification, the outcome of surgery depends on the underlying diagnosis. Functional improvement after surgery is not equal among diagnostic subgroups. Since surgery is a highly technical treatment modality, it is imperative to clearly define the pathological condition causing the symptoms rather than relying on simply a 'diagnosis' of low back pain, being a symptom rather than a clinical diagnosis or disease. Identifying a pathological condition allows surgeons to determine whether a surgical intervention can correct the problem and, in turn, improve the symptoms.

The aim of the chapter is to discuss current options of surgical treatment of degenerative spinal disorders presenting with predominantly axial low back pain. Although the perception of the benefits of surgery for axial spine pain stemming from degenerative changes remains controversial, our aim is to discuss the current literature on the relative merits of surgery for selected patient groups.

2. Differential diagnosis

There are many causes of axial low back pain. Generally, the history and physical can play a paramount role in illuminating the etiology. Patients exhibiting constitutional signs such as fevers and chills can insinuate infectious etiologies, whereas weight loss, night sweats, and personal or family history of cancer can imply malignancy. Obviously, any recent trauma warrants imaging to rule out fracture.

Other causes of axial low back pain are divided into non-structural and structural entities relating to the vertebral column. Non-structural causes, sometimes referred to as non-specific low back pain, are due to strain or sprain around the vertebral column, whereas structural reasons involve abnormalities within the vertebral column identified on imaging and can be considered as stable or unstable conditions. For this chapter, we will focus on structural degenerative causes of axial low back pain. Stable conditions include degenerative disc disease (DDD) and facet arthropathy, while degenerative and isthmic spondylolistheses and degenerative scoliosis are more unstable conditions. Before delving into specific causes, general surgical outcomes for low back pain will be discussed.

3. Surgical outcomes for low back pain

Most people with low back pain are successfully treated non-surgically through medications, modified activities, physical therapy, localized injections, and alternative therapies that are well described in other chapters of this book. However, there is a minority with persistent or increased pain, needing further workup and possibly surgery. There is a wealth of studies gauging the efficacy of surgical treatment for non-neurogenic axial low back pain. Systematic reviews of randomized controlled trials provide a strong level of evidence by setting inclusion and exclusion criteria when looking at the study methodologies, participants, interventions, and outcome measures. Yet, when it comes to comparing operative and non-operative results for axial low back pain, there can be conflicting results due to the lack of specificity in describing the cause of back pain since the outcomes from surgery differ between diagnostic subgroups. The shortcoming of these systematic reviews is their broad categorization of causes of back pain by combining the aforementioned causes as just degenerative disease. Also, differences in patient inclusion criteria, fusion technique, non-surgical treatment, and outcome measures make it hard to draw conclusions.

The general view of surgery for axial low back pain is met with skepticism. Mirza and Deyo (2007) systematically reviewed surgical compared to non-surgical treatment of discogenic back pain and concluded that surgery may not be more efficacious than structured cognitive behavior therapy. However, careful analysis of this study shows that the specific diagnostic indications for surgery are poorly defined. The population in this review was deemed to have low back pain for 12 months or longer without a specific diagnosis, and there was no established way of diagnosing discogenic pain. Furthermore, there was no uniform surgical technique, but rather an inclusion of a myriad of interventions, including one group that received flexible stabilization without fusion. Thus, due to its limitations, this study fails to accurately measure the effectiveness of surgery for axial low back pain.

Similarly, Chou et al (2009) systematically reviewed the benefits and harms of surgery for non-radicular back pain compared with non-operative measurements. They looked at different trials addressing fusion for chronic back pain mostly due to DDD, but not exclusively limited to DDD. Their conclusion is that fusion is no better than intensive rehabilitation with cognitive behavior emphasis, but slightly better than non-intensive non-surgical therapy. Again, a fault of the analysis is the inability to specify the specific causes of low back pain. It also combined various surgical techniques for undefined chronic back pain. This leads to the presumption that there is no role for surgery in axial low back pain when, in fact, the success of surgery depends on specific causes.

Glassman et al (2009) demonstrated that it is possible to stratify by specific diagnostic indication when looking at lumbar fusion for different diagnostic subgroups. In contrast to prior systematic reviews, they prospectively collected clinical outcome measures and reported on the impact of lumbar posterolateral fusion on different subgroups. Outcome measures such as the Oswestry Disability Index (ODI), Short Form-36 (SF-36), and numeric rating scales for back pain were used. Their findings showed that outcomes scores were not equal among diagnostic subgroups. In fact, the diagnostic subgroup that demonstrated the most significant improvement in ODI scores were the patients with spondylolisthesis, followed in decreasing order by scoliosis, disc pathology (i.e. DDD), postdiscectomy revisions, stenosis, and adjacent level degeneration. This was based on the percentage of patients in each subgroup to reach minimum clinically important difference, or an improvement of at least 10 points in ODI scores, during a 2-year follow-up.

Carreon et al (2008) provided another study looking at outcome measures while stratifying subgroups. They also used ODI and SF-36 to compare surgery with no surgery. The mean improvement in ODI in the surgical group was higher than the non-surgical group. Within this group, patients with spondylolisthesis had the greatest improvement, followed by those with DDD, then patients with chronic non-structural non-specific low back pain. This implies that non-specific etiologies, as displayed by the chronic low back pain group, can portend less success with surgery. On the other hand, more specific structural etiologies such as spondylolisthesis can benefit from surgical interventions.

4. Stable degenerative conditions

4.1 Degenerative disc disease

Degenerative disc disease (DDD) (Figure 1) stems from structural changes of the disc, which eventually leads to disc space narrowing, endplate osteophyte formation and sclerosis, and gas formation within the disc space. This is in contrast with internal disc disruption (IDD), which displays only abnormal discal properties without loss of disc height or endplate changes. The exact pathophysiology of DDD is not fully understood, and so its natural history is still unknown. Kirkaldy-Willis et al (1978) proposed a pathoetiology for this condition. They viewed each level of the lumbar vertebra as a three-joint complex consisting of a disc and two posterior joints. Stresses to one joint can affect the others. The process of degenerative disc disease starts with internal disruption, followed by resorption of the disc and endplate changes. With a degenerative disc and therefore more strain on the posterior facet joints, this eventually leads to advanced facet arthropathy and spondylosis. With minor repeated trauma, the degenerative interaction between the three-joint complex leads to more stresses on the adjacent levels, thus, potentially leading to a multilevel degenerative spine. In 10 to 39 percent of chronic low back pain cases, the intervertebral disc is suggested to be the source of pain (Schwarzer et al, 1995; Manchikanti et al, 2001). Despite this, it is still controversial as to how much DDD correlates to low back pain. As a result, there is a debate regarding its treatment. Clinical examination may show midline spinal tenderness and reduced range of motion, typically in flexion.

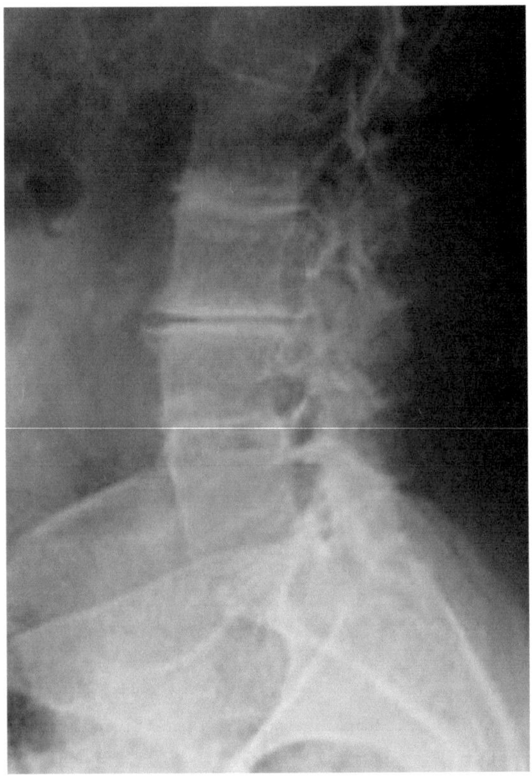

Fig. 1. Degenerative disc disease
Lateral radiograph of the lumbosacral spine depicting marked loss of disc spaces from L2 to S1. There are anterior osteophytes and endplate subchondral sclerosis.

Non-surgical management may include such modalities as physical therapy, medications, and interventional injection treatments. According to Cochrane reviews, long-term bedrest and back braces are not recommended (Hagen et al, 2004; van Duijvenbode et al, 2008). Interventional modalities such as epidural injections and intervertebral disc injections and manipulation have yet to be proven effective by randomized controlled trials. Acetaminophen and non-steroidal anti-inflammatory drugs are the most commonly used medications. Non-surgical options for this condition are discussed elsewhere in this book.

Surgical options for DDD may include fusion or motion-preservation strategies, such as artificial disc replacement. There have been only a handful of high quality randomized controlled studies assessing the effectiveness of surgery for DDD. Because the diagnosis of DDD is still controversial, these studies are still non-specific in terms of diagnostic categorization. In a meta-analysis of randomized trials comparing fusion to conservative treatment for DDD, surgery led to improved functional scores compared to non-surgical treatment (Ibrahim et al, 2008). However, the difference in functional improvement was not statistically significant. Meanwhile, disc replacement can be an option for isolated disc pathology, without arthrosis of the facet joints or spinal instability. There has been little research comparing total disc replacement versus conservative care, however, some studies have shown non-inferiority of disc replacement to fusion (Blumenthal et al, 2005; Zigler et al, 2007). Surgical options for these conditions will be discussed in more detail later in the chapter.

4.2 Facet arthropathy

Along the spectrum of degenerative changes, facet arthrosis results from increased load to the posterior elements due to abnormal load sharing from disc derangement and repetitive minor trauma over time as discussed by Kirkaldy-Willis. Similar to other synovial joints, like the hip, knee, and shoulder, degenerative arthropathy of the facet joints can lead to joint space narrowing, osteophyte and cyst formation, joint effusions, and mechanical pain. Facetogenic pain is typically worse with extension and may be relieved with rest. Patients often get relief of their back pain with leaning on a walker or shopping cart. It is the same degenerative process that may be implicated in spinal stenosis whereby the osteophyte formation, cyst formation, disc bulging, and redundancy of the ligamentum flavum from disc height loss all cause encroachment on the neural elements.

Conservative management, in addition to physical therapy and medications, consists of intra-articular facet injection and medial branch block. Medial branches of the dorsal rami are usually blocked at the junction between the superior articular facet and transverse process. In patients who respond to medial branch blocks, medial branch neurotomy via radiofrequency ablation presents as an effective non-surgical treatment for facet arthropathy (Dreyfuss et al, 2000). Once the diagnosis is confirmed with positive blocks, but the pain still recurs, then surgical options include posterior fusion and facet replacement (discussed later in the chapter).

5. Unstable degenerative conditions

5.1 Degenerative scoliosis

5.1.1 Introduction

Scoliosis is defined as an abnormal curvature of the spine of more than 10 degrees (Figure 2). In adults, scoliosis can be a result of untreated scoliosis that existed before skeletal maturity or can develop after skeletal maturity, otherwise known as de novo scoliosis. An example of de novo scoliosis is degenerative scoliosis, which is caused by a continuum of degenerative changes as described in DDD and facet arthropathy, leading to central canal and foraminal narrowing. Adult scoliosis has detrimental effects on the health status of the affected person. Berven et al (2003) illustrated that compared with control subjects, adults with scoliosis have more pain, lower self-image, less functional capacity, and lower mental health scores. They also concluded that radiographic parameters do not necessarily correlate well with the patient's self-assessment of health status.

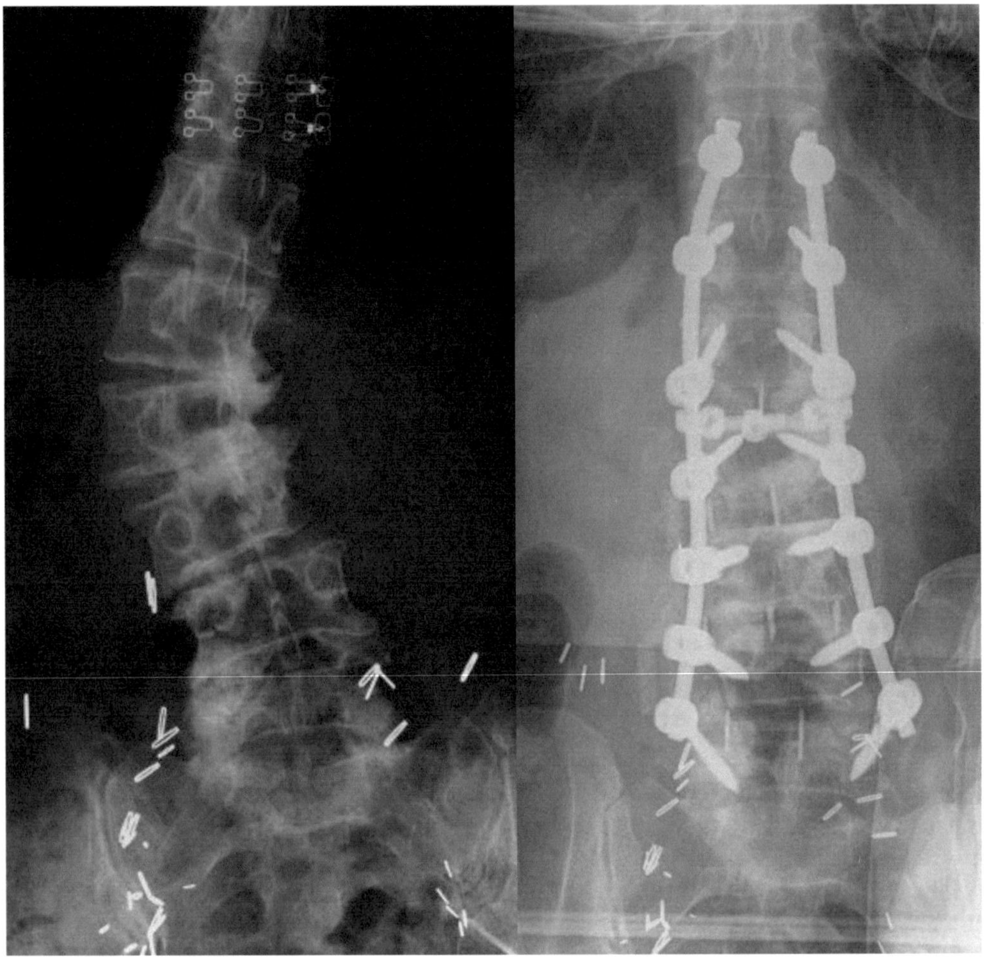

Fig. 2. Degenerative Scoliosis

Figure 2a shows an anteroposterior (AP) radiograph of the lumbosacral spine showing degenerative lumbar scoliosis with the apex at the L2-3 level. Notice the degenerative disc disease from L1-L5 and the rotational deformities of the vertebral bodies based on the asymmetric pedicles. Figure 2b shows postoperative AP radiograph with posterior instrumentation and lateral interbody fusions for deformity correction. Note that the lumbar spine now sits in line with the mid-sacrum.

5.1.2 Natural history

Progression of the curve is common, but the extent of it is unknown so continual observation is important. Chin et al (2009) demonstrated that women older than 69 years of age with levoscoliosis and lateral listhesis of more than 5mm might progress rapidly. Deviren et al (2002) showed that increasing age and curve magnitude correlate to decreased curve flexibility. Also, the degree of axial back pain associates with increasing age. Pritchett and Bortel (1993) studied 200 patients older than 50 years of age with adult scoliosis and found that certain factors might predict curve progression. In general, those with significant curves with rotation can progress rapidly.

5.1.3 Diagnostic imaging

Full-length standing radiographs are required to fully assess the overall spinal balance. Cobb angles are determined for the structural curve, which is usually the largest curve, and any compensatory curve. Coronal balance is determined with a plumb line from the middle of C7 vertebral body on posteroanterior (PA) radiograph. This should intersect the midsacrum. Sagittal balance is measured using a plumb line from the center of C7 on the lateral radiograph. This line should typically fall within 2 to 4 centimeters from the the posterior margin of the lumbosacral disc. If it falls anterior to the posterior margin of the disc then positive sagittal imbalance is present, while negative sagittal imbalance is when the line falls posterior to the disc. Global sagittal imbalance has the most significant impact on pain and function compared to other radiographic parameters (Glassman et al, 2005). The evaluation of the flexibility of the main structural curve and its compensatory curve is done through various specialized radiographs. Obliquities such as pelvic tilt and shoulder asymmetry should also be noted. Flexibility and obliquity assessment dictate which level to fuse and instrument when performing surgery. Any radicular or neurogenic pain should merit obtaining magnetic resonance imaging (MRI).

5.1.4 Clinical diagnosis

While axial back pain is common in scoliosis, Smith et al (2008) showed that neurological symptoms and deficits are also frequently found in these patients. The incidences of back pain and radiculopathy were found to be 99 percent and 85 percent, respectively. Neurogenic symptoms typically arise from the concave side of the curve from asymmetric disc collapse and resultant neural foraminal stenosis. In addition to axial pain and neurogenic symptoms, spinal imbalance can manifest in late presentation. With the patient standing up without hip or knee bending, coronal balance can be evaluated with a plumb line from the C7 spinous process. Normally this plumb line should intersect the gluteal cleft in a balanced spine. Gross sagittal balance can be determined by evaluating the relationship of the pinna of the ear to the greater trochanter of the femur.

5.1.5 Treatment

Non-operative treatment aims to control pain and function. This includes modalities such as medications, physical therapy, activity modification, orthotics, and injections. Numerous studies have shown that conservative treatment does not lead to improved pain and function compared with surgery (Bridwell et al, 2009; Smith et al, 2009; Glassman et al,

2010). In a non-randomized, prospective study looking at 123 patients with a 2-year follow-up, Glassman et al (2010) questioned the cost-effectiveness of non-operative treatment when the average cost over 2 years was $10,815 US.

Thus, surgical management has been preferred. Like other spinal conditions, surgery is only entertained after conservative treatment has failed and the patient is presenting with refractory pain limiting function, progressive deformity or neurologic deficits. Smith et al (2009) have shown that surgery can lead to better outcomes in back pain, leg pain, disability, and health status after 2 years compared to non-operatively treated patients. This is in the face of greater pre-operative back pain, leg pain, and functional disability. Grubb et al (1994) also showed that pain relief was associated with a solid fusion, and surgically managed patients demonstrated improved standing and walking.

While the goal of adolescent idiopathic scoliosis is to prevent progression of deformity and subsequent sequelae such as pain and neurological symptoms, the objectives of surgical care for degenerative scoliosis are to improve current pain and neurologic symptoms, and to restore normal spinal balance, particularly in the sagittal plane; all while maintaining as many mobile segments as possible. Although surgery has been shown to be more effective than non-surgical treatment, there is no consensus regarding the optimal approach and the levels to be included. This is due to the variety of clinical presentations and extent of disease and lack of clear evidence-based literature on approaches. The number of levels to involve in a fusion has been debated. Cho et al (2008) demonstrated that short fusion is a viable method in patients with small Cobb angles and good global balance. In their study, the average Cobb angle in patients receiving short fusion (average was 3 levels) was 16 degrees. Careful assessment of the global alignment must be performed to prevent progression of deformity prior to performing instrumented fusion. The inclusion of L5-S1 is also debatable. While stopping the fusion at L5 reduces perioperative complications and chances of pseudoarthrosis, the theoretical advantages of including this segment include complete sagittal balance correction and obviating future revisions due to degenerative changes at the L5-S1 level (Bridwell et al, 2003). Some clear indications to extend fixation to the sacrum include spondylolisthesis, stenosis requiring decompression, and degenerative disc changes at the L5-S1 level.

Combined anterior and posterior approaches provide presumed circumferential fusion and generous sagittal correction. However, they are associated with high perioperative complication rates as shown in Berven et al's retrospective study (2003). Despite achieving good sagittal correction, 32 percent of the patients developed perioperative complications including infections, dural tears, pneumonia, and acute renal failure. Overall, 40 percent needed repeat surgery for various causes including revision of fusion, hardware complications, and infection. With recent advances in instrumentations and techniques, circumferential fusion and deformity correction can be performed through a posterior-based approach. Interbody fusion can be done through the posterior lumbar interbody fusion (PLIF) and transforaminal lumbar interbody fusion (TLIF) techniques via a posterior-based approach. Crandall and Revella (2009) demonstrated equivalent results between these posterior-based interbody fusion approaches and anterior-based interbody fusion technique. Another recent technique to lessen the perioperative complication rates is the lateral lumbar interbody fusion through a minimally invasive trans-psoas approach (Figure 2). Although it has a steep learning curve, this method can provide lower complication rates, lower blood loss, and shorter hospital stay (Mundis et al, 2010).

In a recent retrospective study comparing surgical outcomes between decompression alone, decompression with limited fusion, and decompression with full curve fusion, Transfeldt et al (2010) showed that decompression alone had the lowest rate of complications followed by decompression and limited fusion, while the decompression and full curve fusion group had the highest rate of complications. On the other hand, post-surgical satisfaction questionnaire showed that the group with the full curve correction had the highest satisfaction rate, while the decompression alone group had the lowest satisfaction rate. Therefore, in spite of higher complication rates associated with full curve correction, patients subjectively prefer global curve balance through full curve correction.

5.2 Isthmic spondylolisthesis

5.2.1 Introduction

Spondylolisthesis is the slippage of one vertebral body on another. Isthmic spondylolisthesis is a common condition encountered in adolescents and adults and involves a defect of the pars interarticularis, which is the junction where the lamina and inferior facet meet with the pedicle and superior facet (Figure 3). This leads to a disconnect between the anterior and posterior elements of the vertebra, leading to slippage (olisthesis). The most common level is at the L5-S1 level (Figure 3), with the pars defect commonly discovered on L5 (Figure 4). The reason for this could be due to the fact that as one goes caudad on the lumbar spine, the pars

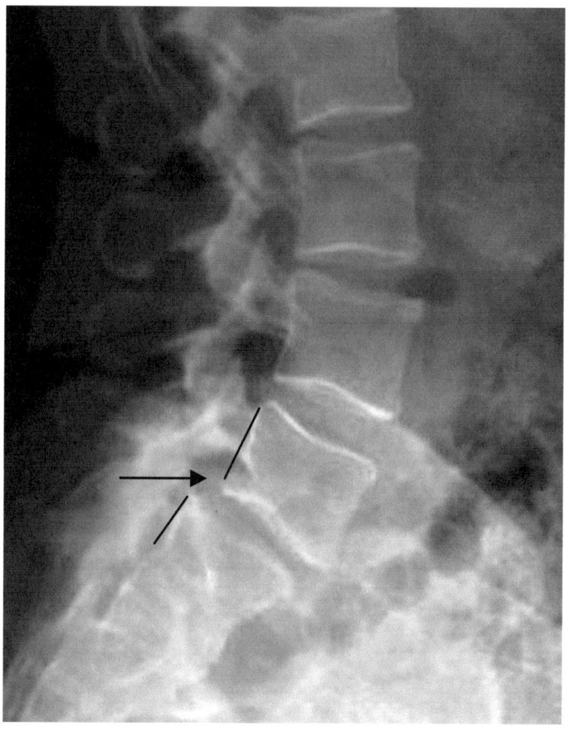

Fig. 3. Isthmic spondylolisthesis.
Lateral radiograph of the lumbar spine showing less than 25% spondylolisthesis at the L5-S1 level.. Notice the posterior cortices of L5 and S1 vertebral bodies do not line up.

get thinner. This, coupled with the fact that the L5-S1 junction endures a lot of stresses, places this level at a high risk for isthmic spondylolisthesis. Contrary to degenerative spondylolisthesis (DS), the isthmic subtype is more common in males. The incidence of pars defect is estimated to be 4 to 6 percent in the general population (Meyerding, 1932; Boxall et al, 1979; Taillard, 1976). In their prospective study, Frederickson et al (1984) reported an incidence of 4.4 percent of pars defect and 2.6 percent of spondylolisthesis at the age of 6. At adulthood, the incidence of pars defect is 5.4 percent while spondylolisthesis is 4 percent.

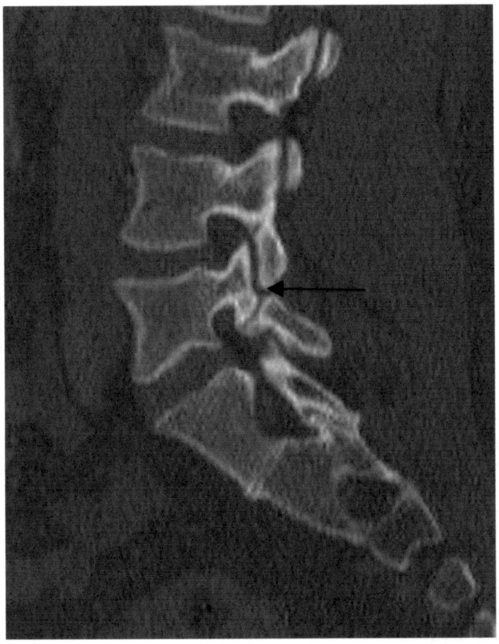

Fig. 4. Pars defect.
Sagittal reformat cut on computed tomography showing disruption of L5 pars interarticularis.

5.2.2 Natural history

In a 45-year follow up, Beutler el al (2003) showed that subjects with unilateral pars defects did not develop slippage. In those with bilateral pars defects without initial slippage, half showed no further slippage while the other half slipped a mean of 24 percent. Also, progression of the spondylolisthesis slowed with each decade and there was no association of slip progression and low back pain. Saraste (1987) showed that risk factors for low back symptoms were slippage greater than 25 percent, pars defect at the L4 level, and early disc degeneration.

5.2.3 Diagnostic imaging

Just like in DS, the lateral standing radiographs can depict spondylolisthesis and often the pars defect. If the pars defect cannot be seen on the lateral view, 30-degree oblique lateral views can be obtained. Computed tomography (CT) can provide the best bony details if still suspecting pars defect (Figure 4). Bone scan can aid in detecting stress fracture or reaction.

5.2.4 Clinical diagnosis

Most people with isthmic spondylolisthesis are asymptomatic. Back pain is generally worsened by activities and relieved with rest. This pain can be caused by lumbar hyperlordosis, which is associated with tight hamstrings. Occasionally, a step-off deformity of the spinous processes can be palpated adjacent to the level of the spondylolisthesis. Neurologic symptoms are usually in a radicular and dermatomal distribution due to impingement of the exiting nerve root, which is frequently L5 for the L5-S1 level. The site of impingement is at the site of the pars defect where the body forms hypertrophic fibrocartilaginous tissue or Gill lesion in an attempt to heal the defect.

5.2.5 Treatment

Most people with symptomatic isthmic spondylolisthesis improve with non-surgical treatment. This includes nonsteroidal anti-inflammatory drugs, activity modification (not including prolonged bedrest), and physical therapy. Radicular symptoms can be treated with epidural or transforaminal injections. Indications for surgery include failure of conservative therapy, progressive instability and/or neurological function, and intractable back or leg pain specific to the spondylolisthetic level. Surgical management of isthmic spondylolisthesis shows favorable outcomes compared to non-surgical treatment. In a prospective, randomized study comparing posterolateral fusion with an exercise program, Moller and Hedlund (2000) demonstrated that the surgical group had better functional outcome based on the Disability Rating Index and pain reduction.

The general basis of surgery for this condition is stabilization of the spondylolisthesis with or without decompression of affected neural structures. Since decompression alone fails to stabilize the spondylolisthesis, the options include decompression and non-instrumented posterior fusion, decompression and instrumented posterior fusion, decompression with posterior fusion augmented with anterior column support in the form of interbody fusion, and direct pars repair.

Controversy exists about non-instrumentated versus instrumentated posterior fusion. In a 5-year prospective randomized study comparing the two techniques, Bjarke et al (2002) showed that patients with non-instrumented posterior fusion had better clinical outcomes than their counterparts, and there was no difference in fusion rates between the two groups. Moller and Hedlund (2000) also echoed similar findings in that instrumentation does not add to the fusion rate nor improve clinical outcomes. Proponents of instrumentation claim that it can attain slip reduction and can restore sagittal alignment. Pertaining to reduction, Poussa et al (2006) showed that patients receiving in situ fusion had better outcome scores compared to the group that had reduction and fusion. Moreover, the reduction group had more neurologic complications and pseudoarthroses than the in situ fusion group. Hence, instrumentation and slip reduction have not been shown to have clear superiority over non-instrumentation and in situ fusion.

The addition of anterior support with interbody fusion theoretically provides circumferential fusion sites. Multiple studies have shown positive effects of anterior support with interbody fusion in high-grade spondylolisthesis (Helenius, 2006; Molinari, 1999, Shufflebarger, 2005). These include better functional outcomes and fusion rates. On the other hand, the use of interbody fusion is debatable for low-grade spondylolisthesis. Stand-alone interbody fusion without posterior instrumentation is discouraged in this condition due to high rates of failure such as cage migration (Button et al, 2005).

The theoretical advantage of direct repair of the pars defect relates to its ability to preserve motion compared with fusion, possibly leading to decreased degeneration in the adjacent segment. Although direct repair has been proven to be successful with low-grade spondylolisthesis in the short-term period (Morelos, 2004), it has not been shown to be as effective in the long-term period as initial improvement in functional outcomes declined with time and the adjacent segment degeneration phenomenon was comparable to those who received posterior fusion (Schlenzka et al, 2006). However, the method of direct repair shown in Schlenzka et al's study involved cerclage wiring, whereas today's fixation typically involves screws/hooks and/or rods (Figure 8). As a result, it is unknown whether today's technology could prove otherwise and long term follow up studies are needed.

5.3 Degenerative spondylolisthesis

5.3.1 Introduction

Degenerative spondylolisthesis (DS) is a condition generally found in females older than 40 years of age. The usual level of involvement is L4-L5, with L4 slipping anterior to L5 (Figure 5). The cause of this is presumed to be a result of structural degenerative changes in disc and ligaments, more importantly the facet capsules. In a review of magnetic resonance imaging (MRI) in 140 subjects, Boden et al (1996) suggested that more sagitally oriented facets might be the cause of DS.

5.3.2 Natural history

Matsunaga et al (1990) studied the natural course of DS by observing 40 patients from 5 to 14 years. Slip progression was seen in 12 (30 percent) of the patients, but this did not correlate well with clinical symptoms. Meanwhile, 4 of the 28 patients who did not show progressive slip displayed clinical deterioration. Therefore, there is a lack of correlation between progressive slip and clinical symptoms. Also, the study infers that there is no correlation between degenerative changes, such as intervertebral disc narrowing, spur formation, subcartilaginous sclerosis, or ossification of ligaments, and slip progression, hence, suggesting that these anatomic changes may act to stabilize the spine.

5.3.3 Diagnostic imaging

Since DS is a dynamic condition involving instability of the spine, the preferred radiological imaging study is a lateral radiograph, in the standing position. Dynamic flexion and extension views can be added for further inspection of the instability. In a study by Boden and Wiesel (1990) looking at dynamic flexion and extension views, 90 percent of asymptomatic volunteers had 1 to 3mm of translation, therefore, it was considered that anything more than 4mm is abnormal. Slippage is graded based on the percentage of antero-posterior displacement on the vertebral body. Grade 1 equates to less than 25 percent of displacement on the caudad vertebral body; grade 2 is up to 50 percent; grade 3 is up to 75 percent; and grade 4 is up to 100 percent. Additionally, supine views are not helpful

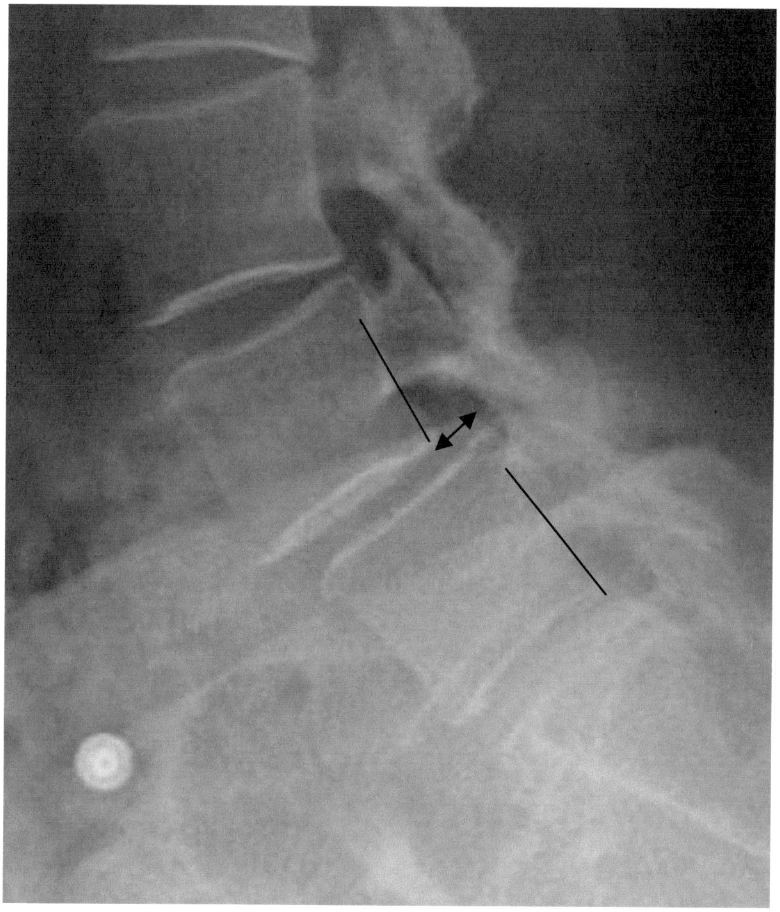

Fig. 5. Degenerative spondylolisthesis.
Lateral radiograph of the lumbosacral spine showing grade 1 spondylolisthesis at the L4-5 level. Notice the posterior cortices of L4 and L5 vertebral bodies do not line up. The percentage of displacement is approximately 20-25 percent of the vertebral body of L5.

since this position may reduce the slippage. Although MRI portrays a static condition, a study by Chaput et al (2007) showed that large (>1.5mm) facet effusions are highly predictive of DS at L4-L5.

5.3.4 Clinical diagnosis

Axial back pain in DS is frequently associated with back extension, whereas back pain in discogenic back pain is classically related to sitting and flexion. Other features of the condition can mimic spinal stenosis and lead to neurogenic claudication. The predominant symptom is pain, radiating from the buttock to the legs, and commonly involves bilateral legs. The neurogenic symptoms do not resemble radicular symptoms in affecting a specific dermatome, but may be diffuse in nature. If there are associated radicular signs, L5 is the most commonly involved root. Also, neurogenic claudication must be differentiated with vascular claudication when diagnosing DS.

5.3.5 Treatment

Generally, a comprehensive course of non-surgical treatment is the first line unless the patient exhibits any sign of neurological deterioration. This is defended by Matsunaga et al's (2000) study showing that 76 percent of his sample size remained without neurological deficit at the 10 year follow up. Those who have failed conservative treatment and display increased or persistent pain, with or without neurologic symptoms, may be considered for surgery. The Spine Patient Outcomes Research Trial (SPORT) depicts the benefits of surgical treatment in patients with DS associated spinal stenosis. They followed 607 subjects for 4 years and rated their progress with outcome measures including, SF-36 and ODI. Despite their high cross over rate between surgical and non-surgical treatment groups, their conclusion was that patients with DS treated with surgery showed better improvement in pain and function during the 4 year follow up. Another shortcoming of this study was that it did not compare different types of surgical techniques. However, there are numerous studies that offer insights into the optimal surgical treatment.

The surgical options include decompression alone, decompression with posterior non-instrumented fusion, decompression with posterior instrumented fusion, and decompression with posterior fusion and anterior column support. Several papers have clearly shown that posterior non-instrumented fusion in conjunction with decompression leads to better clinical outcome than decompression alone in DS patients (Herkowitz, 1991; Mardjetko, 1994). As far as whether or not to add instrumentation to the fusion is still debatable. Fischgrund et al (1997) demonstrated in a prospective, randomized study comparing instrumented fusion with non-instrumented fusion, that fusion rate at 2 years was better in the instrumented group compared to the non-instrumented group. In spite of this, clinical outcome was similar for both groups. As a result, it is up to the physician's discretion to determine when it is appropriate to place instrumentation in this setting of spinal instability. Similarly, there is no convincing data to support the routine use of anterior column support, such as interbody fusion, in addition to posterior fusion. The purported advantages of this would be restoration of disc height and neuroforaminal space, circumferential fusion leading to higher likelihood to fuse, and better sagittal alignment restoration.

6. Surgical methods

Surgical treatment for degenerative lumbar conditions causing axial low back pain can be considered in two broad categories: fusion procedures and motion-preservation techniques. For stable conditions causing low back pain, fusing two vertebrae together will eliminate the pain arising from their articulation. In an attempt to preserve motion, like in the hip or knee, and prevent accelerated degeneration at the adjacent level, motion-preservation strategies have been developed. For more unstable conditions, such as spondylolisthesis or scoliosis, fusion surgery with or without correction of the deformity, is considered the best surgical option.

6.1 Lumbar fusion

Spinal fusion is the surgical attempt at bonding two vertebrae together to stop the motion between them and restore the normal anatomical relationships. Fusion procedures are most commonly performed for those who are considered candidates for surgery. There are a variety of fusion techniques that may include the use of instrumentation, the location of fusion (interbody, intertransverse, interspinous, etc.), the approach (posterior, anterior, lateral), and the type of graft material used (e.g. autograft, allograft, osteogenic biologics) or a combination. A detailed account of all of these techniques is beyond the scope of this chapter.

The most commonly employed fusion technique is the posterior approach using pedicle screw-rod instrumentation and fusion across the transverse processes or facet joints (Figure 6). Pedicle screw placement is a technically demanding procedure, but it is the most commonly used technique to stabilize the spine. A retrospective study showed that the rate of screw misplacement can reach 6.7 percent, but no major neurological compromise was observed (Jutte and Castelein, 2002). Therefore, pedicle screw fixation is safe and has an acceptable complication rate despite pedicle breach. Spinal fixation can also be performed with a variety of other instrumentation, such as screws alone, hooks, plates, or wires. Non-instrumented fusions remain a viable option, however, they fail to stabilize the spine during the healing process and are associated with higher rates of failure of fusion (pseudarthrosis).

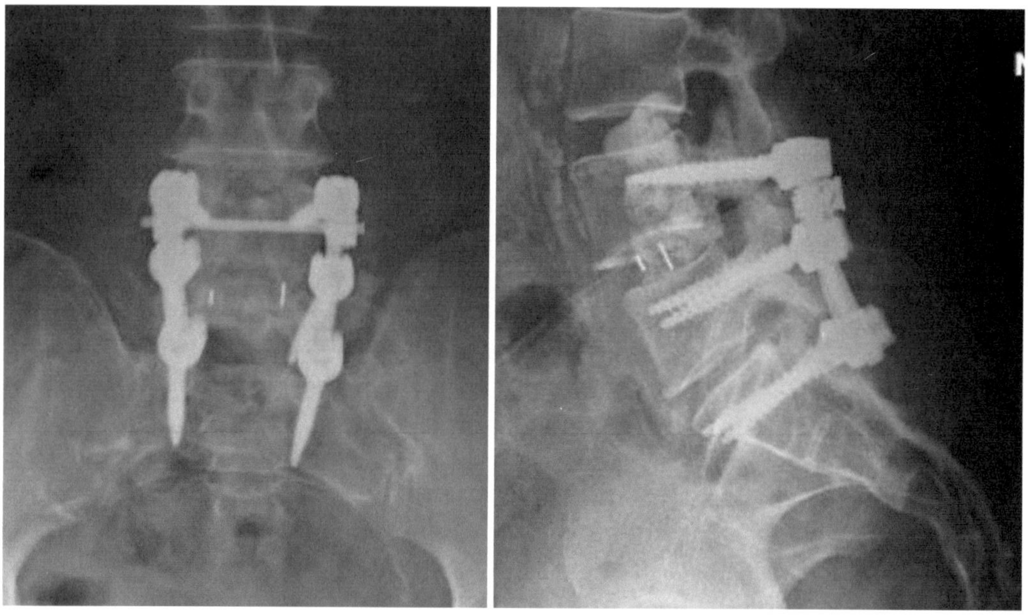

Fig. 6. Posterior and anterior fusion through posterior-based approach.
AP and lateral radiographs of two-level fusion with posterior pedicle screw-rod construct and TLIF at L4-5. (identified by radio-opaque vertical lines).

Anterior fusions through the disc spaces improve our ability to restore the normal anatomy of the anterior column of the spine by restoring normal disc height and curvature. Generally accepted indications for interbody fusions include degenerative disc disease, disc collapse with resultant neuroforaminal stenosis, and the need to restore sagittal and coronal balance. Interbody fusion creates a bond between two vertebral bodies through the disc space and can be done in combination with posterior fusion or as a stand-alone technique. Anterior fusion can be approached via several different routes: posterior, lateral or directly anterior. The posterior approach, most commonly done in association with a posterior fusion and/or decompression, is performed through a posterolateral approach into the disc space similar to removing a herniated disc fragment. There are two commonly used methods for interbody fusion done through a posterior approach: the posterior lumbar interbody fusion (PLIF) and the transforaminal lumbar interbody fusion (TLIF). PLIF is performed bilaterally and uses the same approach as disc fragment removal. A laminotomy or laminectomy is created to allow exposure of the nerve roots, which are carefully retracted and mobilized. Once the disc is identified, a window is created in the disc, the disc material is removed and the vertebral endplates are denuded of cartilage until there is bleeding bone. A prosthetic cage or structural allograft bone filled with bone graft is inserted into the disc space on both sides. TLIF involves resection of the facet and unroofing the neuroforamen on one side only to get to the posterolateral corner of the intervertebral disc. The traversing nerve root requires less retraction with the TLIF since the approach is slightly more lateral than PLIF. Once inside the disc, it is prepared in a similar way as PLIF. A prosthetic cage or structural allograft filled with bone graft is inserted into the disc space only from one side and placed in a central position inside the disc space (Figure 6). The difference between the two is that TLIF entails less neural manipulation to get to the vertebral disc and is done with a unilateral approach so it is more widely practiced. They both take advantage of the commonly used posterior approach to establish access to the anterior column of the spine.

Anterior lumbar interbody fusion (ALIF) approaches the spine directly anteriorly through the abdomen either through a trans-peritoneal or retroperitoneal approach. The rectus abdominus is retracted laterally which makes this approach truly muscle-preserving. The advantage over a posterior interbody approach (i.e. PLIF or TLIF) is ease of clearing out the disc for fusion, the ability to place a large graft for better restoration of normal anatomical height and better fusion rates, and obviating the need to retract the thecal sac or nerve roots. The potential risks include vascular injury, ileus, and retrograde ejaculation in males.

The lateral trans-psoas approach, is a relatively new procedure that has been gaining in popularity (Figure 7). The patient is placed in the lateral position, and with the use of fluoroscopy and nerve monitoring, a safe corridor through the retroperitoneum and psoas muscle is created to access the disc. While the obvious advantages are that it avoids the need for a posterior approach and can correct spinal instabilities or deformities, it cannot be used to access the L5-S1 disc space.

6.2 Motion-preservation techniques

The technology for motion-preservation techniques are developing at an exponential rate and include a wide range of options such as simple as direct pars repair (Figure 8) (for isthmic spondylolisthesis), interspinous spacers, to more complex devices such as disc replacement, facet replacement, and posterior dynamic stabilizations (Figure 9). Because they are relatively novel concepts, there is a lack of long-term clinical studies demonstrating their effectiveness and safety. While disc replacement is indicated primarily for discogenic pathology, facet replacement aims to treat posterior degeneration and dynamic stabilization intends to limit, but not abolish motion in an unstable spine. The purported benefits of

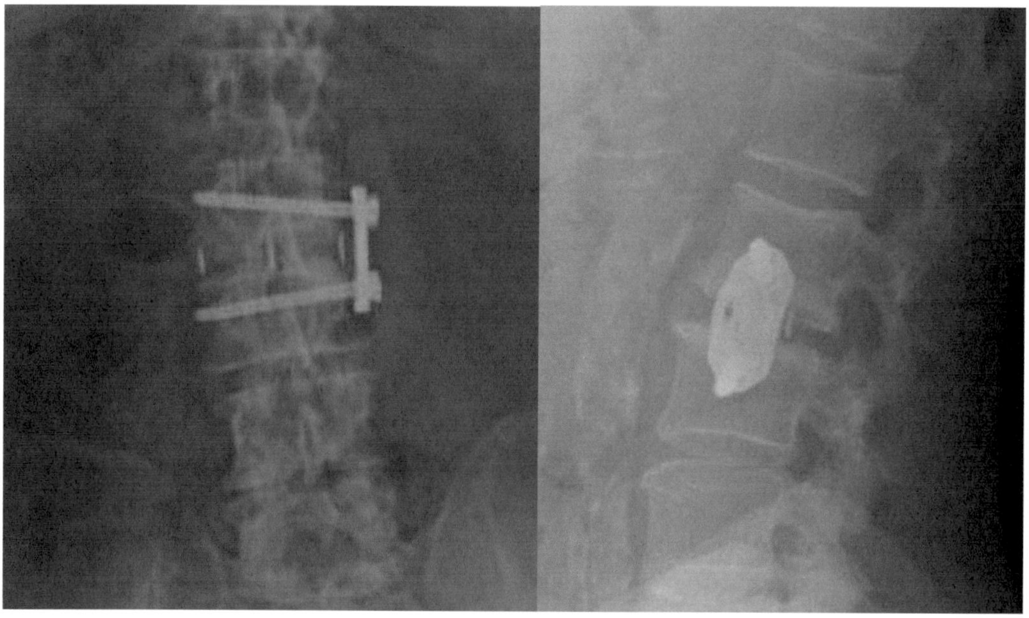

Fig. 7. Lateral lumbar interbody fusion.
AP and lateral radiographs of the lumbar spine showing lateral trans-psoas interbody fusion at the L2-L3 level with a side plate and interbody fusion mass as depicted by the white markers.

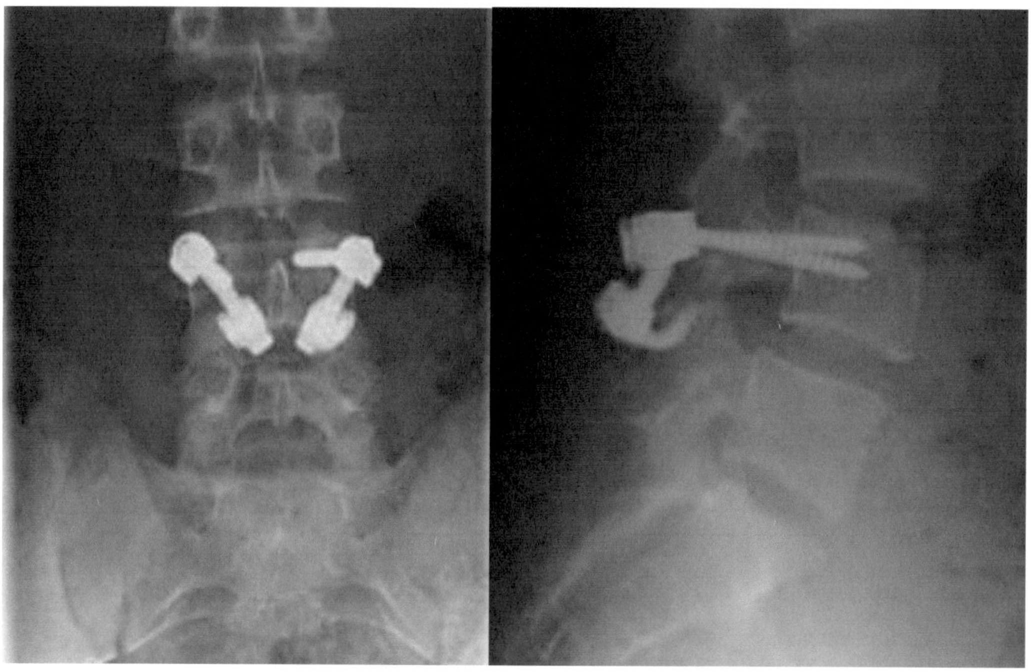

Fig. 8. Pars repair.
AP and lateral radiographs of the lumbar spine showing pars repair of L4 with pedicle screws, hooks, and rods.

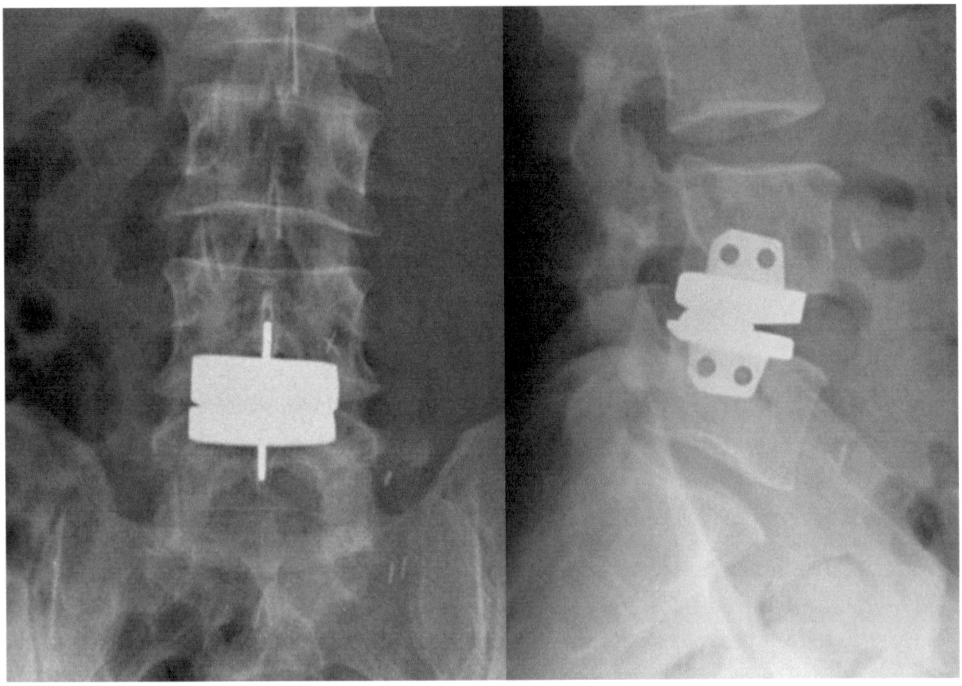

Fig. 9. Artificial total disc replacement.
AP and lateral radiographs of an artificial disc replacement at L4-5.

lumbar disc replacement, facet replacement, and dynamic stabilization are to maintain normal motion of the lumbar spinal segment and therefore to potentially decreasing the risk of degeneration at the adjacent segments. Mid-term outcomes of single level total disc replacement showed sustained improved outcome measures at an average follow up of 44.9 months in the treatment of DDD (Scott-Young et al, 2011). However, complications reported in literature such as implant subsidence, loosening, early wear, displacement, malposition, and the difficulty with revision surgery, have limited its widespread use.

7. Conclusion

Surgical treatment for low back pain remains controversial largely due to confusion in terminology and the inability of literature to stratify the results based on specific diagnostic indication. Low back pain should be viewed as a symptom, not a disease or diagnosis. When considered only as a diagnosis, study results are mixed and confounded due to the many different causes. Therefore, it is imperative to elucidate the conditions causing low back pain whether structural or non-structural. When stratified into diagnostic subgroups, results of surgery differ. For example, surgery is beneficial for more structural abnormalities, in particular those with more instability such as spondylolisthesis and degenerative scoliosis, as opposed to non-structural conditions which are better treated with non-surgical modalities. While the preferred method of treatment for these degenerative conditions is a non-surgical approach, there are many patients who are candidates for surgery. Although the traditional surgical strategy for structural degenerative conditions is fusion, motion-sparing techniques are showing promise, however, long-term studies are needed. More unstable degenerative conditions benefit more from fusion procedures with correction of deformities. Only with a more refined diagnostic ontology and a better understanding of the pathomechanical processes, can we hope to determine the best treatments available for patients suffering from these conditions.

8. References

Andersson, G. (1997). The epidemiology of spinal disorders, In: The Adult Spine: Principles and Practice, Frymoyer J et al, pp93-141, Raven Press, ISNB 978-0781703291, New York, USA

Andersson, G. (1999). Epidemiologic features of chronic low back pain. *Lancet*, 354, 1999, pp. 581-585

Atlas, SJ; Chang, Y; Kammann, E; Keller, RB; Deyo, RA & Singer, DE. (2000). Long-term disability and return to work among patients who have a herniated lumbar disc: the effect of disability compensation. *J Bone Joint Surg Am*, 82(1), Jan 2000, pp. 4-15

Berven, SH; Deviren, V; Smith, JA; Hu, S & Bradford, D. (2003). Management of fixed sagittal plane deformity: outcome of combined anterior and posterior surgery. *Spine*, 28, 2003, pp. 1710-1715

Berven, S; Deviren, V; Demir-Deviren, S; Hu, S & Bradford, D. (2003). Studies in the modified Scoliosis Research Society Outcomes Instrument in adults: validation, reliability, and discriminatory capacity. *Spine*, 28, 2003, pp. 2164-2169

Beutler, WJ; Fredrickson, BE; Murtland, A; Sweeney, CA; Grant, WD & Baker, D. (2003). The natural history of spondylolysis and spondylolisthesis: 45-year follow-up evaluation. *Spine*, 28. 2003, pp. 1027-1035

Bjarke Christensen, F; Stender Hansen, E; Laursen, M; Thomsen, K & Bünger, CE. (2002). Long-term functional outcome of pedicle screw instrumentation as a support for posterolateral spinal fusion: randomized clinical study with a 5-year follow-up. *Spine*, 27, 2002, pp. 1269-1277

Blumenthal, S; McAfee, PC; Guyer, RD, Hochschuler, S; Geisler, F; Holt, T; Garcia, R; Regan, J & Ohnmeiss, D. (2005). A prospective, randomized, multicenter Food and Drug Administration investigational device exemptions study of lumbar total disc replacement with the CHARITE artificial disc versus lumbar fusion: Part I. Evaluation of clinical outcomes. *Spine*, 30, 2005, pp. 1565-1575

Boden, SD; Riew, KD; Yamaguchi, K; Branch, T; Schellinger, D & Wiesel, S. (1996) Orientation of the lumbar facet joints: Association with degenerative disc disease. J Bone Joint Surg Am. 78, 1996, pp. 403-411

Boden, SD & Wiesel, SW. (1990). Lumbosacral segmental motion in normal individuals: Have we been measuring instability properly?. *Spine*, 5, 1990, pp. 571-576

Bone and Joint Decade. US Bone and Joint Decade Web site. (2005). The Burden of Musculoskeletal Diseases in the United States. Accessed December 28, 2005. Available from: http://www.usbjd.org/index.cfm

Boxall, D; Bradford, DS & Winter RB. (1979). Management of severe spondylolisthesis in children and adolescents. *J Bone Joint Surg*, 61A, 1979, pp. 479-495

Bridwell, KH; Edwards, CC & Lenke, LG. (2003). The pros and cons to saving the L5-S1 motion segment in a long scoliosis fusion construct. *Spine*, 28, 2003, pp. S234-S242

Bridwell, KH; Glassman, S; Horton, W; Shaffrey, C; Schwab, F; Zebala, L; Lenke, L; Hilton, J; Shainline, M; Baldus, C & Wooten, D. (2009). Does treatment (nonoperative and operative) improve the two-year quality of life in patients with adult symptomatic lumbar scoliosis: A prospective multicenter evidence-based medicine study. *Spine*, 34, 2009, pp. 2171-2178

Brox, J; Reikeras, O; Nygaard, O; Sorensen, R; Indahl, A; Holm, I; Keller, A; Ingebrigtsen, T; Grundnes, O; Lange, J & Friis, A. (2006). Lumbar instrumented fusion compared with cognitive intervention and exercises in patients with chronic back pain after previous surgery for disc herniation: a prospective randomized controlled study. *Pain*, 122, 2006, pp. 145-155

Button, G; Gupta, M; Barrett, C; Cammack, P & Benson, D. (2005). Three- to six-year follow-up of stand-alone BAK cages implanted by a single surgeon. *Spine J*, 5, 2005, pp. 155-160

Carreon, L; Glassman, S & Howard, J. (2008). Fusion and nonsurgical treatment for symptomatic lumbar degenerative disease: a systematic review of Oswestry Disability Index and MOS Short Form-36 outcomes. *Spine J*, 8, 2008, pp. 747-755

Chaput, C; Padon, D; Rush, J; Lenehan, E & Rahm, M. (2007) The significance of increased fluid signal on magnetic resonance imaging in lumbar facets in relationship to degenerative spondylolisthesis. *Spine*, 32, 2007, pp.1883-1887

Chin, K; Furey, C & Bohlman, H. (2009). Risk of progression in de novo low-magnitude degenerative lumbar curves: natural history and literature review. *Am J Orthop*, 38, 2009, pp. 404-409

Cho, K; Suk, S; Park, S; Kim, J; Kim, S; Lee, T; Lee, J & Lee, JM. (2008). Short fusion versus long fusion for degenerative lumbar scoliosis. *Eur Spine J*, 17, 2008, pp. 650-656

Chou, R; Atlas, S; Stanos, S & Rosenquist, R. (2009). Nonsurgical Interventional Therapies for Low Back Pain: A Review of the Evidence for an American Pain Society Clinical Practice Guideline. *Spine*, 34, 2009, pp. 1078-1093

Chou, R; Baisden, J; Carragee, E; Resnick, D; Shaffer, W & Loeser, J. (2009). Surgery for low back pain: a review of the evidence for an American Pain Society Clinical Practice Guideline. *Spine*, 34, 2009, pp. 1094-1109

Crandall, DG & Revella, J. (2009). Transforaminal lumbar interbody fusion versus anterior lumbar interbody fusion as an adjunct to posterior instrumented correction of degenerative lumbar scoliosis: Three year clinical and radiographic outcomes. *Spine*, 34, 2009, pp. 2126-2133

Deviren, V; Berven, S; Kleinstueck, F; Antinnes, J; Smith, JA & Hu, S. (2002). Predictors of flexibility and pain patterns in thoracolumbar and lumbar idiopathic scoliosis. *Spine*, 27, 2002, pp. 2346-2349

Dreyfuss, P; Halbrook, B; Pauza, K; Joshi, A; McLarty, J & Bogduk, N. (2000). Efficacy and validity of radiofrequency neurotomy for chronic lumbar zygapophysial joint pain. *Spine*, 25, 2000, pp. 1270-1277

Fairbank, J; Frost, H; Wilson-MacDonald, J; Yu, L; Barker, K & Collins, R. (2005). Randomised controlled trial to compare surgical stabilisation of the lumbar spine with an intensive rehabilitation programme for patients with chronic low back pain: the MRC spine stabilisation trial. *BMJ*, 330, 2005, pp. 1233

Fanuele, JC; Birkmeyer, NJ; Abdu, WA; Tosteson, TD & Weinstein, JN. (2000). The Impact of Spinal Problems on the Health Status of Patients: Have We Underestimated the Effect?. *Spine*, 25, 2000, pp. 1509-1514

Fischgrund, JS; Mackay, M; Herkowitz, HN; Brower, R; Montgomery, DM & Kurz, LT. (1997). Degenerative lumbar spondylolisthesis with spinal stenosis: a prospective, randomized study comparing decompressive laminectomy and arthrodesis with and without spinal instrumentation. *Spine*, 22, 1997, pp. 2807-2812

Fredrickson, BE; Baker, D; McHolick, WJ; Yuan, H & Lubicky, J. (1984). The natural history of spondylolysis and spondylolisthesis. J *Bone Joint Surg*, 66A, 1984, pp. 699-707

Fritzell, P; Hagg, O; Wessberg, P & Nordwall, A. (2001). Swedish Lumbar Spine Study Group. 2001 Volvo Award Winner in Clinical Studies: lumbar fusion versus nonsurgical treatment for chronic low back pain: A multicenter randomized controlled trial from the Swedish Lumbar Spine Study Group. *Spine*, 26, 2001, pp. 2521–2532

Glassman, SD; Berven, S; Bridwell, KH; Horton, W & Dimar, J. (2005). Correlation of radiographic parameters and clinical symptoms in adult scoliosis. *Spine*, 30, 2005, pp. 682-688

Glassman, S; Carreon, L; Djurasovic, M; Dimar, J; Johnson, J; Puno, R & Campbell, M. (2009). Lumbar fusion outcomes stratified by specific diagnostic indication. *Spine J*, 9, 2009, pp. 13-21

Glassman, SD; Carreon, LY; Shaffrey, CI; Polly, D; Ondra, S; Berven, S & Bridwell, K. (2010). The costs and benefits of nonoperative management for adult scoliosis. *Spine*, 35, 2010, pp. 578-582

Grubb, S; Lipscomb, H & Suh, P. (1994). Results of surgical treatment of painful adult scoliosis. *Spine*, 19, 1994, pp. 1619-1627

Hagen, K; Hilde, G & Jamtvedt, G. (2004). Bed rest for acute low-back pain and sciatica. Cochrane Database Syst Rev CD001254

Hart, LG; Deyo, RA & Cherkin DC.(1995). Physician office visits for low back pain. Frequency, clinical evaluation, and treatment patterns from a U.S. national survey. *Spine*, 20, 1995, pp.11–19

Hanson, DS; Bridwell, KH; Rhee, JM & Lenke, LG. (2002). Correlation of pelvic incidence with low- and high-grade isthmic spondylolisthesis. *Spine*, 27, 2002, pp. 2026-202

Helenius, I; Lamberg, T & Osterman, K. (2006). Posterolateral, anterior, or circumferential fusion in situ for high-grade spondylolisthesis in young patients: A long-term evaluation using the Scoliosis Research Society questionnaire. *Spine*, 31, 2006, pp. 190-196

Herkowitz, H & Kurz, L. 1991. Degenerative lumbar spondylolisthesis with spinal stenosis: A prospective study comparing decompression with decompression and intertransverse process arthrodesis. *J Bone Joint Surg Am*, 73, 1991, pp. 802-808

Hicks, J; Singla, A; Shen, F & Arlet, V. (2010) Complications of pedicle screw fixation in scoliosis surgery; a systematic review. *Spine*, 35, 2010, pp. E465-470

Ibrahim, T; Tleyjeh, IM & Gabbar O. (2008) Surgical versus non-surgical treatment of chronic low back pain: A meta-analysis of randomised trials. *Int Orthop,* 32, 2008, pp. 107-113

Jutte, P & Castelein, R. (2002). Complications of pedicle screws in lumbar and lumbosacral fusions in 105 consecutive primary operations. *Eur Spine J,* 11, 2002, pp. 594-598

Kirkaldy-Willis, WH; Wedge, JH; Yong-Hing, K & Reilly, J. (1978). Pathology and pathogenesis of lumbar spondylosis and stenosis. *Spine,* 3, 1978, pp. 319-328

Manchikanti, L; Singh, V; Pampati, V; Damron, K; Barnhill, R; Beyer, C & Cash, K. (2001). Evaluation of the relative contributions of various structures in chronic low back pain. *Pain Physician,* 4, 2001, pp. 308-316

Mardjetko, S; Connolly, P & Shott, S. (1994). Degenerative lumbar spondylolisthesis: A meta-analysis of literature, 1970-1993. *Spine,* 19(20 Suppl), 1994, pp. 2256S-2265S

Matsunaga, S; Sakou, T; Morizono, Y; Masuda, A & Demirtas, A. (1990). Natural history of degenerative spondylolisthesis: Pathogenesis and natural course of the slippage. *Spine,* 15, 1990, pp. 1204-1210

Matsunaga, S; Ijiri, K & Hayashi, K. (2000). Nonsurgically managed patients with degenerative spondylolisthesis: A 10- to 18-year follow-up study. *J Neurosurg,* 93, 2000, pp.194-198

Meyerding, H. (1932) Spondylolisthesis: surgical treatments and results. *Surg Gyn Obstet,* 54, 1932, pp. 371-377

Mirza, SK & Deyo, RA. (2007). Systematic review of randomized trials comparing lumbar fusion surgery to nonoperative care for treatment of chronic back pain. *Spine,* 32, 2007, pp. 816-823

Molinari, RW; Bridwell, KH; Lenke, LG; Ungacta, FF & Riew, KD. (1999). Complications in the surgical treatment of pediatric high-grade, isthmic dysplastic spondylolisthesis: A comparison of three surgical approaches. *Spine,* 24, 1999, pp. 1701-1711

Möller, H & Hedlund, R. (2000). Surgery versus conservative management in adult isthmic spondylolisthesis: A prospective randomized study. Part 1. *Spine,* 25, 2000, pp. 1711-1715

Möller, H & Hedlund, R. (2000) Instrumented and noninstrumented posterolateral fusion in adult spondylolisthesis: A prospective randomized study. Part 2. *Spine,* 25, 2000, pp. 1716-1721

Morelos, O & Pozzo, AO. (2004). Selective instrumentation, reduction and repair in low-grade isthmic spondylolisthesis. *Int Orthop,* 28, 2004, pp. 180-182

Mundis, G; Akbarnia, B & Phillips, F. (2010). Adult deformity correction through minimally invasive lateral approach techniques. *Spine,* 35, 2010, pp. S312-321

Poussa, M; Remes, V; Lamberg, T; Tervahartiala, P; Schlenzka, D; Yrjonen, T; Osterman, K; Seitsalo, S & Helenius, I. (2006). Treatment of severe spondylolisthesis in adolescence with reduction or fusion in situ: long-term clinical, radiologic, and functional outcome. *Spine,* 31, 2006, pp. 583-590

Pritchett, JW & Bortel, DT. (1993). Degenerative symptomatic lumbar scoliosis. *Spine,* 18, 1993, pp. 700–703

Saraste, H. (1987). Long-term clinical and radiological follow-up of spondylolysis and spondylolisthesis. *J Pediatr Orthop,* 7, 1987, pp. 631-638

Schlenzka, D; Remes, V; Helenius, I; Lamberg, T; Tervahartiaia, P, Yrjonen, T; Tallroth, K; Osterman, K; Seitsalo, S & Poussa, M. (2006). Direct repair for treatment of symptomatic spondylolisthesis in young patients: no benefit in comparison to segmental fusion after a mean follow-up of 14.8 years. *Eur Spine J*, 15, 2006, pp. 1437-1447

Schwarzer, AC; Aprill, CN; Derby, R; Fortin, J; Kine, G & Bogduk, N. (1995). The prevalence and clinical features of internal disc disruption in patients with chronic low back pain. *Spine*, 20, 1995, pp. 1878-1883

Scott-Young, M; Lee, M; Nielsen, D; Magno, C; Kimlin, K & Mitchell, E. (2011). Clinical and Radiological Mid-Term Outcomes of Lumbar Single-Level Total Disc Replacement. *Spine*, September, 2011

Shufflebarger, HL & Geck, MJ. (2005). High-grade isthmic dysplastic spondylolisthesis: Monosegmental surgical treatment. *Spine*, 30, 2005, pp. S42-S48

Shufflebarger, H; Suk, SI & Mardjetko, S. (2006). Debate: Determining the upper instrumented vertebra in the management of adult degenerative scoliosis: stopping at T10 versus L1. *Spine*, 31, 2006, pp. S185-S194

Smith, JS; Fu, KM; Urban, P & Shaffrey, C. (2008) Neurological symptoms and deficits in adults with scoliosis who present to a surgical clinic: Incidence and association with the choice of operative versus nonoperative management. *J Neurosurg Spine*, 9, 2008, pp. 326-331

Smith, JS; Shaffrey, CI; Berven, S; Glassman, S; Hamill, C; Horton, W; Ondra, S; Schwab, F; Shainline, M; Fu, K & Bridwell, K. (2009). Spinal Deformity Study Group: Improvement of back pain with operative and nonoperative treatment in adults with scoliosis. *Neurosurgery*, 65, 2009, pp. 86-93

Smith, JS; Shaffrey, CI; Berven, S; Glassman, S; Hamill, C; Horton, W; Ondra, S; Schwab, F; Shainline, M; Fu, K & Bridwell, K. (2009). Spinal Deformity Study Group: Operative versus nonoperative treatment of leg pain in adults with scoliosis: A retrospective review of a prospective multicenter database with two-year follow-up. *Spine*, 34, 2009, pp. 1693-1698

Swank, S; Lonstein, JE; Moe, JH; Winter, RB & Bradford, DS. (1981). Surgical treatment of adult scoliosis. A review of two hundred and twenty-two cases. *J Bone Joint Surg Am*, 63, 1981, pp. 268-287

Taillard, W. (1976). Etiology of spondylolisthesis. *Clin Orthop*, 115, 1976, pp. 30-39

Transfeldt, E; Topp, R; Mehbod, A & Winter, R. (2010). Surgical Outcomes of Decompression, Decompression With Limited Fusion, and Decompression With Full Curve Fusion for Degenerative Scoliosis With Radiculopathy. *Spine*. 35, 2010, pp. 1872-1875

van Duijvenbode, IC; Jellema, P & van Poppel, MN. (2008). Lumbar supports for prevention and treatment of low back pain. Cochrane Database Syst Rev CD001823

van Tulder, M; Koes, B & Bombardier, C. (2002). Low back pain. *Best Pract Res Clin Rheumatol*, 16, 2002, pp. 761–775

Weinstein, JN; Lurie, JD; Tosteson, TD; Zhao, W; Blood, E; Tosteson, A; Birkmeyer, N; Herkowitz,H; Longley, M; Lenke, L; Emery, S & Hu, S. (2009). Surgical compared with nonoperative treatment for lumbar degenerative spondylolisthesis. *J Bone Joint Surg*, 91-A, 2009, pp. 1295-1304

Zigler, J; Delamarter, R; Spivak, JM; Linovitz, R; Danielson, G; Haider, T; Cammisa, F; Zuchermann, J; Balderston, R; Kitchel, S; Foley, K; Watkins, R; Bradford, D; Yue, J; Yuan, H; Herkowitz, H; Geiger, D; Bendo, J; Peppers, T; Sachs, B; Girardi, F; Kropf, M & Goldstein, J. (2007). Results of the prospective, randomized, multicenter Food and Drug Administration investigational device exemption study of the ProDisc-L total disc replacement versus circumferential fusion for the treatment of 1-level degenerative disc disease. *Spine,* 32, 2007, pp. 1155-1162